Handbook of Fiber Science and Technology: Volume I
Chemical Processing of Fibers and Fabrics

FUNDAMENTALS AND PREPARATION

Part B

INTERNATIONAL FIBER SCIENCE AND TECHNOLOGY SERIES

Handbook of Fiber Science and Technology: Volume I

Chemical Processing of Fibers and Fabrics
FUNDAMENTALS AND PREPARATION Part A
edited by Menachem Lewin and Stephen B. Sello

Chemical Processing of Fibers and Fabrics
FUNDAMENTALS AND PREPARATION Part B
edited by Menachem Lewin and Stephen B. Sello

Handbook of Fiber Science and Technology: Volume II

Chemical Processing of Fibers and Fabrics
FUNCTIONAL FINISHES Part A
edited by Menachem Lewin and Stephen B. Sello

Chemical Processing of Fibers and Fabrics
FUNCTIONAL FINISHES Part B
edited by Menachem Lewin and Stephen B. Sello

Other volumes in preparation

Handbook of Fiber Science and Technology: Volume I
Chemical Processing of Fibers and Fabrics

FUNDAMENTALS AND PREPARATION

Part B

edited by

Menachem Lewin
*Israel Fiber Institute
and Hebrew University
Jerusalem, Israel*

Stephen B. Sello
*J. P. Stevens & Co., Inc.
New York, New York
and Greenville, South Carolina*

CRC Press
Taylor & Francis Group
Boca Raton London New York

CRC Press is an imprint of the
Taylor & Francis Group, an **informa** business
A TAYLOR & FRANCIS BOOK

First published 1984 by Taylor & Francis

Published 2019 by CRC Press
Taylor & Francis Group
6000 Broken Sound Parkway NW, Suite 300
Boca Raton, FL 33487-2742

First issued in paperback 2019

No claim to original U.S. Government works

ISBN-13: 978-0-367-45182-0 (pbk)
ISBN-13: 978-0-8247-7117-1 (hbk)

Visit the Taylor & Francis Web site at
http://www.taylorandfrancis.com

and the CRC Press Web site at
http://www.crcpress.com

Library of Congress Cataloging in Publication Data
(Revised for volume 1, part B)
Main entry under title:

Handbook of fiber science and technology.

(International fiber science and technology series ;
1-   )
Includes bibliographical references and indexes.
Contents: v. 1. Chemical processing of fibers and fabrics--fundamentals and preparation -- v. 2. Chemical processing of fibers and fabrics--functional finishes.
1. Textile finishing. I. Lewin, Menachem, [date].
II. Sello, Stephen B., [date]. III. Series: Inter-national fiber science and technology series ; 1, etc.
TS1510.H3   1983   677'.02825   83-7685
ISBN 0-8247-7010-2 (v. 1, pt. A)
ISBN 0-8247-7117-6 (v. 1, pt. B)

Publisher's Note
The publisher has gone to great lengths to ensure the quality of this reprint but points out that some imperfections in the original may be apparent.

ABOUT THE SERIES

When human life began on this earth *food* and *shelter* were the two most important necessities. Immediately thereafter, however, came *clothing*. The first materials used for it were fur, hide, skin, and leaves—all of them sheetlike, two-dimensional structures not too abundantly available and somewhat awkward to handle. It was then-quite a few thousand years ago—that a very important invention was made: to *manufacture* two-dimensional systems—fabrics—from simple mono-dimensional elements—fibers; it was the birth of textile industry based on fiber science and technology. Fibers were readily available everywhere; they came from animals (wool, hair, and silk) or from plants (cotton, flax, hemp, and reeds). Even though their chemical composition and mechanical properties were very different, yarns were made of the fibers by spinning and fabrics were produced from the yarns by weaving and knitting. An elaborate, widespread, and highly sophisticated art developed in the course of many centuries at locations all over the globe virtually independent from each other. The fibers had to be gained from their natural sources, purified and extracted, drawn out into yarns of uniform diameter and texture, and converted into textile goods of many kinds. It was all done by hand using rather simple and self-made equipment and it was all based on empirical craftsmanship using only the most necessary quantitative measurements. It was also performed with no knowledge of the chemical composition, let alone the molecular structure of the individual fibers. Yet by ingenuity, taste, and patience, myriads of products of breathtaking beauty, remarkable utility, and surprising durability were obtained in many cases. *This first era* started at the very beginning of civilization and extended into the twentieth century when steam-driven machinery invaded the mechanical operations and some empirical procedures—mercerization of cotton, moth-proofing of wool and loading of silk—started to introduce some chemistry into the processing.

The second phase in the utilization of materials for the preparation and production of fibers and textiles was ushered in by an accidental discovery which Christian Friedrich Schoenbein, chemistry professor at the University of Basel in Switzerland, made in 1846. He observed that cotton may be converted into a soluble and plastic substance by the action of a mixture of nitric and sulfuric acid; this substance or its solution was extruded into fine filaments by Hilaire de Chardonnet in 1884.

Organic chemistry, which was a highly developed scientific discipline by that time, gave the correct interpretation of this phenomenon: the action of the acids on cellulose—a natural fiber former—converted it into a *derivative*, in this case into a cellulose nitrate, which was soluble and, therefore, spinnable. The intriguing possibility of manipulating natural products (cellulose, proteins, chitin, and others) by chemical action and thereby rendering them soluble, resulted in additional efforts which led to the discovery and preparation of several cellulose esters, notably the cellulose xanthate and cellulose acetate. Early in the twentieth century each compound became the basis of a large industry: viscose rayon and acetate rayon. In each case special processes had to be designed for the conversion of these two compounds into a fiber, but once this was done, the entire mechanical technology of yarn and fabric production which had been developed for the natural fibers was available for the use of the new ones. In this manner new textile goods of remarkable quality were produced, ranging from very shear and beautiful dresses to tough and durable tire cords and transport belts. Fundamentally these materials were not truly "synthetic" because a known natural fiber former—cellulose or protein—was used as a base; the new products were "artificial" or "man-made." In the 1920s, when viscose and acetate rayon became important commercial items polymer science had started to emerge from its infancy and now provided the chance to make *new fiber formers* directly by the polymerization of the respective monomers. Fibers made out of these polymers would therefore be "truly synthetic" and represent additional, extremely numerous ways to arrive at new textile goods. Now started the *third era* of fiber science and technology. First the basic characteristics of a good synthetic fiber former had to be established. They were: ready spinnability from melt or solution; resistance against standard organic solvents, acids, and bases; high softening range (preferably above 220°C); and the capacity to be drawn into molecularly oriented fine filaments of high strength and great resilience. There exist literally many hundreds of polymers or copolymers which, to a certain extent, fulfill the above requirements. The first commercially successful class was the *polyamides*, simultaneously developed in the United States by W. H. Carothers of duPont and by Paul Schlack of I. G. Farben in Germany. The *nylons*, as they are called commercially, are still a very important class of textile fibers covering a remarkably wide range of properties and uses. They were soon (in the 1940s) followed by the *polyesters*, *polyacrylics*, and *polyvinyls*, and somewhat later (in the 1950s) there were added the *polyolefins* and *polyurethanes*. Naturally, the existence of so many fiber formers of different chemical composition initiated successful research on the molecular and supermolecular structure of these systems and on the dependence of the ultimate technical properties on such structures.

As time went on (in the 1960s), a large body of sound knowledge on structure-property relationships was accumulated. It permitted embarkation on the reverse approach: "tell me what properties you want and I shall *tailor-make* you the fiber former." Many different techniques exist for the "tailor-making": graft and block copolymers, surface treatments, polyblends, two-component fiber spinning, and cross-section modification. The systematic use of this "macromolecular engineering" has led to a very large number of *specialty fibers* in each of the main classes; in some cases they have properties which none of the prior materials—natural and "man-made"—had, such as high elasticity, heat setting, and moisture repellency. An important result was that the new fibers were not content to fit into the existing textile machinery, but they suggested and introduced substantial modifications and innovations such as modern high-speed spinning, weaving and knitting, and several new technologies of texturing and crimping fibers and yarns.

This third phase of fiber science and engineering is presently far from being complete, but already a *fourth era* has begun to make its appearance, namely in fibers for uses *outside* the domain of the classical textile industry. Such new applications involve fibers for the reinforcement of thermoplastics and duroplastics to be used in the construction of spacecraft, airplanes, buses, trucks, cars, boats, and buildings; optical fibers for light telephony; and fibrous materials for a large array of applications in medicine and hygiene. This phase is still in its infancy but offers many opportunities to create entirely new polymer systems adapted by their structure to the novel applications outside the textile fields.

This series on fiber science and technology intends to present, review, and summarize the present state in this vast area of human activities and give a balanced picture of it. The emphasis will have to be properly distributed on synthesis, characterization, structure, properties, and applications.

It is hoped that this series will serve the scientific and technical community by presenting a new source of organized information, by focusing attention to the various aspects of the fascinating field of fiber science and technology, and by facilitating interaction and mutual fertilization between this field and other disciplines, thus paving the way to new creative developments.

Herman F. Mark

INTRODUCTION TO THE HANDBOOK

The Handbook of Fiber Science and Technology is composed of five volumes: chemical processing of fibers and fabrics; fiber chemistry; specialty fibers; physics and mechanics of fibers and fiber assemblies; and fiber structure. It summarizes distinct parts of the body of knowledge in a vast field of human endeavor, and brings a coherent picture of developments, particularly in the last three decades.

It is mainly during these three decades that the development of polymer science took place and opened the way to the understanding of the fiber structure, which in turn enabled the creation of a variety of fibers from natural and artificial polymeric molecules. During this period far-reaching changes in chemical processing of fabrics and fibers were developed and new processes for fabric preparation as well as for functional finishing were invented, designed, and introduced. Light was thrown on the complex nature of fiber assemblies and their dependence on the original properties of the individual fibers. The better understanding of the behavior of these assemblies enabled spectacular developments in the field of nonwovens and felts. Lately, a new array of sophisticated specialty fibers, sometimes tailor-made to specific end-uses, has emerged and is ever-expanding into the area of high technology.

The handbook is necessarily limited to the above areas. It will not deal with conventional textile processing, such as spinning, weaving, knitting, and production of nonwovens. These fields of technology are vast, diversified, and highly innovative and deserve a specialized treatment. The same applies to dyeing, which will be treated in separate volumes. The handbook is designed to create an understanding of the fundamentals, principles, mechanisms, and processes involved in the field of fiber science and technology; its objective is not to provide all detailed procedures on the formation, processing, and modification of the various fibers and fabrics.

Menachem Lewin

INTRODUCTION TO VOLUMES I AND II

Textiles have undergone wet chemical processing since time immemorial. Human ingenuity and imagination, craftmanship and resourcefulness are evident in textile products throughout the ages; we are to this day awed by the beauty and sophistication of textiles sometimes found in archaeological excavations.

The objectives of the chemical processing, while basically unchanged over the centuries, have in recent times been diversified and expanded. Comfort and esthetics, durability and functionality, safety from fire and health hazards, easy care performance, such as washability, soil release, water and oil repellency, and stability against biological attack are examples of the objectives of chemical treatments of fibers and fabrics. Before these treatments can be applied, the textile materials have to be prepared by appropriate chemical procedures such as sizing, desizing, scouring, bleaching, and mercerization.

The array of fibers used at present is highly diversified. The advent of polyester, nylon, acrylic, and polyolefin fibers in recent years has greatly increased the complexity of the treatments as well as the range of the chemicals used. It became clear that approaches such as those practiced until 3 decades ago cannot continue to serve the solution to the wide range of problems facing chemists and technologists in the industry today. This realization coincided with rapid developments in polymer science and technology and brought about a surge in research and development activities in textile chemistry.

The studies carried out in the last 3 decades yielded a staggering amount of new data and not only a better understanding of the fibers and fiber assemblies and of the chemical interactions and structural changes, but also a large number of innovative ideas were created and put forward. Many of these ideas were developed into new processes, machines, and instruments, and culminated in a remarkable reshaping of the textile industry.

In these books an attempt is made to review and summarize the most important developments in this field. The emphasis is placed on the chemical aspects of the problems discussed. While technological aspects as well as industrial applications of the processes are being dealt with, only a brief treatment is given to factory layouts and to the machinery used.

Chemical Processing of Fibers and Fabrics is divided into two major areas. The first area, the fundamentals underlying the chemical treatments of fibers and fabrics and the preparation processes, are presented in Vol. I, Parts A and B. The second area, the functional finishes of textiles, are discussed in Vol. II, Parts A and B.

The need for a new comprehensive book in the field of chemical processing of fibers and fabrics has been felt for a long time. The vast amount of information accumulated in recent years in this field necessitated the preparation of the present books. They are intended for scientists and technologists both in the field of textiles and polymers as well as for students and researchers in other fields of human endeavor.

It is hoped that these books will not only further the knowledge and understanding of the complex field of textile chemistry, but will also bring about an interaction between people dealing in this field and people of other disciplines and will trigger off new and innovative developments for the benefit of all humanity.

Menachem Lewin
Stephen B. Sello

PREFACE

This is the second of two parts on fundamentals and preparation. It reviews the chemistry and technology of the various phases of preparation in detail.

Warp sizing--especially of cellulosics with starch--has been carried out in the textile industry for centuries. This field is still of great interest to scientists and technologists and in recent years many efforts have been made to replace the empirical approaches previously used by techniques developed after systematic scientific studies. A relationship is being established between the characteristics of films cast from sizing polymers, the properties of sizing warp yarns, and weaving efficiency. The introduction of new fibers, fiber blends, and modern high speed looms made it necessary to develop new sizing polymers and slashing technologies. Desizing of textiles, including modern size recovery systems, is also discussed.

Subsequent chapters review the bleaching of cellulosic, wool and synthetic textiles and the application of fluorescent whitening agents. It is an objective of this book to lead the reader to a better understanding of the mechanism underlying the preparation procedures and to discuss the important recent developments in this field.

The editors wish to thank the editorial advisory board of the International Fiber Science and Technology Series, the contributors, and the editorial staff of Marcel Dekker for their cooperation and their contributions to this book.

Menachem Lewin
Stephen B. Sello

CONTRIBUTORS

Peter G. Drexler Research and Development Division, Chem-Mark, Inc., Middlesex, New Jersey

Raphael Levene Israel Fiber Institute, Jerusalem, Israel

Menachem Lewin Israel Fiber Institute, and School of Applied Science and Technology, Hebrew University, Jerusalem, Israel

Giuliana C. Tesoro* Department of Mechanical Engineering, Massachusetts Institute of Technology, Cambridge, Massachusetts

---

*Present affiliation: Department of Chemistry, Polytechnic Institute of New York, Brooklyn, New York

CONTENTS

CONTENTS OF OTHER VOLUMES

Handbook of Fiber Science and Technology: Volume I
Chemical Processing of Fibers and Fabrics

FUNDAMENTALS
AND PREPARATION

Part B

1

MATERIALS AND PROCESSES
FOR TEXTILE WARP SIZING

PETER G. DREXLER / Chem-Mark, Inc., Middlesex, New Jersey

GIULIANA C. TESORO* / Department of Mechanical Engineering,
Massachusetts Institute of Technology, Cambridge, Massachusetts

---

*Present affiliation: Department of Chemistry, Polytechnic Institute of
New York, Brooklyn, New York.

1. INTRODUCTION AND GENERAL CONSIDERATIONS

1.1 Purpose of Sizing

The primary purpose of sizing—an important step in preparation for weaving known as warp slashing—is to attain optimum weaving efficiency. The process requires selection of sizing materials that are appropriate for the specific warp yarns, of formulations, and of processing conditions which are suitable for the equipment used.

The evolution of sizing technology in the textile industry has a long history. When mechanical weaving gained momentum in the second half of the nineteenth century, attention was focused on breakage of warp yarns made from natural fibers (silk, cotton, and wool), which at the time were used without protective lubricants. In later years, soap solutions, gelatine, tallow, linseed oil, and starch dispersion were applied to the yarns to minimize damage by abrasion, and about sixty years ago the processes were investigated in some depth at the Shirley Institute in Great Britain. The flow properties of the applied materials were examined, and the fundamental principles of rheological behavior were recognized, although the term "rheology" had not yet been coined [1-5].

The current concepts of sizing are thus based on half a century of pragmatic technological advances in the mills, coupled with new knowledge and developments in the fields of polymer science, fiber science, and chemical manufacturing.

A warp size may be defined as a film-forming polymeric material which is applied to a sheet of yarns for the purpose of protecting it during the weaving process, in which warp yarns are subjected to high abrasion (e.g., at the sites of drop wires, split rods, heddles, reed, shuttle). In the case of staple yarns like cotton, the size partially encapsulates the yarn and shields defects such as weak spots, knots, or crossed ends which occur in normal yarn production. The size also envelops protruding fibrils, lessening friction. When filament yarns like acetate or nylon are sized, the filaments are aligned to prevent chafing. The degree of penetration of the yarn by the size depends on the system, but in all instances, a protective coating is formed on yarn surfaces.

The size is generally applied from solution to a sheet of yarns which travels at a speed of 45-300 m/min and is then dried to very low moisture content on heated rolls. After drying, the yarns in the warp sheet are separated ("split") by an assortment of bars. No yarn breakage occurs if the adhesive strength of the size is sufficiently high, and no shedding of size if essentially complete encapsulation of the yarn is attained.

The size formulation is generally prepared and kept for several hours at elevated temperature, and it should be stable under these conditions. In the weaving room, humidities are usually high, and the dried film or protective coating of size should not become tacky. After weaving, easy removal of size (desizing) and freedom from residual deposits in the fabric are important requirements.

Many sizes are applicable to yarns made from a broad spectrum of fibers, both natural and synthetic. However, no universal size which can exhibit all the desired properties on all substrates is available, and each sizing system has specific shortcomings or limitations. The requirements for sizing yarns made from natural and from synthetic fibers are different, and only a few polymeric systems can bridge this gap. Furthermore, for each sizing system, consideration must be given to environmental effects of materials released into public streams, and the need for completely biodegradable sizes is considered acute at this time. On the other hand, energy constraints suggest that biodegradability of sizing materials may not be the ultimate answer and that recoverability and recycling of sizes may best serve the commercial interests of the textile manufacturing industries in the future.

1.2 Critical Requirements of the Size

With a focus on the film-forming polymer which is the dominant constituent of all size formulations, a discussion of the critical requirements of sizing materials should distinguish several categories of properties that are determined by different system parameters and that have important effects on yarn performance in the sizing and desizing processes. These may be described, for example, as follows:

1. Properties that depend primarily on the molecular structure of the film-forming polymer, namely, solubility/dispersibility,

tensile strength, elongation, abrasion resistance, and moisture response.

2. Properties that determine the behavior of sized yarn, namely, adhesion, flexibility, lubricity, transparency, and mildew resistance.

3. Properties related to ease of processing and to conditions of application, namely, compatibility (in the formulation), lack of corrosiveness of solutions, viscosity and rheology of solutions, penetration, foaming, skinning of solutions, fiber lay, uniformity, and ease of removal.

Even with such distinctions, it is difficult to discuss clearly this complex array of principles and pragmatic definitions of performance. Some of the most important requirements of sizing formulations for optimal warp preparation and weaving efficiency are briefly reviewed below as background for the overview which follows.

Adhesion

The adhesion of the size to the yarn substrate is critically important. Yarn surfaces vary in structure and geometry, depending on the fiber composition. For example, natural fibers are polar and hydrophilic, with numerous hydrogen bonding sites, and have a rough surface. Man-made fibers, on the other hand, generally exhibit lower polarity and hydrophilicity and smoother surfaces. A sizing formulation must wet fiber surfaces thoroughly within the yarn, and its ability to do so depends, in part, on the surface tension of the applied liquid. Thus, the values of surface tension for aqueous size compositions containing different polymers have considerable significance. The role of wetting in the adhesion of polymer films to solid surfaces has been discussed in a review of the cohesive and adhesive strength of polymers [6]. Swelling of the fibers by the applied size formulation is an additional consideration in this context. Natural fibers swell when in contact with aqueous size solutions, and this behavior leads to better adhesion than in the case of nonswelling (hydrophobic) synthetic fibers. Sizes based on highly polar systems, such as those derived from acrylates, vinyl acetate, or starch, tend to provide strong adhesive bonding to natural fibers and weaker, although adequate, adhesion on synthetics.

Tensile Strength

The tensile strength of a size film depends primarily on the molecular and supramolecular properties of the film-forming polymer. The chemical structure and molecular weight of the polymer, its orientation and crystallinity affect the tensile strength of the film. However, as a practical matter, the strength of the size film is lower than the value theoretically predicted from polymer structure, in part because of

stresses and anisotropy developed during processing of sized yarns.
For example, overcuring of size films on drying cans can cause significant tensile loss; recovered size polymers may yield films of lower
tensile strength than virgin polymers because of contamination by
foreign material in the course of the recycling process.

Hardness

The hardness of a size may be defined as resistance to localized de-
formations. It is a complex property which depends primarily on the
structure, molecular weight, and elastic modulus of the material. The
wide range of hardness in the sizing systems used makes it imprac-
tical to classify component polymers according to hardness, but as a
first approximation, film hardness may be assumed to decrease in the
order acrylic > styrene copolymers > polyvinylacetate > polyvinyl alcohol (PVA) > carboxymethyl cellulose (CMC) > starch > polyester
dispersions > polyurethane. It must be emphasized that film hardness is not a valid criterion for predicting performance of the size
but can only provide a pragmatic indication of other properties, such
as flexibility of sized yarns and adhesion.

Flexibility

A good size film must be supple and bend with the yarn with minimal changes of energy levels at stress points at the size/yarn inter-
face. This condition is approached when the stress-relaxation behavior is similar for size and yarn, for example, in the case of sizes
made from polyester dispersions, applied to polyester fibers. Differences in stress-relaxation behavior are greater for polyester dispersions on polyester-cotton blends, or on regenerated cellulose,
acrylics, wool, glass, or olefin fibers.

Starch, starch derivatives, and CMC have highly compatible ex-
pansion/contraction coefficient values with cotton, rayon, or acetate
fibers, but the stress values of starch films differ from those of polyamide fibers, acrylics, and olefin fibers.

Polyvinyl alcohol sizes show a behavior similar to that of starch,
starch derivatives, and CMC sizes. Acrylate size films exhibit stress-
relaxation behavior similar to that of polyamide fibers, and to a lesser degree, of polyesters. It should be noted that the flexibility of
commercially useful sizes should remain adequate over a wide range
of relative humidities (55-85% RH), and should not change appreciably
as a consequence of aging.

Lubricity

In a size film, lubricity is essential in order to minimize the effects
of friction and wear between surfaces (yarn to yarn, and yarn to
processing equipment).

Many polymers used in sizing (e.g., starches, CMC, PVA) do not yield films of adequate lubricity, and sizes formulated with these polymers require the addition of a lubricant in the formulation. Suitable additives have been identified largely on the basis of their utility in other textile processes, for example, modified vegetable oils (used in fiber processing), mineral oil systems (used in texturizing), and synthetic lubricants (used as finishes and in knitting). The latter are generally nonpolar compounds containing hydrocarbon chains, for example, polyethylene waxes, hydrogenated fatty acid derivatives, and paraffins.

The lubricant, being hydrophobic, orients itself on the surface of the size film during the drying process. A good lubricant should not be appreciably soluble in the fiber and should not adversely affect the strength, hardness, and adhesion/cohesion properties of the size film. In addition to the use of lubricant additives, it is also possible to improve lubricity of size films by chemical modification of the film-forming polymer, but this internal lubrication approach is more complex, because it requires consideration of concomitant changes in the film-forming properties of the polymer.

Moisture Response and Elongation

The response of the film to changes in the relative humidity of the environment, and in the moisture content, are of critical importance in processing, and elongation of the film should thus be evaluated over a broad range of relative humidities. When the warp leaves the drying cans, the moisture content is very low and the film may become brittle; as moisture is regained in the weaving room, flexibility and elongation increase, while tensile strength may decrease significantly as absorbed water plasticizes the film. The effects of humidity and moisture content on film properties depend on the specific polymer used; starch and acrylic sizes are particularly sensitive to moisture, and appropriate adjustments in the size formulation must be made in order to compensate for changes in tensile strength and elongation which result from variations in the humidity of the environment.

The phenomenon of "wet tack" is related to the moisture response of the size film. Tackiness is observed when the film adheres to itself in preference to the substrate on application of a light pressure. Tackiness is frequently caused by increased moisture content ("wet tack"); it is detrimental to adhesion and it may render the size unacceptable.

Abrasion Resistance

This property of sized yarns may be defined as the ability to withstand progressive removal of polymer and yarn particles from the surface by metallic machine parts and by adjacent yarns throughout the weaving process. Abrasion in the repetitive mechanical action of the loom has a serious detrimental effect on weaving. Some advantage in abrasion resistance may be derived from yarn construction: for example, a high-twist yarn exhibits better abrasion resistance and requires a lower percentage of size than a low-twist yarn [7].

Under most conditions, abrasion of a polymeric size film proceeds by several simultaneous wear mechanisms, and it is difficult to present a satisfactory theory of wear in brief. According to Glaeser [8], under comparable conditions of surface nonhomogeneity and surface temperature, the wear rate of polymeric compositions is directly proportional to the specific tensile strength of the material. Empirical correlations of abrasion loss to surface roughness, to work of abrasion, and to tensile strength have also been proposed. The following generalizations regarding abrasion resistance have been supported by practical experience and experimental observations.

Polyurethane films have excellent abrasion resistance, and the behavior has been attributed to the presence of polar groups that enhance hydrogen bonding along the polymer chain. High abrasion resistance in styrene-maleic acid size compositions has been attributed to the regularity of the macromolecule. In acrylic copolymers, the addition of acrylonitrile or acrylamide in low concentration as comonomer has been reported to yield improvements in abrasion resistance, the magnitude depending on the specific system. More generally, for starch, PVA, CMC, and polyester dispersion sizes, abrasion resistance has been reported to depend on the energy the films can absorb without breaking when subjected to stress.

The properties of importance in determining the abrasion resistance of size films are elastic modulus, breaking elongation (dry and wet), torsional rigidity, shear tenacity, elastic recovery (instantaneous and delayed), and residual energy to rupture.

It may be possible to evaluate and rank different size films by measuring these properties in systematic experiments. However, data on films prepared under controlled conditions from each important group of sizing materials and formulations are nòt available.

Viscosity and Penetration

The viscosity of the size formulation applied determines the uptake of solution by the yarn, and thus the amount of size available for film formation and fiber/yarn encapsulation on drying. Under dynamic conditions of processing (warp speeds of 70–300 m/min), typical aqueous size solutions (75-95°C; 9-11% solids) have a viscosity in the range of 50-370 CPS. For most substrates, this results in a solution uptake (pick-up) by the yarn of about 100-115%, and thus a size pickup of about 9-12% based on yarn weight. If the viscosity is too high, penetration is low and excessive amounts of size are deposited on yarn surfaces, resulting in poor efficiency and in shedding of the size. Lower viscosity, on the other hand, increases penetration. The viscosity of the size solution should not be too sensitive to temperature changes, because exact control of the solution temperature is difficult and the distribution of size within the yarn is viscosity dependent. Penetration of the yarn by the size is critically important. In the case of spun yarns, strength is dependent on twist, and optimal effect of sizing is obtained when good penetration of the size into the yarn allows retention of the twist. This does not necessarily entail penetration to the core of the yarn. For filament yarns, interfiber bonding by the size does not play a role, the penetration requirements are different, and the protective function of the size film is based on other factors.

Requirements in Processing

Low Foam: The size liquor should not contain entrapped air because films formed by application of foaming solutions cause localized stress points on the warp around the micro bubbles, shedding of the size, and impaired adhesion. If agitation leads to foaming, defoamers may be added.

Compatibility: Auxiliary chemicals such as emulsifiers, lubricants, defoamers, surfactants and softeners, are frequently included in the size formulation. A high level of compatibility of the size components is essential because nonhomogeneities severely impair the cohesive and adhesive strength of the size film formed.

Skinning Tendency: This phenomenon is an indication of excessive moisture loss from the surface of a liquid size system prior to application. Skinning is most severe in starch compositions, but it may occur in acrylates, PVA, and polyester dispersions as well. A balanced blend of humectants, a lower temperature in the size box, and constant agitation can overcome the problem.

Resolubility (Insolubilizing Effects): Resolubilization of dried size films may occur by scission of intermolecular bonds (hydrogen bonds) and removal from the substrate. Chemical and structural changes in the polymer during film formation and drying may alter the solubility and removability of the size. For example, overcuring of sizes during heat setting of fabric may cause a drastic reduction in solubility because of cross-linking, which can pose severe problems in desizing. The phenomenon has been observed especially with starch sizes and, to a lesser degree, with PVA, CMC, and acrylate polymers.

Noncorrosiveness: A size should not be corrosive to or attacked by the common metals of mill equipment and should preferably possess rust-inhibiting properties. Most size formulations are noncorrosive and are applied at a pH of 5.8-9.5. However, several good polyacrylic acid filament size systems are applied at lower pH (2.6-3.4), and thus require the addition of corrosion inhibitors.

Mildew Resistance: Many polymeric size components are natural nutrients for microorganisms. In order to avoid bacterial contamination and degradation over a prolonged period of time, the addition of bacteriocides (0.02-0.05%) is commonly practiced.

Among the critical requirements of the size that have been briefly outlined, adhesion to the substrate and resistance to abrasion may be pinpointed as the most important. However, selection and evaluation of materials, optimum processing conditions, and practical experience must interact in order to obtain satisfactory results in the mill.

2. SUBSTRATES

The selection of film-forming polymer and size formulation is dictated, in part, by the fiber content of the warp yarns. The fiber properties, yarn size, twist, and other parameters of yarn manufacture and design also play a significant role. A detailed discussion of fiber properties is beyond the scope of this review, and the reader is referred to textbooks and technical publications on the subject [9-11] for data and for detailed discussion. A brief summary of relevant properties for fibers of major commercial importance and volume is presented here as background for the review of sizing materials which follows.

2.1 Natural Fibers

Cotton

Cotton contains about 93-94% cellulose (see Fig. 1.1), the balance consisting of protein, pectin, waxes, and inorganic material. Cotton fiber (staple) varies in length from 1 to 5 cm, and has a convoluted surface. The fiber is hydrophilic and readily reaches its equilibrium moisture content. The specific gravity of the fiber under standard conditions of temperature and humidity is 1.54. At moderate temperatures, cotton is resistant to degradation by heat but is discolored on exposure to 120°C for prolonged periods, and it burns readily when ignited in air. The fiber swells without degradation in caustic soda solutions; it disintegrates in hot dilute or cold concentrated acids; it is resistant to organic solvents; and it is attacked by mildew.

The range of physical properties of cotton fibers (at 65% RH and 20°C) is as follows:

Tenacity, dry (g/denier)	3.0-4.9
Wet tenacity (% of dry)	100-110
Tensile strength (PSI × 10^3)	60-120
Elongation at break (%)	3-11

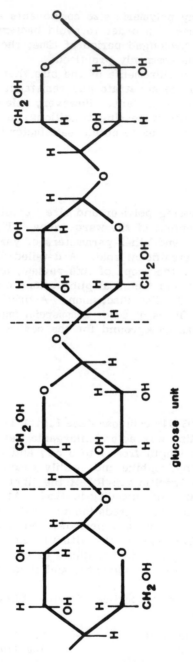

Figure 1.1 Structure of cellulose.

Elastic recovery (% recovery)	75-45
Stiffness (g/denier)	57-60
Moisture regain (%)	7.0-8.5

In the context of sizing cotton warp yarns, the hydrophilic properties of the fiber and its capability for hydrogen bonding are important. Hydrogen bonds may be formed between hydroxyl groups of the cellulose and polar groups of the polymer used for sizing. Water-soluble, film-forming polymers that form hydrogen bonds are thus particularly suitable for yarns containing 30% or more cotton fiber; these are starch and starch derivatives, gelatine, carboxymethyl cellulose (CMC), and fully hydrolyzed polyvinyl alcohol (PVA).

Wool

Wool is a protein staple fiber exhibiting high elongation and high elastic recovery. It has the highest moisture regain among natural fibers. It is not heat stable; it loses its strength when heated at 125°C. When wool burns, it leaves ash particles which appear as shiny brittle beads and emanates an odor similar to that of burning hair. Wool is destroyed by hot sulfuric acid, which cleaves the peptide chain hydrolytically. Strong alkalies attack the disulfide bonds in wool, and dilute alkalies cause significant strength loss. Wool has good resistance toward organic solvents and mildew. The specific gravity of the fiber is 1.32. The range of physical properties of wool fibers at 65% RH and 20°C is summarized below:

Tenacity, dry (g/denier)	1.0-2.0
Wet tenacity (% of dry)	78-90
Tensile strength (PSI $\times 10^3$)	17-29
Elongation at break (%)	20-40
Elastic recovery (% recovery)	99-65
Stiffness (g/denier)	3-9
Moisture regain (%)	11-17

When wool is treated with a hot aqueous size solution, it absorbs water and swells, its volume often increasing by about 10%. This effect requires that sizes used on wool exhibit a high degree of elasticity so that they may expand and coat the surface of the swollen wool yarn without subsequent shedding. Sizes based on natural gums and on starch derivatives have excellent adhesion to wool yarns, but synthetic polymers, including partially hydrolyzed polyvinyl alcohol, and acrylics may also be used.

Silk

Silk is a protein filament fiber consisting mainly of sericin and fibroin, the remainder being fatty matter and wax. Silk is readily dissolved in oxidizing agents, in zinc chloride, and in calcium chloride solutions.

The heat stability of silk is higher than that of wool (142°C), but the fiber decomposes at 176°C. Mild alkaline solutions do not attack silk, which is also resistant to organic solvents. The specific gravity is 1.25.

Physical properties of silk filament fibers at 20°C, 65% RH are in the following range.

Tenacity, dry (g/denier)	3-5
Wet tenacity (% of dry)	80-85
Elongation at break (%)	25-35
Elastic recovery (% recovery)	76-52
Moisture regain (%)	∿11

Silk absorbs moisture readily, and it can take up to a third of its weight in water, with considerable swelling. Appropriate sizing materials for silk include alginates, and their mixtures with starch, natural gums, and partially hydrolyzed polyvinyl alcohol of low viscosity.

2.2 Man-Made Fibers

Regenerated Cellulose (Rayon)

Regenerated cellulose fibers made by different processes encompass a broad range of properties, but several features are common to rayon fibers generally. They are hydrophilic; they lose strength on heating to about 177°C, decompose at about 240°C, and burn readily when ignited in air, without melting and leaving little ash. Rayon is attacked by cold concentrated and by hot dilute acids. Its resistance to alkali depends on the specific fiber type. Resistance to organic solvents is excellent, and mildew resistance is moderate. Specific gravity ranges from 1.42 to 1.55, depending on fiber type. Some physical properties of several types of rayon filament fibers at 20°C and 65% RH are summarized below [11].

	Medium tenacity	High tenacity	High wet modulus
Tenacity, dry (g/denier)	0.7-3.2	3.0-5.0	2.5-5.0
Wet tenacity (% of dry)	44-72	44-72	68-75
Tensile strength (PSI × 10^3)	28-47	58-88	66
Elongation at break (%)	15-30	9-26	9-18
Elastic recovery (% recovery)	82-90	70-100	95
Stiffness (g/denier)	6-16.6	13-50	15-35
Moisture regain (%)	11.5-16.5	11.5-16.5	11-13

All rayons consist entirely of cellulose, and are highly hydrophilic, but structural differences between standard viscose rayon and polynosics suggest somewhat different requirements for sizing. This is illustrated by the data below [11].

	Viscose rayon	Polynosic rayon
Cross section	serrated	round, smooth
Microstructure	none	fibrillar
Degree of polymerization (DP)	270-300	500-550
Dry elongation (%)	17	9
Wet elongation (%)	22	12
Water imbibition (%)	100	66
Increase in diameter on wetting (%)	26	15
Recovery from stretch	poor	good

As a rule, size add-on can be much lower on the more dimensionally stable polynosics, but the type of size used is similar and includes gelatin (with lubricant added), starch, modified starches, carboxymethyl cellulose, and partially hydrolyzed polyvinyl alcohols.

Cellulose Acetates

Cellulose acetate is thermoplastic, melts around 230°C, and decomposes at 260°C, forming a charred hard ash. Secondary cellulose acetate fibers lose strength on prolonged exposure to sunlight. They are soluble in many organic solvents, including acetone and dioxane, and they swell in alcohols and in chlorinated solvents. Cellulose acetate is essentially hydrophobic and is resistant to mildew. It is attacked by acids and saponified by strong alkali. The specific gravity is 1.32. Properties of acetate filaments at 20°C and 65% RH are in the following range [10].

Tenacity, dry (g/denier)	1.2-1.5
Wet tenacity (g/denier)	0.8-1.2
Tensile strength (PSI × 10^3)	22-30
Elongation at break (%)	22-42
Elastic recovery (% recovery)	94-22
Moisture regain (%)	6.0

Comparison of the properties of secondary cellulose acetate with those of rayon clearly shows the effect of the esterification of the hydroxyl groups: conversion of a hydrophilic polymer to a hydrophobic

and thermoplastic macromolecule. This effect is even more significant in triacetate fibers, which do not contain residual hydroxyl
groups. Triacetate fibers (completely acetylated cellulose) have a
moisture regain of only 3.5%; they are soluble in methylene chloride-
alcohol mixtures; they have no sites for hydrogen bonding, and consequently, adhesion of size to triacetate fibers depends on other
factors. Sizing polymers for secondary acetate and for triacetate fibers include starch acetate esters, carboxymethyl cellulose in conjunction with acrylic binders, polyvinyl alcohol, styrene-maleic anhydride copolymers, and vinyl acetate copolymers. It is evident from
this list that the synthetic polymers have a major role in the sizing
of yarns made from acetate fibers.

Nylon (Aliphatic Polyamides)

The important commercial fibers in this class are nylon 6 (polycap-
rolactam) and nylon 66 (condensation polymer of adipic acid and hex-
amethylenediamine). The major difference between the two fibers is
the somewhat lower melting point of nylon 6 (215°C vs. 250°C for
nylon 66). The moisture regain is about 4.2% for both; resistance
to alkalis is excellent. Nylon fibers are decomposed by strong mineral acids but have good resistance to weak acids. After prolonged
exposure to ultraviolet light, nylons show strength losses of 23-46%.
The fibers are excellent insulators, but because of this property
they are also subject to accumulation of static electricity. Both types
of nylon are soluble in phenols and in concentrated formic acid. Specific gravity is 1.14. Physical properties at 20°C, 65% RH are in the
following range:

	Nylon 6			Nylon 66		
	Filament (regular)	Filament (high ten.)	Staple	Filament (regular)	Filament (high ten.)	Staple
Tenacity, dry g/denier	4.1-5.8	7.4-8.2	2.6-6.0	3.0-6.0	7.4-8.3	4.4-6.6
Wet tenacity (% of dry)	85-90	85-90	85-90	85-90	85-90	85-90
Tensile strength (PSI × 10³)	60-85	110-122	55-86	65-84	110-122	62-68
Elongation at break (%)	25-40	16-22	34-60	25-40	16-22	23-58
Elastic recovery (%)	100	—	—	100	100	—

Adhesion of sizes to nylon fibers is governed in part by hydrogen bonding. The extent of wetting and the strength of adhesive bonds obtained between polyamide fibers and acrylic sizes is excellent. The polymers that are predominantly used for sizing yarns made from polyamide filament and staple fibers are polyacrylic acid, with and without the addition of CMC and PVA binders, acrylic copolymers with modified starches, oxidized starches with wetting aids, and PVA (partially hydrolyzed, low molecular weight).

Polyester

Polyester is the leading synthetic fiber in volume. Its popularity may be attributed to the great versatility of the fiber, which can be manufactured in many variants to impart a wide range of properties tailored to specific end uses. Polyester fiber melts at 250°C, and burns slowly when ignited. It has good resistance to organic solvents, but it is dissolved by some compounds (e.g., *meta*-cresol, *ortho*-chlorophenol, trifluoroacetic acid). Polyester has a very low moisture regain and good dielectric properties. The specific gravity is 1.22-1.38.

Physical properties at 20°C and 65% RH are reported as follows:

	Filament (regular)	Filament (high tenacity)	Staple
Tenacity, dry (g/denier)	4.5-5.0	6.4-6.7	5.5-6.4
Wet tenacity (% of dry)	100	100	100
Tensile strength (PSI × 10³)	80-88	104-160	88-118
Elongation at break (%)	19-24	11-13	20-28
Elastic recovery (% recovery)	97-80	100-100	—
Moisture regain (%)	—	0.4-0.8	—

One limitation of polyester fibers is their hydrophobic character, resulting in low absorption and transport of water. In sizing, hydrophobicity causes difficulty with some aqueous size systems which do not uniformly wet fiber surfaces and do not adhere well to the substrate. This problem has been successfully solved with the development of sizes based on aqueous dispersions of polyester film formers. Other types of sizes are also used in applications where requirements for adhesive/cohesive strength are moderate (e.g., filament weaving). Sizes recommended for polyester yarns include polyester dispersions (sulfonated or polyglycol type); mixtures of polyester dispersions with PVA, CMC, or acrylates; acrylic copolymers in mixtures with modified starches; copolymers of ethyl vinyl ether with maleic anhydride; and some polyurethanes.

Olefin Fibers

The most important fibers in this class are polyethylene and polypropylene filaments (mono- and multifilaments) as well as staples. Olefin fibers generally have a high degree of crystallinity; they are chemically inert and have low specific gravity. In the case of syndiotactic polypropylene, commercially the most important olefin fiber, the specific gravity is 0.90, and average physical properties (20°C and 65% RH) may be summarized as follows:

Tenacity, dry (g/denier)	6.5-9.0
Wet tenacity (% of dry)	100
Tensile strength (10^3PSI)	36-80
Elongation at break (%)	18-60
Elastic recovery (% recovery)	100-90
Moisture regain (%)	0.1

Olefin fibers have excellent extensibility and tenacity. They are hydrophobic, and because of their very low moisture regain, they tend to accumulate static charges. The fibers have generally excellent chemical resistance, but they dissolve in hot tetralin or decalin and are attacked by strong oxidizing agents. Olefin fibers have good resistance to abrasion. They do not require high size add-ons, but very few sizing systems are considered satisfactory for olefin yarns; specifically Polyox resins (high-molecular-weight polyethylene oxides) and copolymers of vinyl acetate have been recommended.

Acrylics

Acrylic fibers are polymers of acrylonitrile, with minor amounts of comonomers added to enhance specific properties (e.g., dyeability). Acrylic fibers melt at 255-275°C. When ignited, they burn with a black beadlike ash residue. They are insoluble and do not swell in most common solvents but are dissolved by strong hydrogen bond-breaking solvents such as dimethylformamide and dimethyl sulfone. Acrylic fibers are destroyed by boiling alkaline solutions. Specific gravity is 1.17-1.19, and physical properties (20°C and 65% RH) are in the following range.

Tenacity, dry (g/denier)	2.0-4.0
Wet tenacity (% of dry)	80-100
Tensile strength (10^3PSI)	30-62
Elongation at break (%)	24-50
Elastic recovery (% recovery)	92-99
Moisture regain (%)	1.0-2.4

The abrasion resistance of acrylic fibers is inferior to that of nylon, but better than that of wool or acetate. Many different size compositions have been recommended for acrylics, including starch derivatives in combination with PVA or CMC or polyacrylic acid, PVA and styrene/maleic anhydride copolymers.

Modacrylics

These fibers are copolymers of acrylonitrile (35-85%) with significant amounts of halogenated comonomers (e.g., vinyl chloride). Modacrylics have excellent resistance to sunlight, acids, alkalies, and biological attack. They are unaffected by most solvents, but dissolve in hot acetone. Modacrylic fibers are flame resistant and do not drip when exposed to elevated temperature. Modacrylics have good dimensional stability in hot water below the boil. Specific gravity is 1.28-1.38, depending on the comonomer. A summary of physical properties (20°C and 65% RH) shows:

Tenacity, dry (g/denier)	2.4-3.1
Wet tenacity (% of dry)	100
Tensile strength (10^3PSI)	43-50
Elongation at break (%)	32-41
Elastic recovery (% recovery)	78-97
Moisture regain (%)	0.4-4.4

Size systems recommended for modacrylics are similar to those used on acrylic fibers.

Fiberglas

Glass fiber is manufactured in staple and in filament form. From the point of view of sizing, only the latter is important. Filament glass glows but does not burn. Heating to about 320°C decreases fiber strength, and softening begins at 730°C. Glass fibers have outstanding chemical resistance and very low moisture content. Glass fibers have remarkably high tensile strength, attributed to the completely amorphous nature of the material and to the absence of stresses caused by crystallite imperfections. Specific gravity is high (2.5-2.7) and other properties are, typically (20°C, 65% RH):

Tenacity, dry (g/denier)	6.2-6.9
Wet tenacity (% of dry)	70-100
Tensile strength (10^3PSI)	200-575
Elongation at break (%)	2-4
Elastic recovery (% recovery)	100
Moisture regain (%)	0

Resistance to abrasion is very low for glass fibers, and as filaments abrade each other, a very rough surface is produced. Sizing agents applied to glass yarns in weaving are completely different from the size - lubricant coatings applied in the course of glass fiber formation. In the production of continuous glass filaments, the molten glass continuously flows through a plurality of orifices while concurrently attenuated and wound in package form at speeds of 2500-3500 m/min. During this process, a chemical coating composition must be applied to encapsulate the forming filaments, because of the low abrasion resistance and limited flexibility of glass fibers. These "forming sizes" are generally complex mixtures containing organofunctional silane coupling agents (e.g., [12]). They are documented in a voluminous patent literature that is beyond the scope of the present review. However, they are important in the context of this discussion because they leave a residual layer on the glass filament, and, unless they are decomposed by a coronizing process, these residues have a significant effect on the requirements for the sizing system in subsequent weaving [13-16]. The size polymers suggested for glass filament weaving are starch ethers (with low ash contents), acrylic resin with epoxidized soya oil, dextrins in mixtures with PVA, and polyester dispersions.

Carbon Fibers

Carbon and graphite fibers are advanced materials which even more than glass, present special problems in processing and weaving. Commercially important carbon fibers are manufactured from several types of precursors. Those made from polyacrylonitrile predominate at this time; and illustrative manufacturing processes are based on highly oriented polyacrylonitrile fibers that can yield carbon fibers of high modulus and excellent tensile strength (e.g., [17]).

Carbon/graphite filaments find application in end uses, where their high corrosion and temperature resistance, low density, chemical inertness, high tensile strength, and high modulus of elasticity justify the high cost. Among the most important applications are those in composites for aerospace structural components, deep submergible vehicles, rocket motor casings, and ablative materials for heat shields on space reentry vehicles.

The property ranges for commercial carbon fibers are difficult to summarize because of continually advancing technology in carbon fiber manufacture. Although some of the current applications require weaving, and thus the use of sizing materials, the properties and the microstructure of the fibers pose unique problems. Wetting and penetration of carbon/graphite yarns by conventional sizing formulations is not possible, and the special materials developed for use on carbon fiber substrates are a closely guarded secret.

Elastomeric Fibers

Elastomeric fibers are characterized by an extension at break in excess of 180-200% and a rapid recovery of 94-99% upon release of tension. They include nylons (nylon 610) and, more importantly, polyurethane elastomeric fibers [18-20]. The latter fibers, designated as spandex, are composed of at least 85% of a segmented polyurethane in which high-melting, crystalline "hard" segments are present in a low-melting, amorphous "soft" segment matrix. The hard segments are commonly formed from an aromatic diisocyanate and a diamine, linked by urethane groups to polyglycol soft segments.

Spandex elastomeric fibers burn with melting, leaving fluffy black ash particles. They deteriorate at temperatures above about 150°C, and melt at 230-234°C. They are resistant to dilute alkali, attacked by acid, soluble in some organic solvents (e.g., dimethyl formamide). Specific gravity is 1.20-1.22. Some typical properties of spandex fibers at 20°C and 65% RH are:

Tenacity, dry (g/denier)	0.74-0.92
Wet tenacity (% of dry)	100
Tensile strength (10^3PSI)	9-12
Elongation at break (%)	430-720
Elastic recovery (% recovery)	93.5-96
Moisture regain (%)	0.75-1.30

Due to the elastic nature of the fiber, no truly film-forming polymer sizes can be applied to yarns made from spandex fibers. Some materials used to reduce interfiber friction are talc, some spin finishes, and gum-arabic solutions.

Miscellaneous Fibers

Polyvinyl fibers are specialty items (e.g., Vinyon, American Viscose) used in small quantities, and usually blended with other fibers. Fabrics woven from specialty fibers also include those made from asbestos, teflon, steel, alginate, casein fibers, etc., for special purposes. These require approaches to sizing which must be tailor-made.

Substrate Considerations

The above summary of fiber properties has focused on the fibers of major interest in the industry so as to allow appreciation of the materials in the system. The importance of blends should be pointed out. When blends are used, the selection of sizing formulation must be guided by the particular composition, with awareness of difference in properties (e.g., moisture response) which exist between the component fibers, and of the special problems posed in each case.

Differences between spun fiber yarns and filament yarns, which have been briefly touched on the introductory section, cannot be overemphasized. Because of differences in fiber surface area and interfiber geometry, penetration and distribution of the size are vastly different, and both requirements and effectiveness of the protective size film are affected.

3. POLYMERS FOR SIZING

Proper selection of film-forming polymers as sizes can be a demand-
ing process. A large number of synthetic water-soluble or water-
dispersible polymeric materials compete with natural gums, alginates,
animal glues, gelatins, and starches in the natural and modified forms
for this application. Combinations of natural and synthetic polymers
in the size recipes of the trade seem almost endless [21,22]. The
most important materials from the viewpoint of overall physicochemical
properties and economics may be listed as the following: starches
and chemically modified starches, carboxymethyl cellulose (CMC),
acrylic polymers, polyvinyl acetate, polyvinyl alcohol, polyester dispersions, styrene copolymers, and specialized compositions developed
for hot-melt sizing and for application from organic solvents. These
are reviewed in turn in this section.

3.1 Starch and Starch Derivatives

Chemically, starch is defined as a naturally occurring polysaccharide
(poly-α-glucopyranose). Starch is synthesized in plants, and may
be obtained from many different plants, but only a few species yield
useful commercial quantities of the product. These are corn, wheat,
rice, potato, tapioca, sago, and sorghum grain. Corn is by far the
most important source of starch in the United States: close to 150
million tons of corn are harvested yearly in the United States, and
about 6-7% of this amount is used for the production of adhesives,
binders, coatings, sizes, and food intermediates.
     The textile industry uses starch in many forms for the warp
sizing of yarns made from natural and synthetic fibers and from
blends. In this application, starches are seldom used in unmodified
form, more frequently as derivatives identified as "thick-boiling,"
"thin-boiling," oxidized, acetylated, hydroxyethyl, cationic, high
amylose, etc., to cite but a few, which are water-soluble, film-
forming polymers meeting the requirements of sizing compositions.
There are many diverse formulations based on starch derivatives.
For natural fibers, conventional cooking of pearl, compacted pellet,
or grit starches is still used, with modifications depending on the
requirements and on the type of equipment available in the chemical
manufacturers' plants. Specific formulations are determined by many
factors including yarn count, number of warp ends sized, type of
loom used in weaving, and construction of the fabric being woven.
Cooked starches are usually formulated with softeners, lubricants,
and other additives such as humectants, defoamers, antistatic agents,
etc. Since starches in unmodified form are not effective as sizes for
yarns made from synthetic fibers, numerous derivatives of starch
have been developed for several systems. Among the advantages of
starches and derivatives are compatibility with many chemicals, easy

removal from finished fabric by conventional desizing procedures and the absence of serious problems of biodegradability.

A schematic of starch conversion processes related to products used in sizing is shown in Fig. 1.2.

Chemical modification of starch to introduce specific substituents and/or functional groups involves more complex chemical reactions and processes. Some examples of the effects of chemical modification on the properties and effectiveness of starch in sizing formulations are discussed below.

Conversion of a small percentage of the hydroxyl groups in starch to hydroxyethyl ether groups ($-OH \rightarrow -OCH_2CH_2OH$) yields a polymer that is claimed to be especially suitable for sizing of acrylic, modacrylic, and cotton blend yarns [23], with a significant improvement in viscosity stability over a broad range of temperatures, a decrease in the temperature of gelatinization, and enhanced film-forming properties as compared to unmodified starch.

Dialdehyde starch is reported to yield exceptionally strong films [24]. The derivative is prepared by utilizing the selectivity of periodate as an oxidizing agent for 1,2-glycols as shown in the equation below.

Recovery of the periodate by electrolytic techniques [25,26] makes the process economically feasible.

Two distinct sizing polymers, namely amylose and amylopectin, may be obtained by fractionation of starch. Amylose (22-26%) consists of D-glucopyranose units (average DP, 800-1000), linked primarily by $\alpha(1-4)$ bonds. The amylopectin fraction (74-78%) consists of D-glucopyranose units (DP, 6000-9000), with highly branched 1-6 linkages [27]. The linearity of the polymer in the amylose fraction contributes to the excellent film strength obtained, whereas the branched structure of the amylopectin yields nongelling polymer solutions of stable viscosity at high concentrations. The amylose and amylopectin fractions can be further modified by reaction with ethylene oxide. The hydroxyethyl amylose derivative obtained is used for sizing of continuous glass filament warps.

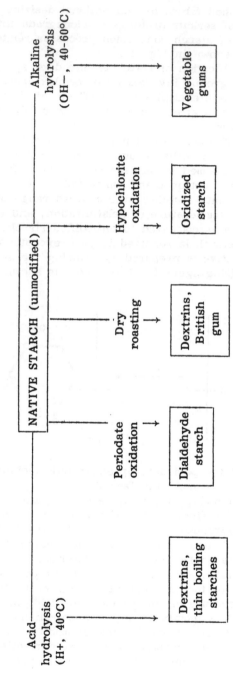

Figure 1.2 Schematic of starch conversion processes.

Table 1.1 Effect of Heating Corn Starch with Gluten

	Heating time (hr)				
	0	1	2	4	6
Solubility (%, 40°C)	Negligible	9.5	41.5	89.5	42.5
Reducing value	0.33	6.3	7.6	20.7	35.0
Intrinsic viscosity	1.08	0.22	0.18	0.08	0.11
β-Amylolysis (%)	62.0	—	51.	24.	5.

Source: Ref. 30.

The amylopectin fraction may be oxidized, or esterified, to lower the viscosity of solutions, yielding polymers that are particularly effective for sizing of fine count cotton and viscose yarns. The lower viscosity allows a significant reduction in the size add-on as compared with an unmodified amylopectin size [28]. Complex derivatives of high amylose starches have been claimed to be particularly suitable for the sizing of yarns made from synthetic fibers and their blends. These derivatives are reportedly obtained by sequential reactions of the starch with an alkylene oxide, with 2-chloroethyl diethylamine hydrochloride, and with propane sultone [29]. Clearly, the derivatives are complex mixtures that defy accurate description by structural formulae. Other approaches to modification of starch have been described to obtain increased water solubility, reduce the temperature of gelatination, and lower the viscosity of solutions. For example, starch has been modified by addition of gluten [30]. The gluten, which contains 60-65% protein, inhibits the dextrinization of the starch. When corn starch and corn gluten are heated together (in the proportion of 95:5) for 6 hr with stirring (samples withdrawn at hourly intervals), the properties are as shown in Table 1.1.

These results indicate that, while solubility reaches a maximum after 4 hr, the reducing power continues to increase as heating is continued. The intrinsic viscosity and β-amylolysis values decrease considerably more than in the absence of gluten.

Several other amino acids and proteins have been recommended, with or without added catalysts, for the controlled modification of starch [31], the objective being to modify the properties and increase the utility of starch products for the sizing of natural and synthetic fibers [32].

Amine salts of starch half esters with dicarboxylic acids have been claimed as exceptionally useful derivatives, providing improved film elongation in the sizing of hydrophobic yarns [33]. The products obtained from starch, maleic anhydride, and tertiary amines (e.g., triethylamine) are reported to be particularly suitable for sizing of polyester/cotton and polyester/rayon yarns. A typical preparation [33] is based on reaction of 2 parts of maleic anhydride for 100 parts of partially hydrolyzed corn starch, maintaining the pH of the mixture at 6.0-6.5 with triethylamine.

Dextrin phosphates have been recommended to obtain lower size pickup and allow faster loom speeds [34]. The preparation of these derivatives from starch powder slurries with alkali metal phosphates and phosphoric acid at 85-90°C is carried out under reduced pressure, and the phosphorylated dextrin product is then vacuum dried, cooled, and powdered [34]. The products are claimed to form strong films from low-viscosity solutions which allow deep penetration into the yarn at low add-ons.

Starch acetates are important derivatives for use in mixtures with many water-soluble natural and synthetic film formers (e.g., CMC, PVA, proteins, etc.). These acetates have a low degree of substitution, and they are available commercially under trade names such as Miraloid, Mirafilm (products of A. E. Staley Manufacturing Co.), and Kofilm (product of National Starch and Chemical Co.). They are supplied in a wide range of viscosities. When cooked, they yield noncongealing sols. When blended with other film-formers, they provide excellent viscosity stability, hydrating capacity, and flow rate, with a low gelatinization temperature.

Very high molecular weight starches can be modified and used as good film-formers by "thinning" them in enzymatic conversion with α-amylase. The conversion takes place in the presence of calcium ions, in a sulfite-bisulfite buffer. The process entails dry blending of the following ingredients in the proportions indicated [35]:

Sodium hypochlorite bleached, unmodified starch	100 pts (dry)
Sodium metabisulfite	0.6
Sodium sulfite	0.2
Calcium hydroxide	0.1

For example, 360 parts of water and 0.4 parts of a 1% solution of α-amylase (e.g., Vanzyme 31, Vanderbilt Chemical Co.) are added to 43 parts of the dry mix, and the slurry formed (at pH 6.9) is heated above the pasting temperature of 80°C for 25 min. Under these conditions hydrolysis takes place. When the desired level of hydrolysis is reached, the enzyme is deactivated by rapid heating to 100°C. The resulting product is a good coagent for sizing hydrophobic/hydrophilic yarn blends.

It is evident from the above illustrative examples that many diverse starch derivatives and products have been developed over the years and play important roles in sizing formulations.

To illustrate the versatility and blending adaptability of the large number of starch compositions investigated, some examples of formulations currently in industrial usage are shown below:

1. Style: 3600-4000 ends cotton
   Charge: 610-630 liter water
             280-320 kg corn starch slurry (104-110 kg solids)
             10-14 kg polyethylene-based softener
             4-6 kg emulsifiable wax

   Solids = 9-11%   Size add-on = 11-13%

2. Style: 3600-3800 ends 50/50 cotton/polyester
   Charge: 675-725 liter water
             40-50 kg corn starch
             28-34 kg CMC/PVA blend (60/40 ratio)
             1-3 kg emulsifiable wax

   Solids = 8-10%   Size add-on = 10-12%

3. Style: 3600-3800 ends 50/50 cotton/polyester
   Charge: 400-420 liter water
             90-100 kg corn starch
             4-6 kg PVA
             3-5 kg emulsifiable wax
             1-2 kg bloc wax
             .065-.095 liter liquid enzyme

   Solids = 13-15%   Size add-on = 15-17%

4. Style: 10,000-12,000 ends 50/50 polyester/cotton (sheeting)
   Charge: 500-520 liter water
             85-95 kg dialdehyde starch
             20-24 kg CMC/PVA blend (25/75 ratio)
             8-10 kg emulsifiable wax
         Mix finished to 700-775 liter with water

   Solids = 14-16%   Size add-on = 17-19%

5. Style: 3300-3500 ends cotton (corduroy)
   Charge: 560-580 liter water
             110-130 kg starch ester
             6-8 kg PVA
             5-8 kg emulsifiable wax
         Mix finished to 750-800 liter with water

   Solids = 13-15%   Size add-on = 15-17%

3.2 Carboxymethyl Cellulose

Cellulose undergoes reaction with monochloroacetic acid in the presence of sodium hydroxide to yield the sodium salt of carboxymethyl

cellulose (CMC). From aqueous solutions, CMC forms clear, cello-
phane-like films with outstanding characteristics. CMC is produced
in different grades and variations for many industries, including the
pharmaceutical, food, oil, paper, and textile industries.

The advantages of CMC as a warp size, particularly suitable for
natural fiber yarns, are derived from its water solubility, its ability
to form tough, flexible, clear films, and its resistance to thermal and
bacterial degradation [36]. Size formulations, applications, and properties are detailed in technical information literature of manufacturers
(e.g., [37]), which includes comparisons with some other film-formers
used in sizing. Warp size grade CMC is carboxymethyl cellulose

of low degree of substitution (DS 0.38-1.4). The solution properties depend on the molecular weight, as well as on the carboxymethyl
content (DS) (a DP of 950-1050 corresponds to an average molecular
weight of 185,000 ± 20,000).

The practical advantages of CMC in processing may be stated
as stable viscosity of solutions on prolonged heating, high rate of
water binding, resistance to bacterial degradation, relatively low
Biological Oxygen Demand (BOD of 11,000-17,000 ppm, compared to
800,000 ppm for starch after 5 days of incubation), and ease of desizing without the need for enzymes.

It has also been suggested [36] that CMC can be reclaimed by
ultrafiltration technology and reused. Illustrative formulations based
on CMC are:

1. Style: (65% polyester, 35% cotton) 2000-2400 ends
   Charge: 60-70 kg CMC
               10-14 kg acrylic resin binder
                3-4 kg emulsifiable wax
               Mix finished to 1100-1300 kg with water
   Solids = 5-7%   Size add-on = 6-8%

2. Style: Warp composition: 65% polyester, 35% cotton
   Charge: 22-28 kg CMC
            2-4 kg PVA or acrylic binder
            1-2 kg emulsifiable wax
            Mix finished to 280-320 kg with water
   Solids = 7-9%   Size add-on = 8.5-10%

Other representative formulations are detailed in CMC manufacturers' literature [37].

3.3 Acrylic Polymers

Polymerization and copolymerization of acrylic monomers offer great flexibility for the synthesis of polymers varying in structure and properties. A broad range of sizing materials based on acrylic polymers has been developed. The monomers employed for these products are primarily

$$CH_2=CH-COOH, \quad CH_2=CH-COONa, \quad CH_2=CH-COOR,$$

Acrylic acid   Salts (K^+, NH_4^+)   Esters ($R = -CH_3$, $-CH_2CH_3$, etc.)

$$CH_2=CH-CONH_2, \quad CH_2=\underset{\underset{CH_3}{|}}{C}-COOH, \quad CH_2=CH-C\equiv N$$

Acrylamide   Methacrylic acid   Acrylonitrile

A schematic formula of acrylate copolymers is shown below:

$$\text{+}(CH_2-\underset{\underset{COOCH_3}{|}}{CH})_X \text{———} (CH_2-\underset{\underset{COONa}{|}}{CH})_Y \text{———} (CH_2-\underset{\underset{COOH}{|}}{CH})_Z\text{+}$$

Each of the functional groups in the acrylate polymer contributes to the performance properties of the film former, and this is illustrated by the following generalizations:

A high carboxyl content (acrylic acid ratio) in acrylic copolymers enhances film toughness, adhesion to the substrate, and removability. The incorporation of acrylonitrile, or acrylamide comonomers, provides improved hardness of the film (lower flexibility) and improved compatibility with hard water cations. The ester content is an important variable in determining elasticity and cohesiveness of the polymer film. The use of nonacrylic comonomers (e.g., ethylene, styrene, vinyl pyrrolidone, etc.) allows even further variation in the specific properties of the polymers, and of the films formed from them in sizing.

Polyacrylic acid (homopolymer) is a useful and important sizing material, of value for continuous filament yarns made from polyester, nylon, and acrylic fibers. Polymers prepared from acrylic acid and acrylate salts (Na, K, NH_4) have been claimed to be components of "improved anionic sizes" [38]. More generally, corrosiveness of the polyacrylic acid solutions has prompted the use of sizing systems in which partial neutralization of the carboxyl groups of polyacrylic acid yields the Na, K, or NH_4 salt. By varying the mole ratio of neutralized to unneutralized carboxyl groups, it is possible to control the pH and viscosity of solutions, as well as the flexibility and strength of the film and its adhesion to specific substrates. For example, for a polyacrylic acid:sodium polyacrylate (1:6) size solution the following concentrations of size pickup have proven satisfactory: for cotton, 7-9%; for linen, 8-11%; for viscose/rayon, 2.5-4.5%; and for polyester staple, 9-12%. On synthetic filament yarns, the recommended solid pickup for acrylic sizes is approximately 2.5-4.0%. Special interest in acrylic sizes is attributed to the fact that they adhere to the substrate more firmly than other polymers and that elongation properties of the films under tension exceed those of most other size classes. In the case of filament nylon, it is also postulated that a reaction occurs between end groups of the polyamide fibers and functional groups in the acrylic polymer.

In early developmental work on acrylate sizes, starches were utilized as coreactants [39]. When acrylic or methacrylic esters are polymerized in the presence of starch dextrins, the products contain a certain percentage of graft copolymers, in which acrylic chains are attached to the polysaccharide. Evidence of grafting has been obtained by paper chromatography and by precipitation techniques. The reaction yields highly viscous dispersions from which clear, water-sensitive films, utilized in filament sizing, may be obtained [39]. Complex combinations of methyl methacrylate (0.5-3.0%), acrylic acid (0.8-4.0%), methyl acrylate (17-21%), and butyl acrylate (0.3-1.0%) have been copolymerized and marketed for high-speed slashers and water jet looms. In the latter application, special precautions must be taken to decrease the water-sensitivity of the film.

Anionic, water-soluble acrylic copolymers are an important class of materials. An example is a copolymer [40] comprising 5-30% of an unsaturated dicarboxylic acid (maleic anhydride half ester), 6-40% of a monocarboxylic acid (acrylic or methacrylic acid), and 40-85% of an aromatic vinyl compound (styrene). The preparation of the copolymer is described as follows [40]: 1000 parts of isopropanol and 600 parts of maleic anhydride are reacted at 120°C to form the half ester. A mixture of 1350 pts of styrene, 250 pts of acrylic acid, 10 pts of t-butyl hydroperoxide, and 12 pts of azodiisobutyronitrile are added in the course of 3 hr. The composition of the resulting copolymer is 37.8% by weight of maleic acid/isopropyl half ester, 4.5% of acrylic

acid, and 52.7% of styrene. The product is heated at 125°C for
2 hr and then at 150°C for 3 hr. The mixture is treated with 1000
pts of 25% aqueous ammonia to yield a clear, 27% active copolymer
solution which can be used directly as a size after appropriate dilution.

The use of acrylamide for sizes has been considered since acrylamide became a commercial commodity in the mid-1950s. Polyacryla-
mide itself has excellent properties but cannot compete successfully
as a sizing polymer because of its relatively high cost. Acrylamide,
however, has become an important comonomer in the synthesis of film
formers with acrylonitrile, acrylic acid, butyl methacrylate, meth-
acrylamide, styrene, vinyl acetate, etc. The water-soluble copolymers have good properties and, depending on the ratio of acrylamide, their solutions exhibit non-Newtonian or pseudoplastic behavior.
Solutions of sizes from copolymers of acrylamide have excellent tolerance of electrolytes and for other ingredients present in the size
bath. At the low concentrations of acrylamide comonomer used in
acrylic size polymers, the compound serves primarily to improve cohesive strength and to increase the flexibility of the film. Flexibility
may be further increased via controlled addition (0.5-1.0%) of a plas-
ticizer such as polypropylene glycol, sorbitan monooleate, or tridecyl
alcohol ethylene oxide adducts.

A special class of acrylate sizes is represented by alkali salts
of ethylene-acrylic acid copolymers, which yield recoverable warp
sizes [41]. According to the description in the patent [41], these
ethylene-acrylic acid "interpolymer salts" may be categorized into
three types:

1. Structural. Insoluble in polar solvents, with a weight percent of sodium acrylate of up to 10%. Products have very
   high melt index values.
2. Transitional. Swellable in polar solvents; exhibit high vapor transmission. Sodium acrylate content in the weight
   percent range of 9-18%. Intermediate melt index.
3. Water-soluble. Soluble in polar solvents, with a 16-35% sodium acrylate content, and very low melt index value.

The interpolymers described [41] dissolve in hot water and remain
in solution on cooling but do not dissolve in cold water initially.
This behavior may result from the fact that at room temperature ionic
sites are hindered by hydrophobic coils and are not accessible to the
solvent. At high temperatures, uncoiling of ionic sites and solubility ensue. The copolymers are good sizing materials for cotton and
for polyester-cotton blends. A feature of the products is the recoverability of polymer from the waste water of desizing by changing the pH (from 8-8.5 to 4.0-5.0) and precipitating the acidified
polymer from solution. The precipitate can be reconverted to the
soluble salt form and applied again.

For sizing of very fine polyester filament and nylon filament yarns, a similar water-dispersible ethylene-acrylic acid derivative has been suggested [42]. This copolymer contains 18-35% of acrylic or methacrylic acid by weight, 55-82% of ethylene, and also a hydrolyzable organofunctional silane of the general formula XR'SiY$_3$ (in which X represents a functional group capable of reaction with carboxyl groups; R' is a saturated hydrocarbon radical containing a minimum of three carbon atoms separating X from Si, and bonded to both; and Y is a hydrolyzable group). The silane (for example, γ-aminopropyl-triethoxy silane, $H_2NCH_2CH_2CH_2-Si(OCH_2CH_3)_3$) is hydrolyzed in situ during drying of the ethylene-acrylic copolymer on the substrate. This silane-modified size is reported to provide good resistance to abrasion, and good fiber-to-fiber, and fiber-to-metal lubricity.

Some examples of typical size compositions based on acrylic polymers are shown below:

1. Low viscosity solutions of polyacrylic acid (4-6% solids) are widely used for nylon filament warps because of their outstanding adhesion to polyamide, attributed to the formation of ionic bonds between size and substrate. For example:

    Style: 100% nylon
    Charge: 100-120 kg high MW polyacrylic acid (25% active)
             No additives; finish to 360-390 kg with water
    Solids = 5.2-6.5%   Size pick-up = 3.5-4.5%

2. In the sizing of twist or rotoset filament nylon (except Qiana), acrylic acid and methacrylic acid copolymers are materials of choice. These materials form brittle films, allowing an easier split at the lease rods.

    Style: 100% unrelaxed nylon
    Charge: 60-70 kg acrylic acid/methacrylic acid copolymer
             (25% active)
             No additives; finish to 340-380 kg with water
    Solids = 3.5-4.5%   Size add-on = 8-10%

In the case of polyester (filament or staple) yarns, ionic bonds cannot be formed between size and fiber, but secondary valence forces provide sufficient affinity to make acrylic sizes appropriate. For example:

3.   Style: 2800-3200 ends polyester
    Charge: 140-160 kg acrylic copolymer (30% active)
             2-4 kg lubricant wax + antifoam
             Mix finished to 360-400 kg with water
    Solids = 11-13%   Size pick-up = 15-17%

4.   Style: 100% textured polyester
    Charge: 70-90 kg acrylic copolymer binder (30% active)
              3-5 kg lubricant
             2-4 kg wetting aid + antifoam
             Mix finished to 360-400 kg with water
    Solids = 6-8%   Size add-on = 8-10%

Several premium formulations based on acrylates have been developed
and are extensively used for sizing of polyester filament warps. These
combine the film-forming and adhesive characteristics of acrylates
with the solubilizing effect of specific comonomers. Examples of these
copolymers are those prepared from acrylic acid and acrylonitrile.
In some instances, the nitrile groups may be hydrolyzed to amide
and to carboxyl to enhance solubility and adhesion further.

3.4 Polyvinyl Acetate

Copolymers of vinyl acetate form the basis of many size formulations
used for filament yarns, including those made from high-wet-modulus
rayon, cellulose triacetate, nylon, polyester, and polypropylene.
Vinyl acetate polymerizes, to form essentially water-insoluble polymers of the structure:

$$\{CH_2-CH\}_X$$
$$|$$
$$O$$
$$|$$
$$C{=}O$$
$$|$$
$$CH_3$$

When polyvinyl acetate is hydrolyzed by a controlled process, the
acetate groups can be fully or partially saponified to yield various
grades of polyvinyl alcohol (PVA). Both polyvinyl acetate and polyvinyl alcohol are manufactured in low-, medium-, and high-molecular-
weight products, providing great versatility in the choice of polymers suitable for sizing.

Copolymers of vinyl acetate have been developed with the objective of improved water solubility. Unsaturated carboxylic acids,
their alkali metal or ammonium salts, have been used as comonomers.
An early patent [43] describes a copolymer consisting of 97 mol% vinyl
acetate and 3 mol% crotonic acid for the sizing of cellulosic filament
yarns. More recently, copolymer latexes of vinyl acetate, vinyl propionate, and acrylic acid have been claimed [44] as sizing materials
especially suitable for texturized filament polyester yarns. By controlling the ratio of vinyl acetate to vinyl propionate in the terpolymers, variations can be made in the tensile, elongation, and adhesion characteristics of the size to meet critical requirements. The
effects are illustrated by the data in Table 1.2 [44].

Other terpolymers have been suggested as good functional products with increased abrasion resistance for nylon filament sizing [45].
A representative composition consists of 49.5 mol% of vinyl acetate,
48.5 mol% of maleic anhydride, and 2 mol% of styrene. The styrene
comonomer can be replaced by methacrylic acid or ethyl acrylate.
These and other related acid-modified terpolymer latex systems [45-51]
are very versatile and are recommended for acetate and polyester

Table 1.2 Effect of Monomer Ratios on Terpolymer Properties

Vinyl acetate %	65	70	75	80	95
Vinyl propionate %	30	25	20	15	0
Acrylic acid %	5	5	5	5	5
Film tensile strength (PSI at 65% RH)	720	1350	1340	1530	1760
Film elongation (%, 65% RH)	860	370	370	430	560
Adhesion to polyester (lb)	40	57	48	37	27

Source: Ref. 44.

filament warps, as well as for sizing spun yarns such as cotton, ray-on, wool, polyester, and blends. Depending on whether free acid copolymers or their salt forms are used, desizing processes are carried out in aqueous or in organic solvent systems. Some comparative data taken from the patent literature for sizes based on vinyl acetate copolymers are summarized in Tables 1.3-1.6 [46-51].

Table 1.3 Vinyl Acetate Copolymers-Composition[a]

Size	Vinyl acetate (VAc) (%)	Comonomer (I) (%)	Comonomer (II) (%)
A	90.5	(DMM) 5.0	(AA) 4.5
B	91.6	(DBM) 5.0	(AA) 3.4
C	96	(CA) 4.0	—
D	93	(MMM) 7.0	—
E	79	(MIBM) 21	—
F	47	(MA) 53	—

[a]Abbreviations: VAc, vinyl acetate; DMM, dimethyl maleate; DBM, dibutyl maleate; AA, acrylic acid; MMM, monomethyl maleate; MIBM, monoisobutyl maleate; CA, crotonic acid; MA, maleic anhydride.

Table 1.4 Vinyl Acetate Copolymers - Size Properties[a]

Size composition	Tensile (PSI)	Elongation (%)	Toughness ($\times 10^4$)	Adhesion to acetate (lb)	Adhesion to polyester (lb)
A	3060	370	113	44	20
B	1580	300	47	36	12
C	2000	200	40	30	18
D	2200	160	36	30	14
E	1400	400	56	27	16
F	500	500	25	—	30

[a]Sodium salt applied to acetate and to texturized polyester filament yarns. (Tested at 65% RH)

Table 1.5 Vinyl Acetate Copolymers - Size Properties[a]

Size composition	Tensile (PSI)	Elongation (%)	Toughness (X 10^4)	Adhesion to acetate (lb)	Adhesion to nylon (lb)
A	3450	370	128	19	13
B	2060	230	47	40	9
C	1660	130	22	27	11
D	2250	160	36	14	16
E	3830	20	8	14	11

[a]Ammonium salt applied to acetate and to nylon filament yarns as loom finish. (Size remains on the fabric.) (Tested at 65% RH)

Table 1.6 Vinyl Acetate Copolymers - Size Properties[a]

Size composition	Elongation (%)	Toughness (X 10^4)	Adhesion (lb)	Desizing aqueous	Desizing in trichloroethylene
A	360	126	180	yes	yes
B	400	40	100	yes	yes
C	200	30	150	yes	yes
D	200	40	80	yes	no
E	620	40	100	yes	no

[a]Ammonium salt applied to spun polyester yarns. (Tested at 80% RH)

Representative size formulations based on vinyl acetate copolymers are well documented in the patent literature and are illustrated by the examples below:

1. Style: 50/50 polyester/cotton
   Charge: 85-100 kg copolymer (VAc: 94 mol%; AA: 5 mol%)
            in 45% solution
            2-4 kg ammonia (28% active)
            No additives; mix finished to 450-500 kg with water
   Solids = 7.5-9.5%   Size add-on = 9-11%

2. Style: Acetate filament
   Charge: 45-50 kg copolymer (VAc: 90.5 weight %; DMM:
            5 weight %; AA: 4.5 weight %0 in 45% active
            solution
            3-5 kg ammonia (28% active)
            Mix finished to 450-500 kg with water
   Solids = 4.5-6.0%   Size add-on = 1.5-2.5%

3. Style: 100% nylon filament
   Charge: 42-52 kg copolymer (VAc: 46 mol%; MA: 46 mol%;
            methyl methacrylate: 8 mol%) in 44% active
            solution
            1-2 kg lubricant, no other additive
            Mix finished to 450-500 kg with water
   Solids = 4.2-5.5%   Size pick-up = 3.5-5.5%

3.5 Polyvinyl Alcohol

Among the film formers investigated in warp size applications, polyvinyl alcohol (PVA) has received a great deal of attention. Commercial grades differing in molecular weight, viscosity, and in the degree of saponification (or residual acetate groups), are available from several manufacturers. Some typical properties of PVA commonly used in warp sizing are shown in Table 1.7 (65% RH at 70°F).

The major differences in properties between fully hydrolyzed and partially hydrolyzed polymers of comparable molecular weight may be qualitatively described as follows.

Properties	Fully hydrolyzed	Partially hydrolyzed
Film strength	High	Moderate
Moisture sensitivity	High	Moderate
Adhesion to synthetics	Fair	Good
Foaming tendency	High	Moderate
Tack	Low	Moderate
Ease of desizing	Fair	Good

Table 1.7 Properties of Polyvinyl Alcohol Polymers Used in Warp
Sizing

PVA	Viscosity[a] CPS	Hydrolysis (%)	Tensile strength (kg/cm^2)	Elasticity (%)
Low MW (copolymer)	5-7	96-98	290-340	110-125
Medium MW	24-26	86-88	450-530	155-165
Medium MW	28-32	94-96	550-675	140-150
Medium MW (copolymer)	28-34	96-98	600-700	135-145
High MW	60-110	99-100	1000-1500	85-105

[a]Viscosity of a 4% aqueous solution, at 20°C.

As the tensile strength values summarized in Table 1.7 indicate, PVA
can give adequate protection to warp yarns even at low add-ons.
PVA films exhibit good flexibility at low humidity and are relatively
unaffected by changes in atmospheric moisture content over a considerable range of humidities. Because of these desirable properties,
PVA has become established as a versatile sizing polymer in many
applications previously dominated by starches, natural gums, and
CMC. Although PVA is more expensive than the natural polymers,
the possibility of recycling, its low BOD, and the higher weaving
efficiency attained make it economically attractive.

Some combinations of PVA and starches are used in some applications where satisfactory properties can be obtained. Laboratory studies
of films have shown, for example, that when 50% of the PVA is replaced
by an equal amount of starch, the resulting loss of tensile strength is
only 18-20%. Changes of similar magnitude in cohesive strength and
abrasion resistance have been noted. Thus, partial replacement of the
PVA with less costly polymers can be implemented without sacrifice in
weaving efficiency. Amylose starch and amylitol (amylose hydrogenated
at high pressure in the presence of Raney nickel) are particularly compatible with PVA and form homogeneous transparent films [52].

When 94-96 weight % vinyl acetate is copolymerized with 4-6 wt %
of methyl methacrylate, the copolymers obtained can be converted to
size components by base-catalyzed alcoholysis of the acetate groups.
The products exhibit a good balance of gel resistance and water sensitivity, which cannot be easily attained with PVA homopolymers. Preferred use of the products is as warp size for polyester-cotton blends

Table 1.8 Copolymers of Vinyl Alcohol and Methyl Methacrylate

Sizing polymer	Add-on (%)	Desized: % removed by 100-sec dip	Total removal: % removed after 2-hr scour
PVA (99% hydrolyzed)	12.7	7.5	73.6
PVA (88% hydrolyzed)	7.7	34.7	88.0
Copolymer (82% VAc)	8.2	44.0	97.7
Copolymer (90% VAc)	10.4	44.0	96.9
Copolymer (91% VAc)	9.0	60.0	99.2

Source: Ref. 53.

which are heat set in the greige. The copolymers also show better desizing characteristics from heat-treated cotton fabrics when compared with PVA homopolymers, as shown by the data in Table 1.8 [53].

Generally, it has been reported that removal of PVA copolymers in desizing requires less energy than removal of homopolymers. For example, the percent of size removed in 30 sec from 50/50 polyester/ cotton fabric heat set at 195°C is as follows:

Size	At 49°C	At 66°C	At 82°C
PVA	17	20	23
PVA/MMA	22	80	96

The foaming tendency of PVA sizes can be generally corrected through judicious application of defoamers such as isooctyl alcohols, or silicon compounds. The application of PVA sizes can be carried out at relatively low size box temperatures (65°C); this increases the viscosity of the size mix slightly, decreases penetration into the yarn, and reduces the total add-on level.

When PVA or its copolymers are not recovered from mill waste waters, a secondary water treatment is generally required to comply with effluent regulations. Although PVA has a low Biochemical Oxygen Demand for periods of 30 days when exposed to nonacclimated sludge microorganisms, planned degradation of PVA is necessary [54].

The ranges tabulated below provide some representative guidelines for the application of PVA sizes [55] to filament yarns.

Fiber	Size pick-up (%)	Type of size (PVA)
Nylon	3.5-4.5%	Low visc. 88% hydr. PVA
Acrylic	2.5-3.5	High visc. 99% hydr. PVA
Acetate	3.5-5.0	Low visc. 88% hydr. PVA
Polyester	4.0-5.0	Low visc. 88% hydr. PVA + acrylic
Rayon	1.5-2.5	Med. visc. 99% hydr. PVA
Glass	1.5-2.0	Low visc. 88% hydr. PVA

Typical PVA formulations for sizing of spun yarn are shown in the examples below:

1. Style: 100% cotton (poplin)
   Charge: 22-28 kg starch ether
   3-5 kg visc. 99% hydrolyzed PVA
   1.0-2.0 kg bloc wax
   Mixture finished to 350-400 kg with water
   Solids = 7-9%   Size add-on = 8-10%

2. Style: 65/35 polyester/cotton
   Charge: 55-65 kg starch ether (medium fluidity)
   10-14 kg med. visc. 88% hydrolyzed PVA
   2-4 kg bloc wax
   Mixture finished to 350-400 kg with water
   Solids = 18-20%   Size add-on - 20-22%

3. Style: 100% acrylic
   Charge: 25-35 kg med. visc. 99% hydrolyzed PVA
   1-3 kg bloc wax
   Mixture finished to 350-400 kg with water
   Solids - 7.5-9.5%   Size pick-up = 5.5-7.5%

4. Style: Rayon (muslin)
   Charge: 16-20 kg starch gum
   2-4 kg med. visc. 99% hydrolyzed PVA
   1-2 kg emulsifiable wax
   2-4 kg emulsifiable wax
   Mixture finished to 350-400 kg with water
   Solids - 5.2-6.2%   Size add-on = 5-7%

5. Style: 50/50 polypropylene/rayon
   Charge: 25-35 kg high visc. 99% hydrolyzed PVA
   2-3 kg emulsifiable wax
   Finished to 360-390 kg with water
   Solids = 7.5-9.0%   Size add-on = 6.8%

    6. Style: 80 ends worsted wool
        Charge: 18-24 kg med. visc. 88% hydrolyzed PVA
                    1-3 kg sulfonated processing oil
                    Finished to 350-400 kg with water
      Solids = 4.5-6%   Size pick-up = 5-6.5%
    7. Style: 55/45 polyester/wool
        Charge: 9-12 kg high visc. 99% hydrolyzed PVA
                    12-15 kg med. visc. 88% hydrolyzed PVA
                    2-4 kg sulfonated processing oil
                    Finished to 350-400 kg with water
      Solids = 6.2-7.5%   Size add-on = 6-8%

3.6 Polyester Dispersions

Polyester filament yarns, both flat and textured, have enormous commercial importance in modern fabrics. The natural and synthetic polymer sizes used for spun polyester yarns are not satisfactory for filament yarns primarily because of inadequate adhesion. This problem has been essentially solved with the development of products based on aqueous dispersion of polyesters of appropriate chemical structure. The materials exhibit low viscosity, good abrasion resistance, good cohesion properties, and adequate capabilities for desizing. They are also good candidates for reclamation from waste streams and for recycling.

    Initially, the newly developed polyester resins were used as a "bridge" or tie-coat between a traditional starch, CMC, or PVA size and the polyester substrate, binding conventional and economical sizing polymers to the polyethylene terephthalate filament yarns, and providing a measure of added protection, with fewer breaks and reduced shedding during slashing. As the use of the polyester resins in textile sizing increased, their cost decreased, and these dispersions are currently used as sizing materials per se, rather than as extenders. Copolymers suitable for the polyester dispersions for sizes can be manufactured in several ways. A typical procedure described in the patent literature may be summarized as follows [56].

    A dihydric alcohol, neopentyl glycol (NPG), and an aromatic polycarboxylic acid, trimellitic anhydride (TMA), are reacted at various molar ratios (see Table 1.9) at 360°C for 2 hr, to yield a low-molecular weight polyester with a high acid number. The product, when tested in the water-soluble, ammonium-salt form, yields films of good flexibility, hardness, and adhesion. In order to reduce moisture sensitivity and tack, another approach has been described [56]. Isophthalic acid is reacted at 230°C for 16 hr with polyethylene glycol and neopentyl glycol until the acid number of the mixture decreases to 15-14. At this stage, trimethylol propane and trimellitic anhydride are added, and the system is further reacted for 4 hr. The endpoint of the reaction is indicated by a final acid number of

Table 1.9 Polyester Dispersions

Molar ratio NPG/TMA	Acid No.	Tack at 55% RH/75% RH	Adhesion to polyester filament
1.0/1.2	261	None/very low	Excellent
1.0/1.0	221	High/high	Excellent
1.0/1.1	292	High/high	Good
1.0/1.3	256	High/high	Good

Source: Ref. 56.

40-42. The polyester resin obtained is then diluted to 30% activity, yielding a product which is readily dispersible in water, and a high number of free carboxyl groups which can be neutralized to attain water solubility. In the free acid form, the dispersion has excellent hydrolytic stability (up to 2 years). This polyester resin has been used in combination with PVA (8% PE: 92% PVA) to give improved results on 50/50 PE/cotton when compared with an acrylic/PVA combination (8% acrylic: 92% PVA) and with PVA alone.

A water-dispersible size composition reportedly results when 0.25 mole of dimethyl isophthalate, 0.125 mole of dimethyl terephthalate, 0.075 mole of hexahydroisophthalic acid, 0.05 mole of dimethyl sodium-sulfoisophthalate, 0.65 mole of diethylene glycol, and 0.8 ml of a 21% active solution of titanium isopropoxide catalyst are reacted at 20°C under 0.3 mm of vacuum for 1 hr. A rubbery, tough polymer is produced, which can be "dissipated" in hot water and is useful as a polyester filament size [57,58].

A broad range of unsaturated polyesters can be synthesized by reaction of polyols with unsaturated dicarboxylic acids in the presence of a saturated dicarboxylic acid. The products obtained are generally insoluble in water, but water dispersibility can be achieved by subsequent sulfonation of the product [59]. For example, 17.6 g of dimethyl isophthalate, 14.4 g dimethyl maleate, and 110.0 g diethylene glycol are reacted at 220-260°C for 4 hr under 2 mm vacuum in the presence of 0.3 ml tetraisopropyl-O-titanate. The polycondensate is cooled to 100°C and treated with 9.5 g of sodium metabisulfite in 20 ml water, yielding an opaque mixture which is readily water dispersible. This product is recommended for sizing low twist yarns made from cellulose acetate and triacetate filaments.

It is claimed that a high level of film flexibility can be achieved with water-dispersible polyesters containing unsaturation, grafted with vinyl or acrylic monomers [60]. Examples of formulations recommended for polyester sizes are given below.

1. Style: 100% polyester
   Charge: 100-120 kg polyester size (30% active)
               3-5 kg emulsifiable wax
               Finish mix to 350-400 kg with water
     Solids = 7.5-9%   Size add-on = 3-4%
2. Style: 100% polyester
   Charge: 60-70 kg polyester size (30% active)
               No additive; only overwax
               Finish mix to 350-400 kg with water
     Solids = 5-6.5%   Size pick-up = 1.3-2.2%

3.7 Polyurethanes

Polyurethanes are, potentially at least, valuable materials for the
sizing of hydrophobic fibers. They have good compatibility with
polyester yarns, and they may be made to exhibit sufficient hydrophilicity for application from aqueous solutions or dispersions, while
retaining excellent film strength. The linear polyurethanes are obtained by reaction of diisocyanates with polyalkyleneether glycols.
For example, m-phenylene diisocyanate or 2,4-toluene diisocyanate
may be reacted with a polyethylene glycol of average molecular weight
400-6000, in the presence of a catalyst, to produce a polyurethane
suitable for sizing [61]. The reaction is terminated by the addition
of isopropanol as a means of controlling the water solubility of the
product. Other polyurethane sizing polymers can be made using
methylenebis(4-phenyl)isocyanate or naphthalene-1,5-diisocyanate,
with other polyethylene glycols providing the hydrophilic segment
of the product. Commercial use of polyurethanes for sizing is not
well documented, and (probably) limited to specialized situations.

3.8 Styrene Copolymers

The aliphatic half-esters of maleic acid, and copolymers of vinyl acetate with maleic acid, have been used in the preparation of sizes for
polyamide fibers, but these materials have been generally somewhat
deficient in their adhesion to nylon, and in abrasion resistance. Terpolymers were later developed from vinyl acetate, maleic anhydride,
and a third monomer which could be either an alkyl ester of acrylic or
methacrylic acid or styrene [62-64]. According to relevant patents
covering the styrene copolymers, the optimum percent of styrene
ranges from 1.8 to 4.0 mol%. Styrene itself may be used, or a mono-
or di-substituted alkylstyrene or a halogenated styrene comonomer
may be substituted for the styrene component. The alkali salts of
styrene-maleic acid ester copolymers are also useful sizes for application to nylon or acetate yarns. For example, a copolymer in the
form of its sodium or potassium salt forms a hard film and is suitable

for conventional sizing, whereas in the form of the ammonium salt or in the free acid form it may be used for loom-finished fabrics. (When the ammonium salt is applied, the ammonia volatilizes during the drying process, leaving a water-insoluble coating on the yarn.)

3.9 Solvent Systems

A great deal of industrial interest and effort has been focused in recent years on the development of solvent systems for slashing. This work has been complex and, in part, slow because coordination among diverse manufacturing companies has been required to attain a "closed-loop" system. Integrated efforts by chemical companies for the development of solvent-soluble sizes, by equipment engineering firms for appropriate machine design, and by fiber producers were essential to the development of viable continuous processes. The materials flow and the sequence of steps involved in solvent slashing and in solvent desizing processes are discussed in subsequent sections of this review. Considerations pertaining to the sizing polymers required for solvent systems are reviewed below.

In solvent sizing, water is partially or completely replaced by an organic liquid, in which the film-forming polymer should be completely soluble. Many solvents have been evaluated experimentally [65], but only a few have been recommended for commercial application. Solvents investigated have included methylene chloride, trichloroethylene, perchloroethylene, 1,1,1,-trichloroethane, 1,1,2-trichloroethane, 1,2,2-trifluoroethane, and methanol/ethanol mixtures. All of these solvents have exhibited specific advantages or disadvantages, but perchloroethylene, trichloroethylene, and trichloroethane have been selected for the best combination of properties. When compared with water, these organic solvents show important differences as media for application of size, namely:

1. Reduced energy requirements. Comparison of some properties of organic solvents to those of water (e.g., latent heat of vaporization, specific heat, etc.) indicate that only about 1/5 as much energy is needed to heat the organic compounds to the boiling point, and only 1/10 as much energy is consumed to vaporize the solvent in the drying step.

2. Decreased water consumption. The highly alkaline waste effluent of water from desizing is eliminated. The fresh water requirements are about 1/15 of those used in aqueous processes.

3. Fast drying rate. In the aqueous process, intermolecular forces (hydrogen bonding, Van der Waal forces) bind water molecules even at elevated temperatures before appreciable evaporation can occur. Organic solvents do not exhibit strong intermolecular forces, and evaporation occurs more rapidly.

4. Improved quality. Fabric aesthetics are improved. Removal of the solvent by evaporation at temperatures above its boiling

Table 1.10 Physical Properties of Chlorinated Solvents

Property	Water	Perchloroethylene	Trichloroethylene
Molecular weight	18.0	165.9	131.4
Specific gravity	1.0	1.63	1.46
Freezing point (°C)	0.0	-22.0	-86.4
Boiling point (°C)	100.0	121	87.0
Specific heat (cal/g/°C)	1.0	0.21	0.22
Latent heat of vaporization (Cal/g)	545.1	50.1	57.3
Vapor pressure (mm Hg/20°C)	0.25	14	58
Heat for evapora- tion at 20°C (Cal/g)	664.9	71.2	72.5
Evaporation coefficient (ether = 1)	80	10	3.5

point results in a soft hand. Waxes and oils are removed, leaving a more uniform, cleaner surface for dyeing.

Table 1.10 summarizes some physical properties of selected chlorinated solvents in comparison with water [65].

Several hydrophobic fibers like polyester, polyamide, and polypropylene have substantivity for perchloroethylene at or above the Tg of the fiber. Thus the solvent is not easily flashed off in the drying step, and up to 4-6% of it may be retained in the yarn. Fluorinated solvents do not exhibit substantivity for hydrophobic fibers, and recovery approaches 99%, but these solvents are, of course, much more costly. A schematic diagram of materials flow for a solvent slashing - solvent desizing system, reproduced from Ref. 65, is shown in Fig. 1.3.

Many of the conventional sizing agents used in aqueous media are insoluble in chlorinated solvents. However, if the free acid form of the polymer or selected unmodified macromolecules are used, a broad range of polymers becomes available for the solvent process.

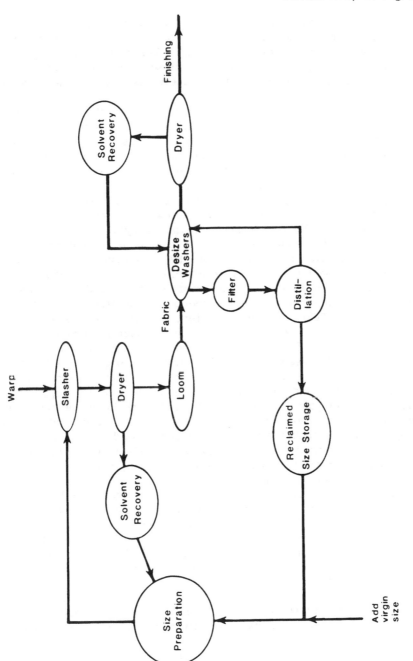

Figure 1.3 Schematic of solvent slashing/solvent desizing system. (From Ref. 65.)

For example, high-molecular-weight ethyl cellulose in perchloroethylene gives generally good performance, comparable with that obtained with PVA or CMC in aqueous medium. Tests of yarn tenacity,
elongation at break, and abrasion have shown that the perchloroethylene solution of ethyl cellulose can be useful as a solvent system for
sizing [65].

A 10% solution of high-molecular-weight (over 300,000) polyethylene oxide in trichloroethylene has been used successfully as a size
on high-density warps of polyester/cotton blends [66]. The yarn
exhibited lower shrinkage than comparable yarn sized from aqueous
medium. Weaving test data also indicated a significant reduction of
warp breakages for the solvent system.

Copolymers of a hydrophobic monomer (e.g., vinyl acetate, styrene, vinyl chloride, acrylate esters, methacrylate esters) and an
unsaturated carboxylic acid monomer (e.g., acrylic acid, crotonic
acid, fumaric acid, maleic acid, etc.) in a chlorinated solvent are
claimed to yield efficient solvent sizes [67]. A copolymer of vinyl
acetate:crotonic acid (97:3 mol%) is soluble in trichloroethylene and
has been suggested as size for polyester, acetate, and nylon filaments. Copolymers derived from styrene, acrylonitrile, or methyl
methacrylate (20-80%), and acrylic esters or methacrylic esters of
higher aliphatic alcohols (80-20%) yield other useful combinations of
solvent soluble sizes [68]. The preferred polymers are claimed to be
those with a Tg between 40 and 80°C, and a Young's modulus of
1×10^3 to 1×10^4 kg/cm^2 [68]. The monomer selected from the
first group is reported to contribute cohesiveness and a degree of
rigidity to the polymer. However, homopolymers have low resistance
to abrasion and a tendency for developing local stresses and shedding.
By copolymerizing with a monomer from the second group, high levels
of flexibility, elasticity, and good abrasion resistance are obtained.
A well-balanced ratio of monomers from two groups has been described as shown in Table 1.11 [68].

The copolymers may be prepared by conventional methods such
as solution, suspension, or emulsion polymerization. The products
are simply dissolved in the halogenated solvent, and applied at 4-10%
concentrations. A systematic study with composition ratios of MMA
and BMA in proportions of 15:85 through 85:15 has demonstrated the
important effects of Tg and Young's modulus on abrasion resistance,
as shown in Table 1.12 [68].

The preferred ratio of MMA and BMA is reported to be between
25/75 and 75/25, but the optimum depends on specific factors including yarn denier, warp density, and other variables of lesser importance.

Chlorinated polymers such as chlorinated polyethylene, chlorinated ethylene-propylene copolymers, and copolymers of vinyl
chloride-vinylidene chloride are film-formers from chlorinated solvents.

Table 1.11 Properties of Copolymers for Solvent Sizing[a]

Hard component monomers			Soft component monomers				Tg (°C)	Young's modulus × 10³ (kg/cm²)
MMA	ST	AN	MA	EA	BA	BMA		
35			65				41.1	3.3
75			25				78.8	10.0
70				30			67.9	6.3
80				20			79.6	9.6
70					30		55.7	4.8
20						80	41.1	4.1
65						35	69.4	8.5
	70		30				70.9	9.5
	60			40			53.2	7.1
	80				20		68.8	9.3
	30					70	47.6	3.3
		50	50				53.5	6.7
		70	30				73.7	9.8
		60			40		40.8	3.4
		70				30	79.8	10.0

[a]Abbreviations: MMA, methyl methacrylate; ST, styrene; AN, acrylonitrile; MA, methylacrylate; EA, ethyl acrylate; BA, butyl acrylate; BMA, butyl methacrylate.
Source: Ref. 68.

However, these polymers have low crystallinity and they tend to be tacky [69,70]. Tack can be minimized by blending with aromatic hydrocarbon polymers, for example, polystyrene of average MW 100,000, or polyvinyl toluene [71]. Typical size compositions for application from chlorinated solvents are outlined below.

1. Style: 50/50 polyester/cotton
   Charge: 22-26 kg vinyl chloride/vinylidene chloride copolymer:polystyrene (4:1)
   1-2 kg wax
   450-480 kg 1,1,1-trichloroethane
   Solids = 4-6%   Size add-on = 8.5-10%

Table 1.12 Effect of Copolymer Composition on Properties

| Copolymer Composition | MMA | 15 | 25 | 35 | 50 | 65 | 75 | 85 |
	BMA	85	75	65	50	35	25	15
Tg (°C)		37.8	48.1	49.2	68.5	69.4	74.6	83.2
Young's modulus ($\times 10^3$ kg/cm^2)		0.38	4.9	6.3	7.9	8.5	8.9	9.9
Abrasion resistance of sized yarn (cotton)		482	465	450	436	412	401	350

Source: Ref. 68.

2. Style: 50/50 polyester/cotton
   Charge: 22-28 kg chlorinated polyethylene: polystyrene
               1-2 kg stearic acid lubricant
               425-480 kg 1,1,1-trichloroethane
   Solids = 4.5-6%   Size add-on = 8.5-9.6%

3. Style: wool/nylon (50/50)
   Charge: 24-28 kg styrene: ethyl acrylate copolymer
               (70:30)
               1-3 kg wax lubricant
               Finished to 480-520 kg with 1,1,1-trichloroethane
   Solids = 4.6-5.5%   Size add-on = 4.8-5.8%

With minor adjustments and variations these formulations can be applied to many commonly used yarns including wool, linen, cotton,
regenerated cellulose, polyester, nylon, acrylic, polypropylene, acetate, and also blends.

3.10 Hot-Melt Sizing Polymers

The concept of eliminating water from the slashing process was first
proposed in the 1940s. Patents for "dry-sizing" were applied for in
1955, and the first machines for dry sizing were exhibited at the
ITMA in Milan, Italy in 1959 [72]. However, these approaches were
not commercialized, because suitable chemical compounds were not
available. The paraffinic and aromatic waxes did not meet the requirements of efficient sizes. The gap was bridged when the high-
molecular-weight polyethylene oxides were introduced commercially
[73]. This was followed by the development of other polymeric systems that could be applied to yarn from the molten state. Hot-melt
sizing compositions essentially consist of thermoplastic polymers that
become free-flowing liquids when exposed to elevated temperature.
They are applied in the fluid state and set by cooling. They contain no solvents, and they shrink less on setting than water- or
solvent-based sizes. Hot-melts form strong, flexible films giving
good adhesion to a great variety of yarns. The advantageous properties of hot-melts also suggest their limitations. It is evident that
they are more heat-sensitive than other sizes; in some instances they
have limited pot stability; they require special equipment with precise control of temperature and volume for consistent performance.
The general requirements for a hot-melt size may be summarized as
resistance to oxidative and thermal degradation, easy removal in
aqueous systems for desizing, low-melt viscosity at the temperature
of application, rapid setting to the nontacky state, and acceptable
economics.

Polymers of high enough molecular weight to be good film-formers
frequently yield melts with excessively high viscosities and slow setting rates. To overcome this problem, mixtures of three components

have been developed [74]. One type of hot-melt size is a blend of a film-forming thermoplastic polymer, a viscosity depressant modifier, and a lubricant which imparts flexibility under dynamic conditions. Water-dispersible copolyesters have been suggested as the polymer components. Meltable vinyl acetate/dibasic acid copolymers have also shown good properties [75,76]. Water- and alkali-soluble phosphonate copolymers are reportedly good candidates. Aliphatic dicarboxylic acids (e.g., adipic, sebacic, succinic acid, etc.) and solid polyhydric alcohols (e.g., sorbitol, mannitol, etc.) are used as viscosity modifiers, and to control setting time. For higher melting ranges (220-250°C), phenolic acids (mono- and polyhydroxybenzoic and naphthoic acid) and polyhydric phenols have been suggested. Selected partial esters of aliphatic acids (e.g., benzyl and substituted benzyl esters) melt in the right temperature range (175-220°C) and are useful. The ratios of polymer:modifier:lubricant in the formulation can vary from 90:8:2 to 50:48:2, with a ratio of about 60:39:1 believed to be optimal [74].

An important polymer for hot-melt systems is the linear copolyester synthesized from isophthalic acid, 5-sulfoisophthalic acid, and diethylene glycol. When this copolymer is blended with adipic acid in a 60:40 ratio, the properties of the hot-melt resin are as follows [75,76]:

Viscosity (at 155°C)	950 cps (Brookfield, #5, 60 sec)
Tg	32°C
Tensile strength (2 mil film)	26 kg/cm^2
Elongation (2 mil film)	26%
Melting range	110-120°C
Solubility in water	∞

Useful water-dispersible hot-melt sizes have been made from linear polyester/amide products [77] in the following manner: A mixture of 73 g of adipic acid, 61.6 g 3,3-ethylene dioxybispropylamine, and 10 ml water is heated under nitrogen for 90 min at 150°C, and then for 10 min at 200°C, removing the water formed by distillation; 100 parts of titanium tetraisopropoxide are then added, and the reaction is continued for 75 min. At this point, 21.2 g of diethylene glycol are introduced, and reaction is continued for 70 min at 200°C under reduced pressure, to remove excess glycol. The mixture is then cooled. This hot-melt polymer product has a melting range of 110-118°C. Similar compositions can be produced with other aliphatic diols and suitable ether diamine components [78]. Ethylene/vinyl ester copolymers blended with paraffin waxes, hydrocarbon resins, and low-molecular-weight polyethylene have reportedly found limited application as hot-melt sizes [79].

The polymer classes possessing the water solubility and melt properties essential for hot-melt sizing are limited because these

properties are somewhat contradictory. Water solubility is provided
by polar groups, but the presence of these groups increases the
melting point and, in some instances, the thermoplastic character
of the resin is lost. The balance between polar and nonpolar con-
stituents is thus critical for a hot-melt size polymer of acceptable
properties. The polymers discussed above generally meet the criteria
for an effective size, i.e., the resins form a protective, flexible film,
which adheres well to the substrate, and they can be removed under
conventional conditions of scouring. Some examples of formulations
for hot-melt sizing are:

1. Style: 65/35 polyester/cotton
   Size: 60/40 linear polyester/adipic acid molded block pressed against the periphery of a rotating grooved cylinder with a surface temperature of 300°C. Speed
   of operation 300 m/min.
   Size add-on = 10.0-10.4%

2. Style: Spun polypropylene
   Size: 60/40 linear copolymer/azealic acid molded block
   pressed against the periphery of a 127°C heated
   grooved roll. Speed of operation 300 m/min.
   Size add-on = 7.9-8.1%

3. Style: 65-35 polyester/cotton
   Size: 60/40 linear copolyester/sorbitol molded block pressed against the surface of a grooved cylinder at
   155°C. Speed 270 m/min.
   Size add-on = 10.6-11.0%

4. Style: Polyester/wool
   Size: 60/40 linear copolyester/mannitol. Same process as
   in formulation 3, above. Roll temp. = 188°C. Speed
   = 250 m/min.
   Size add-on = 5.2-5.5%

A hot-melt sizing composition of current importance is reported to be
the one described in a U.S. patent application recently filed by Burlington Industries [80]. The abstract of this application states, in
part:

A quick-setting non aqueous, water extractable size composition to be applied as a melt to textile yarns consists
of 42-58% 20:80 acrylic acid-ethylene copolymer (I), and
58-42% hydrogenated tallow-type triglyceride wax. Thus,
equal portions of hydrogenated tallow and (I) having
standard melt flow rate 500 were melted together to give
a size with Brookfield viscosity 825-850 cp. Size add-on
on 25/1 - 65-35 polyester-cotton yarn by using a grooved
roll was 15.7%.

4. SIZING MACHINERY AND PROCESSING TECHNOLOGY

The mechanical aspects of sizing are beyond the scope of this discussion, and the process description below is accordingly intended only as a brief summary of the procedures employed in industry.

4.1 Conventional Procedures

Most warp sizing machines, or slashers, in use today apply the size to warp sheets by moving the warp strands from a battery of beam creels through a size box containing an aqueous solution of size polymer, which penetrates the warp yarn. After the warp sheet is wetted, squeezing between hydraulic rollers removes excess solution by a "quetsching" operation, and the sized yarns are then passed around heated drying cylinders, to a take-up mechanism at the head section, where they are wound up on a beam for the weaving process. Although there are many variations in the equipment used, it is helpful to refer to the diagram of a general modular sizing system shown in Fig. 1.4.

The size mix is prepared in insulated cooking kettles under automatic control [22,81,82] and metered continuously to the sizing box at a rate proportional to the consumption in the system [83]. The speed of the warp depends on the type of yarn (spun, filament, denier, etc.). It may vary from 70 to 600 m/min [84]. Either single or double immersion in the size box may be required [85,86], as shown in Fig. 1.5. The pressure of squeeze rolls governs size uptake and controls uniformity of size distribution on the yarn [87]. Drying takes place on 6-20 Teflon-coated cylinders held at temperatures of up to 130°C, aiming for about 4-7% residual moisture in the sized warp [88]. The key variables are instrument-controlled within narrow tolerance limits [89,92]. The critical factors in the process are [90,91]:

> Steam flow (steady steam supply)
> Size box level (same level yields same size pick-up)
> Size box temperature (assures uniformity)
> Size viscosity (assures uniform depositon)
> Squeeze roll pressure (steady size retention)
> Cylinder temperature (governs moisture content)
> Yarn stretch (controls penetration uniformity)
> Slasher speed (optimum uniform production)

4.2 Short Liquor Processes

In order to decrease effluents and potential pollution of public streams, and also to improve the cost-effectiveness of sizing, novel mechanical approaches have been developed in several research institutions around the world [94-96]. These include high-pressure sizing,

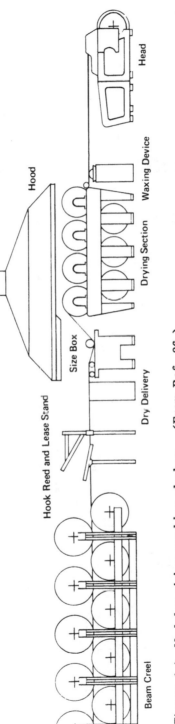

Figure 1.4 Modular sizing machine slasher. (From Ref. 86.)

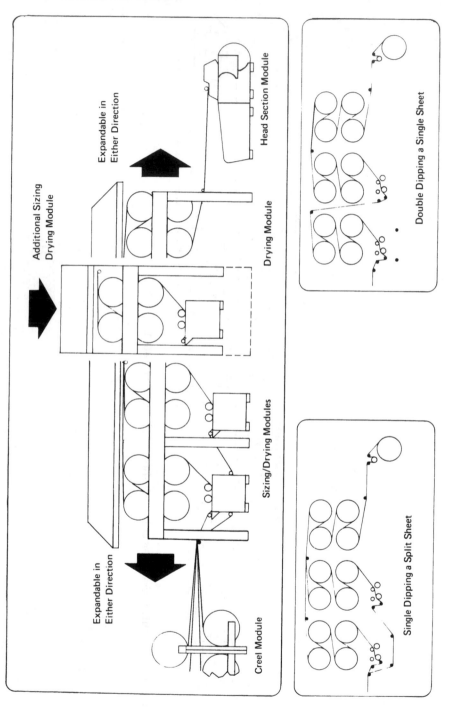

Figure 1.5 Size box with single and double dip. (From Ref. 86.)

sizing in the foamed state (gas dispersion), and dry technology which omits the application of size.

High Pressure Sizing

This technology, which evolved from the conventional slasher, has demonstrated that energy savings, higher production speeds, improved weaving performance, and reductions in chemicals can be achieved with increased squeeze pressure [93,94]. When the 4-10 kg pressure per linear cm on the face length of the squeeze roll is increased to 150-180 kg pressure by means of advanced hydraulic mechanisms, the advantages cited above can be attained [93].

Liquid size systems generally exhibit non-Newtonian rheology, namely, there is a reverse relationship between speed gradients and viscosity coefficients as well as the modulus of elasticity. At higher speeds, a markedly higher shear rate exists for a size formula, and under these conditions a higher solids concentration can be used. This leads to higher size pick-ups with the elimination of 50-60% water. Also, by increasing the squeeze pressure 10- to 30-fold, an average energy saving of 22% can be realized [94]. A comparison of conventional pressure versus high-pressure squeezing indicates that in the latter processes the quantity of water evaporating during drying is reduced to such a degree that heat energy is saved, while a much higher production speed is realized.

Foam Sizing

Foam technology in the treatment of fabrics is not new, but foam systems are relatively new in sizing processes. This approach deals with the controlled application of a stable size formulation in the foamed state, deposited uniformly on a moving warp sheet. This process is also known as "short-liquor" sizing because the water content in the operation is reduced by 60-80% as compared with conventional processes, where elimination of water consumes about 80% of the energy used. In times of increasing cost of energy, scarcity of water, and ever-tightening government regulations, foam sizing is an attractive and potentially viable industrial process. The major advantages of foam sizing are low moisture pickup, accurate control of solids add-on, controlled size penetration into the warp sheet, and reduced energy usage. A brief general review of the background and applications of foam technology in the textile industry [97] includes a useful glossary of terms and reference to several types of equipment used. A more specific recent overview of foam sizing processes and their advantages [98] provides details of the key variables in foam sizing. Table 1.13, taken from this publication [98], summarizes relevant factors in the reduction of energy cost for drying which can be achieved with foamed size slashing.

Table 1.13 Foam Sizing

	Sizing system	
	Conventional	Foam
Size add-on (%)	12.0	12.0
Wet pick-up, obtainable (%)	90.0	48.0
Size concentration in mix (%)	13.3	25.0
Water to be evaporated (lb H_2O/lb yarn)	0.78	0.36
Annual energy cost for drying[a]	$32.011	$14.774

[a]Based on production of 1000 lb of sized yarn per hr on one slasher, energy consumption is 1900 Btu per lb of water evaporated, and energy cost is $3.00 per 1 million Btu.
Source: Ref. 98.

In foam sizing, the most important parameters relating to the length of time an aqueous foamed-size formulation can maintain its initial properties are:

1. Temperature (Change in temperature will alter viscosity as well as surfactant interactions.)
2. Amount of gas (usually air) expressed as "blow ratio," the ratio of the weight of a given volume of liquid before foaming to the weight of the same volume of foam.
3. Viscosity of liquid and foam viscosity.

The foamed size properties are ultimately dependent on the chemical composition and characteristics of the polymers in solution. Size polymer classes such as CMC, PVA, and acrylics are ideally suited for foam application, since with very low levels of foaming auxiliaries, they can yield foams of low air permeability and high surface viscosity.

Size foams are now produced in commercially available machines, which consist of a mechanical agitator capable of mixing metered quantities of air, a liquid size composition, and surfactants, converting these into a foam. In practice it has been found that a 14-18% solids size liquor is optimal with 1-1.5% surfactant added. The foam is generated at a given density at such a rate that the half-life of its cellular structure is assured for 25-35 min. The surfactant provides workable flow characteristics to the foam. The most versatile surfactants are based on C_{12} linear hydrocarbon derivatives such as

sodium laurylsulfate, but others can be used. For the application
of foam size, a conventional slasher machine is used, and the foam
size may be entered into the size box or applied directly to the moving warp sheet. A very low squeeze is applied in order to prevent
foam reversion to liquid state. Operational speeds are governed by
the foam generating capacity. Statistical weaving efficiency tabula-
tions indicate that some foam size applications (PVA) are equal or
slightly superior to conventional liquid size processes [98].

Weaving without Sizing

Preliminary investigations at the Fiber Research Institute (TNO) located in Delft, Holland, have shown [96] that weavable polyester/cot-
ton yarns can be obtained without size by a thermal treatment. When
PE/cotton blends of 65/35, 80/20, and 50/50, in yarn counts varying
between 12 and 45 singles, both open-end as well as ring-spun, are
exposed to a temperature of 250°C for 1.5 sec, a "sizing" effect is
obtained. The heat exposure may be carried out in an infrared Benz
oven, and is performed without tension on the yarn sheet. No loss
of strength or elongation of the yarn should occur. Abrasion resistance and shedding values indicate good weavability.

Similar experiments have been carried out at the Department of
Textile Technology, University of Manchester [99], on 37 tex open-
end-spun cotton, on 37 tex ring-spun cotton, utilizing a Mettler
electric singeing machine: cotton yarn ends were singed singly and
thereafter woven without using sizing agents, with satisfactory results.

4.3 Solvent Slashing

After several years of intensive testing [100-102], processes of sizing and solvent desizing can no longer be considered experimental.
Machinery manufacturers have promoted production size equipment
for the textile industry, claiming that the solvent-sized fabrics are
superior in quality to those treated in aqueous systems, and that
they offer opportunities for lower processing cost.

The pioneer units for solvent processing have been the Derby
Dry-Cleaner machines, which are still functional. In these, the fabric is passed through cold trichloroethylene and squeezed; the residual solvent is removed by drying cans for collection at a distillation
unit. A more complex system is the ICI solvent scouring system
[103]. This equipment is not primarily designed for sizing. It
utilizes boiling trichloroethylene in an enclosed dewaxing chamber.
Clean solvent (with dissolved chemicals) overflows from one compartment to the next, countercurrently to the direction of the warp sheet.
From the last chamber, the material passes through a solvent seal
and high-pressure squeeze for removing entrapped solvent. The

warp sheet then passes through a steam chamber, where the remaining
solvent is released and collected for recycling. With some modifications, the Bohler and Weber (Germany) system and the Dow (United
States) solvent scouring/finishing process equipment have made
great strides toward a complete loop solvent sizing/desizing system.

Currently, the Kanebo (Japan) sizing range is suggested as a
full scale solvent sizing/desizing machine, especially for processing
cellulosic spun yarns with acrylic-methacrylic copolymer size dissolved in a halogenated hydrocarbon solvent. From a beam creel
the warp enters into the size box where the dissolved polymer is
applied at room temperature. The sized warps are air dried (100-
120°C) in multiple drying chambers, under slight vacuum. The solvent rich vapor enters a recovery section where the solvent is reclaimed, while the air passes through a carbon filter unit where residual traces of solvent are captured.

The desizing operation proceeds from a steeping section where
the woven fabric is immersed in solvent at room temperature for
5-10 min. The fabric then passes through a showering and rinsing
section for thorough solvent removal of polymer, and following these
steps, it is dried with hot air (105 ± 5°C). The size and solvent
are fully recovered and purified for reuse. The recovery of size is
claimed to be 95%+. According to the machine manufacturers, this
solvent process has the following advantages: the texture of the
yarns is perfectly retained; cohesion of sized yarn is improved; tensile strength and elongation are not affected by variations in temperature and humidity during the weaving process; desizing is very
rapid; reduced yarn breakages in weaving are observed due to the
high level of yarn uniformity; better quality dyeing is obtained; and
finally, the high initial investment for the equipment is compensated
by low processing cost, good weaving efficiency, water savings, energy savings, and manpower savings.

4.4 Hot-Melt Processing

Several methods and equipment for sizing warps with hot-melt polymers have been disclosed and promoted in recent years [104-106].
Earlier concepts involved the use of sizes in which waxes predomin-
ated. The warp sheet was fed into a sizing chamber and brought
into contact with the top periphery of a sizing roll, while the bottom
periphery of the roll was in the size box touching the molten size.
The sizing roll rotated at a peripheral speed which was slower than
the speed at which the yarn sheet advanced. Leveling rollers removed surplus molten size from the yarn and evenly distributed the
size on the yarn surface, also diffusing it into the interior portions
of the yarn. The size deposition was in the range of 2-6% on the
weight of yarn, and was adjustable by a doctor blade.

From the heated sizing chamber, the sized warp sheet was then fed to an air-cooled, solidifying compartment. Residence time in this area was adjusted so that the enveloping size would be cooled and would solidify to a tack-free state. From this chamber, the warp sheet was then transferred to the take-up beam for weaving.

A more advanced development utilized dicarboxylic acids, diols, and sulfonated dicarboxylic acid transesterified products [107] as hot-melt size polymers. In processing, the solid size in block form is pressed against a heated (100-180°C), grooved rotating applicator cylinder. The viscosity of the molten size is 1000-2000 cps. (This is the right fluidity for transferring the molten polymer to warp yarns moving at 150-300 m/min through the grooves.) The warp sheet contacts the applicator roll where the grooves are already filled with the liquified polymer and are enveloped by the size. Due to the nature of the polymer no special cooling device is needed: Only a few meters from the applicator roll, the yarn sheet is tack-free and can be beamed for weaving. The concepts and current status of hot-melt sizing technology have been reviewed in a recent paper [108].

Generally, the advantages claimed for the hot-melt process are a higher speed of the sizing operation, reduced energy consumption, improved quality of sized warps, and no need for size preparation (cooking). There are still some limitations for hot-melt sizing, but the problems will probably be solved in due course, as technology advances.

4.5 Miscellaneous Innovations

Electrostatic Sizing

The electrostatic deposition of a powdered size on yarn sheets in a continuous operation offers a potential for producing sized warps economically. Savings in the areas of floor space and energy requirements are attainable. In this process, a properly tensioned warp sheet is fed to a padder containing a 0.5% salt solution. The warp is wet-out by dipping in the solution and squeezed to a controlled wet-pick-up of 40-50% by weight. This positively charged, conductive warp sheet obtained travels through a coating chamber where negatively charged, dry size particles are blown on to the warp from opposing directions by electrostatic guns. Depending on the warp speed, the guns can deliver the optimal quantity of powder at the necessary charging energy (1000-2000 g polymeric size at 35-200 Kw power per 70-100 m/min). The coated warp passes through an infrared heater for polymer fusion, then onto the cooling roller and beaming.

Polymers that may find application in electrostatic sizing are primarily the hot-melt systems like polyester resins, but acrylics,

PVA, and styrene-maleic anhydride copolymers are possible candidates as well. At this time, commercial utilization of electrostatic size deposition has not been realized in the United States.

Combined Operations

Sizing machines with added functions have recently become very popular [109]. Machines on which dyeing and sizing take place almost simultaneously and continuously have been used in production [110].

A well-functioning process may be as follows: The warp first passes through a dye bath with dye liquor and auxiliary chemicals for complete dyeing; the warp is partially dried (to 12-25% residual moisture) before it enters the sizing trough; the appropriate final dye-developing catalyst or component is added to the size mix, and the dyeing cycle is concluded during the dwell time of size-drying. Then the warp is beamed as usual.

A method has also been disclosed in which a cellulosic/synthetic blend warp is treated at the slasher by sizing, dyeing, and resin treating in a single pass on conventional equipment [110,111]. The bath contains a starch size, a thermosetting resin and catalysts, textile auxiliaries as lubricants and wetting agents, and pigments.

There are numerous disclosures of experimentation in foreign countries with combined sizing/dyeing operations involving both azoic and direct dyes [112]. Slashers also exist with supporting extensions necessary for impregnation and oxidation procedures during indigo dyeing, as well as for sizing-bleaching preparations.

5. DESIZING

It may be said that a sized warp is both a weaver's necessity and a drawback for the finisher. Except in a few instances, where the size also serves as a permanent hand builder (loom-finished items), the textile processing steps which follow weaving require complete and uniform removal of the size.

Warp size removability depends on the solubility of the film-forming polymer, on the effects of numerous subsequent wet processing steps, and on interactions with added chemicals. Important considerations include rate of dissolution, solvated gel viscosity, concentration of desizing agent, mechanical action, scouring temperature, and the acidic or alkaline conditions required for desizing. Several different methods are used for desizing, depending on the size chemicals used. These may be classified as follows, in the context of their historical evolution:

1. Hydrolytic processes: (a) rot steep, (b) acid steep, (c) enzymatic steep

2. Oxidative processes: (a) chlorine treatment, (b) chlorite treatment, (c) bromite treatment, (d) peroxide treatment
3. Alkaline scour processes
4. Solvent systems
5. Low temperature plasma treatment
6. Durable (permanent) size

5.1 Hydrolytic Processes

Rot Steep

This is an inexpensive old method, without chemical additives. Starch-sized fabric is passed through a padding mangle and saturated with water at 40-45°C to about 90-100% pick-up, then allowed to stand for 24 hr. During this time the microorganisms naturally present multiply and depolymerize the "nutrient" starch. The starch becomes partially soluble in water, to the extent that subsequent washing removes it almost completely. The fermentation process attacks the starch only, but heat buildup and other factors may cause the cellulose fiber itself to be damaged by mildew. The process obviously has many drawbacks: It requires large floor space, it is time consuming, and it does not yield uniform results. Desizing by rot steeping is no longer practiced in industrialized countries.

Acid Steep

In this form of size removal, dilute mineral acid solutions (0.20-0.25% HCl or H_2SO_4) at 25-30°C are used to degrade the starch in the course of an 8- to 12-hr period. This mildly exothermic hydrolysis is uniform, and doesn't degrade the cellulosic fabric unless acid concentration increases by localized evaporation of water. Warm water washing removes the starch which has been converted essentially to glucose.

Enzymatic Desizing

A more advanced method for reducing the molecular weight and removing starch and starch derivative sizes is the enzymatic process. Enzymes are nitrogen-containing, complex proteinaceous substances secreted by the cells of living organisms and have good water solubility. Enzymes are substrate specific: They attack one specific material or chemical linkage and derive their name from the particular substrate on which they exhibit activity. Starch-splitting enzymes that hydrolyze starch (amylum) are called "amylases" [113,114]. The group of amylases that catalyze chain scission in the starch molecule or desizing enzymes may be classified according to their biological source as (a) vegetable - bacterial (fermenting cultures of

microorganisms) or - malt extract (germinated barley concentrates),
and (b) animal - pancreatic (extract from a slaughterhouse waste).
Three principal types of amylases are important, namely α-amylase,
β-amylase, and amyloglucosidase. These enzymes cleave the α-(1—4)-
glucosidic linkage in the starch molecule by different mechanisms.
They work at specific temperature ranges (45-75°C), concentrations
(0.5-15 g/liter), and pH conditions (5.6-7.5). At optimal temperature, the rate of hydrolysis is initially high and gradually approaches
an asymptotic value (1.5-2.0 hr) as low-molecular-weight maltodex-
trines form and the enzyme activity decreases. α-Amylase yields low-
molecular-weight dextrins through a multiple-attack mechanism by
random cleavages, and is the most resistant to thermal attack among
the three amylases. β-Amylase yields predominantly maltose by a
stepwise hydrolysis. Amyloglucosidase cleaves both amylose and amylopectin, yielding predominantly D-glucose. The time of commercial
enzymatic processes is generally long enough for partially depolymer-
izing and liquifying starch, but not for complete conversion of the
starch to D-glucose.

A number of factors influence enzyme activity and must be taken
into account in practical desizing operations. For example, pancreatic amylase works well at low temperatures (40-45°C), and is greatly
activated by the presence of sodium chloride or calcium chloride in
1/100 molar solution. Although pancreatic amylase has low heat-
stability, the fact that its optimum temperature range is low is advantageous in terms of energy requirements.

Surface-active agents are generally added in desizing to aid
wetting of the fabric and penetration of the size film. However, it
has been reported [115] that ionic surfactants inactivate enzymes to
varying extents. Heavy metallic ions such as Fe, Cu, Pb, Zn, and
Co also have retarding or inhibiting effects, as do additives containing phosphate, phenol derivatives, formaldehyde, and cationic softeners. The preferred group of additives are the nonionics. However,
caution must be exercised with these as well. For example, the action of bacterial amylase is retarded by wetting agents based on
ethylene oxide, and malt extracts are totally arrested by nonionic
surfactants.

In normal batch processes, the desizing cycle, using enzymes
from the amylase group, requires about 6-10 hr depending on temperature, enzyme type, concentration, pH, and the thermal deactiva-
tion rate of the enzyme.

The development of continuous scouring and bleaching processes
utilizing J-boxes and steaming equipment for open-width handling of
fabric has caused the slow, traditional desizing practice to become
obsolete. Several continuous desizing methods have been introduced,
based on the higher heat stability of bacterial enzymes. Bacterial
amylases, in the presence of starches, are capable of withstanding

temperatures of 70-120°C for 30-40 min. During this period, the starch is hydrolyzed so that it can be removed by hot water scouring. In practice, preswelling of the starch film is followed by passing the fabric through a hot water bath containing a surfactant and a solution of 6-12 g/liter of a bacterial enzyme; the pH is adjusted to 6.2-6.4 with acetic acid or sodium acetate. The fabric is then passed through a nip at 60-70 m/min and finally through a steamer at a temperature of 85-95°C. Steaming time is 3-5 min, followed by a hot water wash and bleach. This process yields a uniformly desized fabric. Satisfactory desizing of starch products has also been achieved with a 2-4 sec exposure of enzyme-saturated fabric at 120-130°C in open-width pressure-steaming equipment [116].

5.2 Oxidative Processes

Chlorine Treatment

In this oxidative desizing process, water-saturated gray fabric is exposed to chlorine gas. The reaction produces nascent oxygen, which depolymerizes the size film.

$$Cl_2 + H_2O \rightarrow HCl + [O]$$

Great care should be exercised in controlling the conditions of exposure to chlorine (moisture content, dwell time, gas concentration, temperature), because cellulose fibers may be attacked and weakened in the process. Better control can be achieved by replacing chlorine gas with dilute hypochlorite solutions.

Chlorite and Hypochlorite Treatment

The oxidation mechanism of hypochlorite (NaOCl) on polyalcohols (starches, starch derivatives, cellulose, PVA) and on certain gums has been investigated extensively [117-119]. Hypochlorites are nonspecific oxidizing agents. They attack ether bonds and hydroxyl groups and are capable of cleaving C—H bonds as well. Their activity is strongly dependent on the pH of the reaction medium. In chlorite ($NaClO_2$) desizing, several methods of activation can be utilized. Good results have been obtained by ammonium sulfate activation, by photoactivation, and by acid activation of sodium chlorite for starch on cotton yarn as shown below [117].

Efficiency of Desizing with Activated Sodium Chlorite Solutions[a]

	pH of solution		
Process of desizing	Initial	Final	% Size removal
Ammonium sulfate activated:			
10 g/liter sodium chlorite	7.10	6.20	84.6

	pH of solution		
Process of desizing	Initial	Final	Size removal
10 g/liter ammon. sulfate 100°C, 1 hr			
Photoactivated:			
10 g/liter sodium chlorite carbon arc, 1 hr, 45°C	8.94	8.40	61.4
Acid activated:			
10 g/liter sodium chlorite + sodium acetate-acetic acid buffer = pH 4.0 80°C, 1 hr	4.26	4.36	80.4

[a] Size content of yarn = 4.18%
Source: Ref. 117.

Different oxidation reactions with sodium hypochlorite on starches and on PVA have been described as follows [119].

1. In acidic solutions, the chlorine molecule attacks hydroxyl groups to form hypochlorite ester and HCl. Further decomposition of the ester yields a ketone and HCl:

$$H-C-OH + Cl-Cl \rightarrow HC-O-Cl + HCl$$

$$H-C-OCl \rightarrow C=O + HCl$$

2. In alkaline solutions, OCl^- is the oxidant, reacting with the hydroxyl groups of carbohydrates or PVA. The sodium salt of the hydroxyl compound dissociates into alcoholate anion and sodium cation, and the alcoholate ion reacts with OCl^-:

$$H-C-OH + NaOH \rightarrow H-C-ONa + H_2O$$

$$H-C-ONa \rightarrow H-C-O^- + Na^+$$

$$\text{H}-\overset{|}{\underset{|}{\text{C}}}-\text{O}^- + \text{OCl}^- \rightarrow \overset{|}{\underset{|}{\text{C}}}=\text{O} + \text{H}_2\text{O} + \text{Cl}^-$$

3. In neutral solutions, sodium hypochlorite is partially dissociated. The nondissociated fraction reacts with hydroxyl groups to form hypochlorite ester and water. This then is oxidized to ketone and HCl:

$$\text{H}-\overset{|}{\underset{|}{\text{C}}}-\text{OH} + \text{HOCl} \rightarrow \text{H}-\overset{|}{\underset{|}{\text{C}}}-\text{OCl} + \text{H}_2\text{O}$$

$$\text{H}-\overset{|}{\underset{|}{\text{C}}}-\text{OCl} \rightarrow \overset{|}{\underset{|}{\text{C}}}=\text{O} + \text{H}_2\text{O} + \text{HCl}$$

When cotton and cotton blends are treated with hypochlorite solutions in the scouring-desizing-bleaching cycle, it is difficult to prevent some damage to the fabric, but this can be minimized by critical control of processing. It is claimed that this drawback can be eliminated by a hybrid process [121] in which sodium chlorite is recommended for desizing and bleaching in combination with enzymatic agents.

Bromite and Hypobromite Treatment

Desizing of cotton with bromine compounds has been reviewed [120]. Sodium bromite ($NaBrO_2$) is a very active oxidizing agent, combining the effect of bromous acid ($HBrO_2$) and hypobromous acid [117, 120]. When starch is subjected to oxidation by sodium bromite, several reactions are possible, for example, opening of the glucose ring by scission of a C_2-C_3 bond, forming a dialdehyde. The dialdehyde intermediate is not readily soluble in water, but it is further degraded by a hot-water alkaline wash.

Another oxidation path may involve opening the ether linkage of the anhydroglucose by the sodium bromite. The pH of the desizing liquor is very important in the bromite process. Optimal desizing takes place at pH 9.8-10.2. At lower pH, decomposition of bromite is too rapid; at higher pH, oxidative depolymerization of the starch size proceeds too slowly. The process is carried out at about 30°C to minimize fiber degradation. Other advantages are the short immersion periods (5-15 min), the use of low concentrations of reagents, and the low temperature, without need for prolonged storage periods.

Peroxy Treatments

Peroxygens have been extensively studied [122-124] because they are known to react with polyhydric alcohols, including starches and starch derivatives. Desizing with peroxygens [hydrogen peroxide (H_2O_2), potassium peroxydiphosphate ($K_4P_2O_8$), ammonium perdisulfate [$(NH_4)_2S_2O_8$], ammonium hydrogen permonosulfate (NH_4HSO_5), etc.] can be carried out in alkaline solutions following singeing. Continuous processes for desizing, scouring, and bleaching are possible. A convenient method involves saturating the gray fabric with water at 60-65°C after singeing, adding a combination of hydrogen peroxide and potassium peroxydiphosphate (50% active at 3%: 100% active at 0.3%) along with the alkaline scour solution, steaming in a J-box at the boil for 75-90 min, and finally rinsing.

Peroxy-polymer reactions have been shown to involve free radicals and to proceed by homolytic scission of the starch (or PVA) molecules.

A possible mechanism for the reaction using peroxy disulfate-hydrogen peroxide under alkaline conditions can be postulated as follows:

$$S_2O_8^{--} \rightarrow 2\ SO_4 \cdot^-$$

$$SO_4 \cdot^- + H_2O \rightarrow H^+ + SO_4^{--} + OH \cdot$$

$$OH \cdot + H_2O_2 \rightarrow HO_2 \cdot$$

$$HO_2 \cdot + S_2O_8^{--} \rightarrow O_2 + HSO_4^- + SO_4 \cdot^{--}$$

$$HO_2 \cdot + H_2O_2 \rightarrow O_2 + H_2O + OH \cdot^-$$

$$S_2O_8^{--} + H_2O_2 \rightarrow O_2 + 2\ HSO_4^{--}$$

$$HO_2 \cdot \rightleftarrows O_2^{--} + H^+$$

$$O_2^{--} + S_2O_8 \rightarrow O_2 + SO_4^{--} + SO_4 \cdot^-$$

$$O_2^{--} + H_2O_2 \rightarrow O_2 + OH^- + OH \cdot$$

This sequence shows the peroxides to be sources of free radicals, which react instantly with the polymeric substrates. Potassium perphosphate is also an active desizing agent when added to an alkaline scour [122]. It requires a higher activation temperature than the persulfate, and hence is more stable at high temperatures, successfully boosting the scouring effectiveness [125,126]. Even without the synergistic effect of peroxydisulfates or peroxydiphosphates, hydrogen peroxide alone in strongly alkaline solutions is an effective desizing agent for high-molecular-weight PVA size which has become crystalline and partially insoluble in water as a result of heat setting [127]. Acceptable results have been obtained with alkaline hydrogen peroxide, stabilized and buffered with sodium silicate. At elevated temperatures, the oxidative degradation induced by this treatment depolymerizes crystalline PVA to a level where it can be rehydrated and subsequently resolubilized [128]. The optimum conditions for H_2O_2 desizing involve pH 11.4-11.8 (with 1.8-2.2% sodium hydroxide or sodium silicate), and a 0.2-0.4% hydrogen peroxide concentration. Reaction is rapid at 60-65°C, with a preferred dwell time of 10-12 min.

Glass fabric is often sized with a tenacious film-forming dextrinized starch-polyvinyl alcohol mixture. An economical desizing process for such compositions has been patented [129]. This involves the use of sodium carbonate peroxide ($2Na_2CO_3 \cdot 3H_2O_2$): The sized glass fabric is saturated with a solution containing 1.0-18.0 g/liter sodium carbonate peroxide, or about 0.03-0.50% sodium carbonate peroxide on the weight of sized fabric. Subsequent exposure to 75-85°C for 1-3 min, followed by a hot water scour, removes 90% of the sizing polymer, and the residual organic material is eliminated by coronization.

Hydrogen peroxide is a versatile oxidizing agent for continuous scouring, desizing, and bleaching processes, and it can be used in acidic, neutral, and alkaline solutions [130]. Although peroxide treatments are overwhelmingly used for bleaching purposes at acidic and neutral pH, size removal can also take place under these conditions, depending on the temperature, peroxide concentration, and dwell time.

5.3 Alkaline Scour Processes

The best understood and simplest size removal entails an alkaline scour. The principal steps in the process are:

1. Wetting out the greige goods with a suitable low-foaming wetting agent, preferably nonionic or anionic, in a bath of hot water (at 65-95°C) containing 1-2.5% soda ash or sodium hydroxide solution at a liquid-to-fabric ratio of 50-100:1.
2. Steeping, to soften and swell the size film. (Dwell time, depending on the nature of the size, denier, and fabric construction, is 15 min to 2 hr.) This may be done with mechanical agitation, or beaming from end to end several times.
3. Thorough rinsing in cascade-washing or overflowing water at 25-85°C in the roll 3-8 times. The effluent water can be reprocessed through various recycling steps, or discarded, followed up by secondary water treatment.

Although the preferred desizing processes for polysaccharides are based on thorough enzymatic size removal, the periodate-oxidized starches (starch aldehydes) are known to undergo rapid depolymerization in alkali [131]. In the early stages of polysaccharide degradation with sodium hydroxide solutions, the major acidic products are reported to be glycollic acid and α,γ-dihydroxybutyric acid [132, 133].

Alkaline desizing is the dominant choice for removal of sizes based on polyvinyl alcohol. Depending on the residual acetate content, fully hydrolyzed and partially hydrolyzed grades present varying degrees of difficulty. The difference becomes particularly important when the fabric is heat set in the greige state. Removability of partially hydrolyzed polyvinylalcohol is not greatly imparied even after heat treatment at relatively high temperature (170-200°C) because the bulky acetate substituents inhibit molecular orientation and the tendency to form insoluble crystallites. In the case of fully hydrolyzed PVA, similar conditions of heat setting lead to crystallization, reduced solubility, and a high solvated gel viscosity [134]. The partially hydrolyzed resin thus exhibits a significant advantage in ease of removal over the fully hydrolyzed product: desizing steps are similar for both resins but require somewhat more severe conditions in the case of the fully hydrolyzed polymer. The degree of desizing (% size removal) on fabric containing 5% by weight of PVA by different alkaline treatments is shown in Table 1.14.

The effect of an anionic or nonionic wetting agent added to the alkaline solution is demonstrated by these results. More generally, additives that promote swelling, wetting, and diffusion markedly accelerate the removal of sizing polymers, including acrylate, styrene, polyvinyl acetate, and copolymers. These polymers are not attacked chemically by surfactant/alkali solutions, but in some cases, they form water-soluble salts and are thus solubilized by alkali.

Table 1.14 Alakline Scour Desizing

Chemical treatment	Concentration of chemical (%) based on bath weight	Desizing (%)
25% NaOH solution	12	82.0-87.0
25% NaOH solution	4	88.0-92.0
25% NaOH solution	2.5	94.0-96.0
25% NaOH solution + wetting aid	2.5 0.1	99.0-100.0

5.4 Solvent Desizing

Water has been and still is the traditional medium for processing textiles. However, the presence of alkalis, surfactants, polymeric compounds, oxidizing agents, etc., in industrial effluents poses problems of stream pollution. The mandatory treatment of waste waters and increasing limitations in the availability of clean water have motivated development work on solvent processing. The use of recoverable halogenated solvents enables textile mills to size/desize fabrics in solvents. This approach also reduces energy consumption, and little or no water is used. Solvent desizing must be considered in conjunction with solvent sizing (see Sec. 3 and 4). Halogenated solvents selected for sizing/desizing and the special equipment involved have also been discussed in these sections. In this section, the effects of the solvents on fabrics during the desizing steps will be briefly reviewed [135].

The rapid developments in solvent sizing in Europe, Japan, and in the United States have not been followed by extensive industrial application. This may be attributed to declining incentives for investments in new technology at this time, as well as to the growing concern over long-range and largely unknown effects of atmospheric, water, and land pollution by chlorinated hydrocarbons. An ambitious program of physical, chemical, and medical studies on the evaluation of the properties of organic solvents capable of displacing water in textile technology has been implemented. After considering all the relevant factors (ecological, machinery, corrosion, recovery, economy, flammability, personal safety, etc.), it was concluded that halogenated hydrocarbon solvents are the best suited for use in textile "wet" processing.

Compared with water, some organic solvents can "wet" and penetrate sized yarns faster, and to a greater extent. This is because of lower dipole moment, lower viscosity, and dielectric constant of the solvents (Table 1.15). In addition, the low contact angle (θ) of these halogenated solvents allows the liquid to penetrate the intermicellar spaces within the fiber more thoroughly than water. The presence of the chlorinated solvents does not influence the moisture content of the sized yarn, which thus remains "dry." Desizing with solvents is carried out in an enclosed compartment. Close contact with the solvent (dwell time) is usually less than 40 sec at a temperature of 70-120°C, depending on the boiling point of the solvent. There is a rapid removal of the size polymer during the first 10-15 sec of the treatment, and little further polymer solubilization takes place thereafter. Size removal is about 92-97%. The polymer may be separated from the size/solvent mix and recycled, while the solvent is purified by distillation. In continuous operations, the warp is dried in hot air, and residual solvent is removed by activated carbon absorption. When the carbon bed is saturated, the

Table 1.15 Effects of Organic Solvents on Fibers

Solvent	Boiling pt. (°C, 1 atm)	Wetting - surface tension (dynes/cm, 20°C)	Diffusion and swelling		
			Viscosity cps (20°C, 1 atm)	Dielectric constant (20°C, 1 atm)	
Water	100	72.6	1.00	80.3	
Stoddard solvent	185	28	0.90	2.1	
1,1,1-Trichloroethane	74.1	26.5	0.79	7.5	
Trichloroethylene	86.9	32.0	0.54	3.4	
Perchloroethylene	121.1	32.3	0.83	2.3	

solvent is reclaimed by steam distillation, recondensed, and purified in water separators. Some properties of solvent which are relevant to the desizing/recovery process are shown in Table 1.16.

The chlorinated solvents are essentially inert, but if processing temperatures exceed 140-150°C, decomposition by hydrolysis or oxidation may take place, for example:

For perchloroethylene:

$$CCl_2 = CCl_2 \xrightarrow[\Delta]{[O]} CCl_3, COCl, COCl_2$$

$$CCl_2 = CCl_2 \xrightarrow[\Delta]{(H_2O)} CCl_3COOH, HCl$$

For trichloroethylene:

$$CHCl = CCl_2 \xrightarrow[\Delta]{[O]} CHCl_2COCl, COCl_2, HCl, CO$$

To prevent decomposition, small quantities of antioxidants (e.g., organic amines, epoxides, or alkyl phenols) are added to the solvent. After recycling, these additives must be replenished. The present assessment of solvent desizing may be summarized as follows:

Advantages of solvent desizing:

1. It is a very clean operation with no disposal problems; the pollution effect is negligible.
2. The time of desizing and the floor area for the machinery is far less than for conventional aqueous systems.
3. Due to quick and even solvent penetration, the size removal is more uniform, and subsequent dyeing is of higher quality.
4. Acrylics and wool fabrics develop a lustrous, soft hand in solvent, without any felting.
5. In aqueous desizing, polyester and polyester blends may show signs of trimer and oligomer deposits which can cause streaky dyeing. Solvent desizing is capable of removing most of these low-molecular-weight fragments, yielding more uniform dyeing.
6. When solvent desizing is followed by bleaching, the caustic scour can be omitted since removal of the pectin and wax particles is affected by the halogenated hydrocarbons.

Disadvantages of solvent desizing:

1. Although solvent desizing ranges are available commercially, the price of the equipment is high.
2. The majority of size systems used in aqueous processes are not readily applicable in solvent desizing.
3. Cost of processing is higher than for water-based processes.

Table 1.16 Solvent Desizing – Solvent Properties

Solvent	Recovery by condensation	Solubility of:		Heat required to evaporate at 20°C (Cal/g)
	vapor in air (mg/liter)	solvent in water (mg/liter)	water in solvent (mg/liter)	
Water	—	—	—	625
Stoddard solvent	300	100	350	135
1,1,1-Trichloroethane	1400	500	570	73
Trichloroethylene	700	1100	230	73.5
Perchloroethylene	260	100	72	71

4. Most polyester blends and cellulose acetate, when desized in solvents, retain a relatively high concentration of solvents, which requires a high-temperature flash-steam treatment for eliminating it [136].

5. Solvent retention on hydrophobic fibers may cause changes in surface properties.

It is difficult to forecast the future of solvent desizing processes, but it is probable that they will ultimately prove practicable in industry.

5.5 Low-Temperature Plasma Treatment

Cold plasma technology [137] offers an ecologically viable desizing process where solvent treatment or chemical recovery and recycling is not feasible or desirable. Cold plasmas are characterized by the lack of equilibrium between molecules (free radicals) and charged particles (ions and electrons) present. In a "cold plasma," neutral particles have kinetic energies in the 80-100 K range, while charged particles exhibit kinetic energies in the 5,000-60,000 K domain. Plasmas are generated by special equipment, and the substrates must be exposed in an evacuated chamber. Textile materials can survive the moderate operational temperatures and the changes effected by the plasma, since the treatment is restricted to the surface, with a penetration range of 40-100 millimicrons. The degree of size removal depends on variables such as radiofrequency power (W), dwell time, and plasma gas composition.

When an oxygen plasma is used, and the fabric is sized with PVA, CMC, or starch, the process consists of two steps: (a) subjecting the sized fabric to a 100-200 W plasma exposure for 8-10 min. This operation converts the organic sizing materials to innocuous gaseous compounds which are evacuated. This eliminates about 55-65% of sizing materials in gaseous form. (b) Rinsing in water at room temperature. This removes an additional 30-38% of the size, so that the combined steps result in 95-98% size removal. With fine tuning, a higher level of gasification can take place in the first step (a) before fiber loss becomes severe due to deeper plasma penetration. In the case of PVA size, the possible simplified pathways of oxidative decomposition may be formulated as shown in Fig. 1.6 and a flow diagram for the process is shown in Fig. 1.7.

5.6 Desizing by Thermochemical Extraction and Coronization (Glass)

In the case of glass fabrics, the removal of sizing chemicals requires a specialized approach. Suitable sizing formulations for glass may

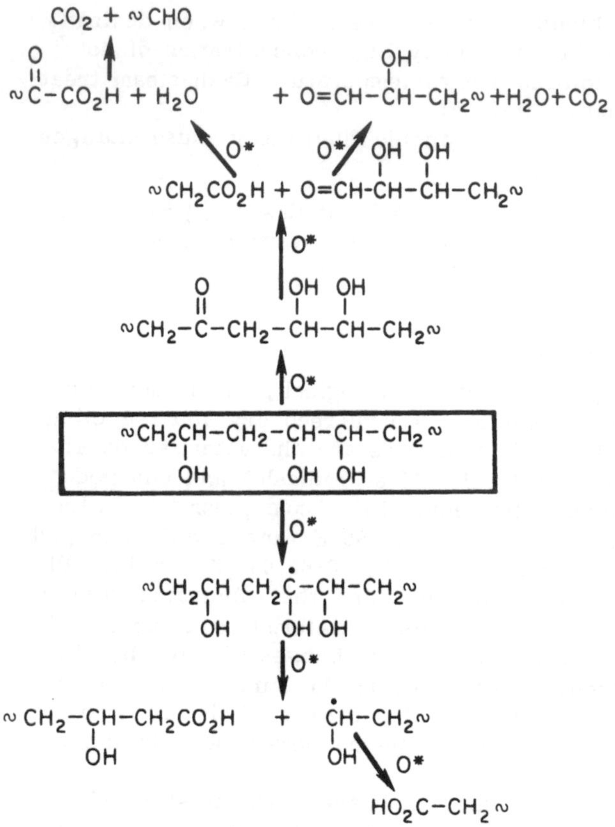

Figure 1.6 Oxidative decomposition of polyvinyl alcohol. (From Ref. 137.)

include PVA, hydrogenated vegetable oils, dextrinized starches, non-ionic and anionic detergents. These agents must be removed after weaving and before coating, resin impregnation, or dyeing can be carried out. The principal method of size removal is a two-step process involving chemical extraction and coronization. The chemicals proposed for the first step include peroxydisulfates [138], peroxydiphosphates, sodium hypochlorite, sodium azide [139], and sodium carbonate peroxide [129]. Treatment with these reagents is carried out in an aqueous solution containing 1.0-18.0 g/liter of chemical and 0.025-2.0% of an anionic or nonionic surfactant at pH 7-9, at 10-80°C, with a residence time of 20-140 sec. This chemical treatment removes about 85-95% of the size film without causing deterioration or tensile loss of the glass fabric. Following washing, the glass fabric is dried and can be weave-set or further desized by coronization at 650-1200°C for 5-25 sec. Although glass fabric has good

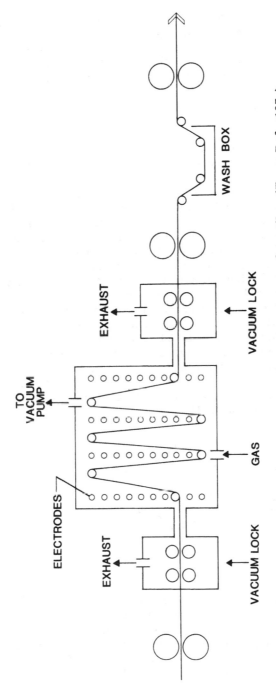

Figure 1.7 Proposed process for continuous plasma treatment of textiles (From Ref. 137.)

resistance to heat exposure over a wide temperature range, a decrease in strength and flexibility may occur at these high temperatures. By employing a very effective wet chemical extraction, the time exposure of the fabric in the coronization step can be reduced, preserving the strength and flexibility of the fabric to a higher degree.

5.7 Durable Sizes

Most sizing compositions are formulated for complete removal by aqueous or solvent processes. However, the loom-finish state is sometimes desirable to enhance hand, wear characteristics, or weight (body). These features can be attained by employing a nonremovable size. Many acrylate polymer sizes can be insolubilized by controlled cross-linking with appropriate catalyst systems without interfering with the primary functions of the size. Other insolubilizing methods for specific polymers include high-temperature heat setting. In the case of styrene-maleic anhydride copolymers, the ammonium salt is a water-soluble system; with high-temperature heat setting (180-220°C for 45-100 sec), the ammonia is driven off, and an insoluble size finish remains [22]. An association complex of polyacrylic acid with polyethylene glycols has also been reported [140]. This complex forms readily at 20-25°C and involves strong hydrogen bonding, yielding films of high strength, considerable hardness, and water insolubility. Other complexes of polyacrylic acid (e.g., with poly-N-vinyl pyrrolidone) have also been disclosed. Insoluble sizes have been obtained by reaction of starch with a solution of a triazine of the general formula [141]:

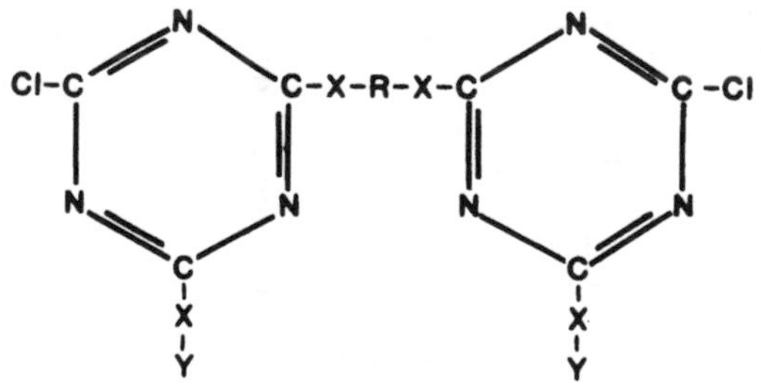

where X is NH; O, or S, or a direct C–C linkage; Y is an aliphatic, aromatic, or heterocyclic radical rendered water-soluble by the presence of at least one sulfo, carboxyl, or polyglycol ether group, and R is a divalent aliphatic, aromatic, or heterocyclic radical that contains at least one substituted triazine group. For example, unmodified starch and a triazine of the formula:

yield an insoluble size suitable for cellulose, polyester, regenerated cellulose, and polyacrylic yarns [141]. Similar results are obtained when aqueous solutions of polyvinyl alcohol are treated with 2-β-naphthoxy-4,6-dichloro-1,3,5-triazine in the presence of sodium bicarbonate [142].

It has also been reported that when methacrylic acid monomer or sebacic acid are heated in the presence of polyvinyl alcohol (20% PVA solution, 90°C, 3-5 hr), clear homogeneous solutions are obtained, which, when deposited on cellulose, polyester, or acrylic yarns and dried at 110-130°C, yield insoluble reaction products of the PVA with the methacrylic or sebacic acid. The clear film formed is insoluble in hot water and in most organic solvents [143]. Generally, PVA can be cross-linked by controlled reaction with polycarboxylic acids (succinic, maleic, adipic, glutaric, diglycolic, etc.) or with aldehydes, yielding films that are acceptable for sizing and that can be insolubilized by heating to form permanent coatings. In a similar manner, a great variety of water-soluble hydroxyl-bearing materials (such as starches, dextrines, galactomannans, etc.) can be insolubilized by reacting them with preformed polymeric acetals (condensation product of diethylene glycol and formaldehyde) followed by curing [144]. A low-viscosity, insoluble permanent size is obtained from PVA and urea-formaldehyde condensation products under conditions of acid catalysis.

6. SIZE RECOVERY AND RECYCLING

About 220 million kg of natural and synthetic sizing agents are annually applied in the United States for weaving (1978 estimate) [145]. These polymers are removed by a scouring process prior to finishing. No other operation in textile finishing offers challenges for energy conservation and pollution control equal to those presented by recovery and recycling of synthetic warp sizes. In the past, the practice has been the simple release of these chemicals into waste-water (public streams), contributing greatly to the pollution load of textile mill effluents. This has not been possible after the Clean Water Act of

1972, requiring zero waste discharge by 1985. Unless treated in wastewater processing facilities, typical warp sizes will decompose in the stream, consuming high levels of dissolved oxygen, and creating a source of Biological Oxygen Demand (BOD). Starches and their derivatives show a high initial BOD, as shown below, but can be effectively degraded in activated sludge treatment.

Material	5-day BOD ppm[a]
Pearl cornstarch	810,000
Methyl cellulose	1,600
Carboxymethyl cellulose	10,000
Hydroxyethyl cellulose	6,500
Polyvinyl alcohol	16,000

[a]Oxygen consumption in 5 days as measured by standard test.

Other materials, however, for example, PVA, CMC, and the acrylates have lower initial BOD, but are more difficult to deal with in wastewater processing. Many approaches to size waste recovery have been explored [146]. Starch and starch derivatives always degrade during desizing (either chemically or enzymatically) and cannot be recycled. PVA, CMC, and some of the acrylate copolymers can be removed from the effluent when the size concentration is above 1.0-1.8%.

A general method for recovery is based on complexing the polymer to form an insoluble salt. Carboxymethyl cellulose responds to this treatment well because it contains reactive sites for complexing. When CMC wastes are treated with 0.1-0.2% aluminum sulfate, a rapid precipitation occurs, with 99% CMC recovery. The process involves:

1. Precipitation with alum:

$$\{C_6H_7O_2(OH)_{2.3}(OCH_2COONa)0.7\}n + \frac{0.7n}{6} Al_2(SO_4)_3$$

$$\{C_6H_7O_2(OH)_{2.3}(OCH_2COO\frac{Al}{3})0.7\}n + \frac{0.7n}{2} Na_2SO_4, \text{ and}$$

2. Resolubilization with alkali:

$$\{C_6H_7O_2(OH)_{2.3}(OCH_2COO\frac{Al}{3})0,7\}n + 0.7n \text{ NaOH}$$

$$\{C_6H_7O_2(OH)_{2.3}(OCH_2COONa)0.7\}n + \frac{0.7n}{3} Al(OH)_3$$

After the sodium salt of CMC is formed, the product can be concentrated and reused as a size, with only slightly impaired properties, for instance, flexibility, elongation, shedding viscosity. Progressively poorer results in subsequent cycles are due to a small but steady buildup of impurities. CMC can also be regained by complexing with ferric chloride, magnesium, or lead salts, but these treatments are not practical because they induce discoloration or they are too costly. Recycled CMC has not yet found extensive application in the industry. Even when blended with 5-8% virgin CMC after each cycle to restore properties, recovered CMC is considered to have only limited industrial value. Ethylene-acrylic acid interpolymers applied from aqueous solution (sodium salt) can be recovered by acidifying the wastewater to pH 5-6, and thus precipitating the polymer almost quantitatively. The polymer can be reconverted to the soluble salt, concentrated, and reused. The coagulant acid may be formic, phosphoric, sulfuric, or hydrochloric acid [147-149]. In one system, the polymer contains acetyl, carboxyl, and sulfate groups in a well-balanced, economically feasible product [149]. The acidic form is an insoluble solid at a pH of 4.6-6.0. At higher pH (6.2-7.5) in a 6-9% aqueous solution, the polymer provides a warp size for staple or filament yarns. After desizing, the polymer is precipitated from wastewater by adding acid to pH 4.0-4.5, and collected. This recovered solid may be treated with an alkali metal base or with NH_4OH, and the soluble salt may be used in a new sizing cycle. A typical reaction sequence involves precipitation of acrylic copolymer at pH 5.0 with Mg salt:

$$\left[(CH-CH_2)_x \text{---} (CH-CH_2)_y \text{---} (CH-CH_2)_z \right]_n + \frac{(y)n}{2} MgSO_4 \longrightarrow$$

with substituents $O-COCH_3$; $O-COCH_2COONH_4$; $O-SO_3Na$

$$\left[(CH-CH_2)_x \text{---} (CH-CH_2)_y \text{---} (CH-CH_2)_z \right]_n + \frac{(y)n}{2} (NH_4)_2 SO_4$$

with substituents $O-COCH_3$; $O-COCH_2COO\frac{Mg}{2}$; $O-SO_3Na$

Changes in polymer properties do occur as a consequence of recycling. The infrared spectrum of the polymers remains unchanged, but the solution viscosity increases steadily in succeeding cycles, and film elongation tends to decrease. There is a loss of chemical of about 3-5% per cycle, and a mechanical loss of 9-11%/per cycle. The chemical loss is due to less than quantitative yield in precipitation. The mechanical deficit is due to the geometry of vessels, faults in collection methods, pipe systems, and abrasion shedding during weaving.

Recyclings of the original polymer without addition of virgin material show accumulation of auxiliary chemicals such as waxes, humectants, inorganics, and foreign matter removed from the warp (e.g., pectins, oligomers of polyester, etc.). At this time, a limit in the number of effective cycles would soon be reached, unless the addition of 7-10% of virgin polymeric material is carried out in each new cycle. With this practice, the limit in the number of cycles is increased, as shown by the data tabulated in Table 1.17 [149].

Hydroxypropylated polysaccharides are not a large class of industrially useful warp sizes, but hydroxypropyl cellulose (HPC) offers a unique and simple method for recycling [150].

HPC:

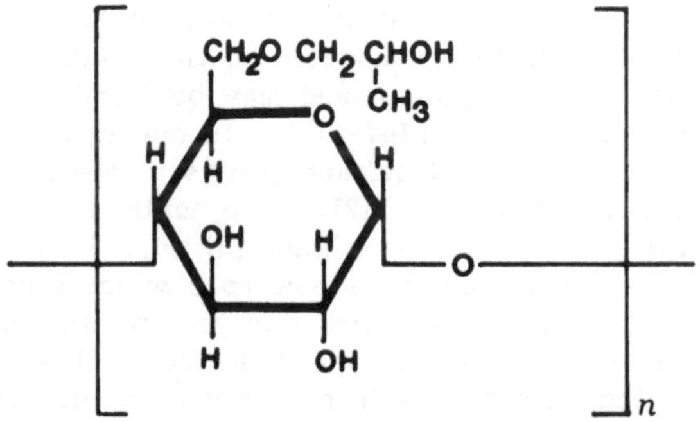

HPC is soluble in cold water, and at a concentration of 8-13% it forms a good warp size. It can be readily precipitated from the desizing waste-water by simply heating the liquor above 45°C, where the poly-mer precipitates quantitatively as a floc. The supernatant liquor may be discarded or reclaimed since it is essentially free of BOD. The concentrated HPC floc can be reutilized as a size without further treatment.

In the case of certain acrylate copolymers, a 70-80% size re-covery can be accomplished by high concentration desizing [151]. When the ammonium salt of an acrylic acid-acrylonitrile (85/15) co-polymer is removed in desizing, this can be done in quantities of water not exceeding 200% on the weight of fabric. The first desize liquor contains 70-80% of the size used, at a solids concentration of 6.5-7.0%. It can be brought to the desired concentration of 8-9% by adding a concentrated solution of the original polymer. The fabric goes through a second desizing step where the remaining (1.5-2.0%) chemicals are removed in water. This portion of the ex-tract is released to the water treatment plant.

Complex formation is not viable with PVA sizes, because reac-tive sites for combination with multivalent metallic ions are not present [146]. Insolubilizers like amine condensates, aldehydes, and trimethylol melamine are not applicable to aqueous solutions of

Table 1.17 Recycling Limitations

Recycle No.:		1	2	3	4	5	6	7	8
A. Same size used for 9 runs									
Polymer content (%)	96	93	89	87	85	82	79	79	77
Inorganics (%)	0	2.2	2.9	3.6	4.2	4.9	5.1	5.3	5.5
Residue (600°C) (%)	13	14.3	14.8	16.0	16.6	17.3	17.9	18.6	19.0
B. 10% virgin polymer added to each run									
Polymer content (%)	96	95	95	94	94	93	93	92	91
Inorganics (%)	0	1.9	2.1	2.4	2.7	2.9	3.0	3.0	3.1
Residue (%) (600°C)	13	13.4	13.7	13.9	14.2	14.2	14.4	14.5	14.5

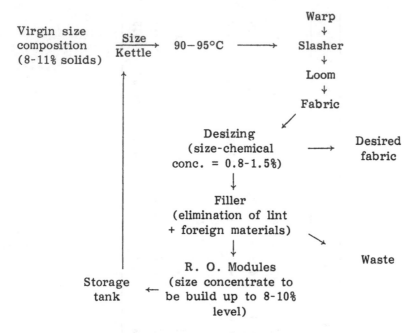

Figure 1.8 Schematic of reverse osmosis size recovery.

PVA [152]. The salt precipitation techniques are not practical because the high dilution of the PVA waste (max 1.4-1.6% PVA) would
require such a high stoichiometric ratio of sodium sulfate or carbonate that the salt itself would be considered a pollution problem.
The conclusion has been reached that since PVA desize water is
thermally stable and resistant to biodegradation, it is ideally suited
for recovery by membrane separation technology or reverse osmosis
processes (R.O.) [153].

By using organic or inorganic membranes in tube modules, the
sizing chemicals can be separated from the waste stream (1.5-1.8%
size polymer) and concentrated to use levels (9.5-11.0% size polymer).
The permeate is discharged from the system with only trace amounts
of the size chemical in it [154-158]. The process can realize significant savings in both water and energy consumption. A flow chart is
shown in Fig. 1.8, and a brief description of the highlights of R.O.
is given below.

In order to obtain pure solvent from a solution through a semi-
permeable membrane, the direction of flow of the solvent must be reversed, hence the process is termed reverse osmosis (R.O.). It is
a useful tool for separation in the ionic and macromolecular range
(10^{-3}-10^{-1} μ). In this range, R.O. overlaps with ultrafiltration, but

it differs from it in that material transport occurs by a diffusive rather than viscosity-flow mechanism.

The membranes used in retaining organic solutions function as molecular sieves. Rejection is dependent on the size and shape of the molecule with respect to the size of the pores. The lower-molecular-weight organics are enriched at the membrane surface. Macromolecules larger than pore size are rejected, and flux rate is dependent [159] on the rate of solute flow (pressure across the membrane), chemical change of membrane (hydrolysis, etc.), bacterial growth on membrane (reduced flow), and temperature of solution.

Commercially useful membrane materials are polyamides, polyimides, cellulose acetate, cellulose nitrate, polytetrafluoroethylene, PVC, and carbon base-inorganics composites. Depending on the pore size, the membranes have a 0% rejection rate for inorganic salt solutions (NaCl, $CuSO_4$, etc.), and a molecular cutoff of 250-1000 MW for organics. This indicates that warp size polymers like PVA, acrylics, or polyester disperse systems are quantitatively retained. The regained polymers can be reused as sizes in slashing. A minimal deterioration is noted in the elongation, tensile, and modulus of the polymers: a shearing effect slowly erodes the average molecular weight of the sizes. This can be rectified by the periodic addition of 5-7% of virgin size to the feed. The quality of the permeate water is also well suited for recycling.

More complex multicomponent warp sizes can also be recovered by the R.O. process, for example:

Type of size	Initial feed concentration	Recovered size concentration	Concentration in water permeate
Vinyl acetate copolymer	2.0%	8.5%	0%
PVA/vinyl acetate copolymer, 70/30 blend	2.9%	3.6%	0%
Acrylic size-binder	5.4%	10.0%	0.2%

The rate of concentration change is related to the viscosity of the recovered size concentrate. In the case of PVA/vinyl acetate copolymer blend, it is significantly slower (1.5 times) than for the vinyl acetate copolymer (3 times) or for the acrylic binder (2 times).

In summary, the successful development and commercial introduction of R.O. filtration processes shows the following advantages:

(a) high-priced polymeric sizes can be used since recovery is high; (b) savings in water, chemicals and energy can amortize the high initial cost of the R.O. installation; and (c) weaving results with recovered size are equal to or better than those obtained with conventional materials.

REFERENCES

1. R. H. Pickard, *J. Text. Inst. 9*, 18 (1918); *10*, 54 (1919).
2. S. M. Neale, *J. Text. Inst. 15*, 443 (1924).
3. F. T. Pierce, *J. Text. Inst. 19*, 237 (1928).
4. F. D. Farrow, *J. Text. Inst. 14*, 414 (1923).
5. R. A. Schutz, Proceedings of the II Internat. Sizing Symposium, Budapest (1974).
6. F. H. Mark, *Adhesive Age*, p. 35-40 (July 1979) and p. 45-50 (September 1979).
7. C. R. Blumenstein, *Text. Ind. 133*, 93-97 (June 1969).
8. W. A. Glaeser, *Wear 6*, 93 (1963).
9. W. E. Morton and J. W. S. Hearle, Physical properties of textile fibers. Second edition, The Textile Inst. (GB), (1975).
10. Properties of Manmade Fibers: Text. Ind., W. R. C. Smith Co., Atlanta, Ga. (yearly publication).
11. R. W. Moncrieff, *Man-Made Fibers*. Fifth Edition. London: Haywood Books, (1970).
12. R. Wong, to Corning Fiber Glass Co. U.S.P. 3,772.870 (1973).
13. J. E. Ward, to Corning Fiber Glass Co. U.S.P. 3,799,981 (1973).
14. A. Marzocchi, to Corning Fiber Glass Co. U.S.P. 3,827,230 (1974).
15. W. Haggerty, to Corning Fiber Glass Co. U.S.P. 3,935.344 (1976).
16. D. M. Fahey, to PPG Ind. U.S.P. 3,940,357 (1976).
17. J. E. Hill, to Great Lakes Carbon Co. U.S.P. 3,837,904 (1974).
18. E. C. Langerak, to DuPont Chem. Co. U.S.P. 2,692,873 (1954).
19. F. S. Martin, to U.S. Rubber Co. U.S.P. 2,751,363 (1956).
20. *Textile Progress 8* (1), 91 (1976).
21. D. A. Bixler, *Amer. Dyestuff Rep. 54*(10), 28-29 (1965).
22. P. V. Seydel, "Textile Warp Sizing" Phoenix Printing, Inc., Atlanta, Ga. (1981).
23. C. C. Kessler, to Penick & Ford U.S.P. 2,516.632-3-4 (1950).
24. Jackson & Hudson, *J. Am. Chem. Soc. 59*, 2049 (1937).
25. Dvonch & Mehltretter, to Sec. Agric. U.S.P. 2,648.629 (1953).
26. Mehltretter, to Sec. Agric. U.S.P. 2,713,553 (1955).
27. Abeve Co., Veendam, Holland. U.S.P. 2,829.987-8-9 (1958).
28. A. N. Edmondson, Development of starch derivatives. Proceedings of III Internat. Sizing Symp., Manchester, England (1977).

29. L. H. Elizer, to Hubinger Co., Keokuk, Iowa U.S.P. 3,793.062 (1974).
30. Ahmedabad Text. Ind., British Patent 1,221.760 (1971).
31. K. P. Satterly, to Standard Brands Inc. U.S.P. 2,831.853 (1958).
32. Ahmedabad Text. Ind. British Patent 1,213.724 (1970).
33. L. P. Hayes and R. L. Drury, to Stanley Manuf. Co. U.S.P. 3,719.664 (1973).
34. V. Schobingen, et al. to Blattmann & Co. Waedenswil, Switzerland U.S.P. 3,719.617 (1973).
35. R. L. Best, to A. E. Staley Manuf. Co. U.S.P. 3,709.788 (1973).
36. G. S. Bairnd and A. L. Griffiths, *Modern Tex. Mag.* 46(6) 34, 59-60. (1965).
37. Hercules Inc., *Technical Service Information, CMC Warp Size* (1981).
38. V. Heap, to Allied Colloids Co. Bradford, England U.S.P. 3,711.323 (1973).
39. D. J. Guest, to I.C.I. Millbank, London, British Patent 809.745 (1959).
40. W. Bonin, to Bayer Akt. Germany. U.S.P. 4,001.193 (1977).
41. A. T. Walter and G. M. Bryant to Union Carbide U.S.P. 3,321.819 (1967).
42. G. C. Johnson, to Union Carbide U.S.P. 3,904.805 (1975).
43. J. R. Fallon and S. H. Foster to Monsanto Chem. Co. U.S.P. 2,859.189 (1958).
44. A. E. Corey and D. D. Donermeyer, to Monsanto Chem. Co. U.S.P. 3,759.858 (1973).
45. Monsanto Chem. Co. Br. Pat. 799.758 (1955).
46. A. E. Corey, and D. D. Donermeyer to Monsanto Chem. Co. U.S. P. 3,716.547 (1973).
47. A. E. Corey and D. D. Donermeyer, to Monsanto Chem. Co. U.S.P. 3,723.581 (1973).
48. A. E. Corey and D. D. Donermeyer, to Monsanto Chem. Co. U.S.P. 3,770.679 (1973).
49. A. E. Corey and D. D. Donermeyer, to Monsanto Chem. Co. U.S.P. 3,817.892 (1974).
50. A. E. Corey and D. D. Donermeyer, to Monsanto Chem. Co. U.S.P. 3,854.990 (1974).
51. A. E. Corey and D. D. Donermeyer, to Monsanto Chem. Co. U.S.P. 3,919.449 (1975).
52. M. Hijiya and M. Hirao, to Hayashibara, Japan. U.S.P. 3,804.785 (1974).
53. H. K. Inskip, to E. I. duPont U.S.P. 3,689.469 (1972).
54. *E. I. duPont Technical Bull.: Biodegradation of Elvanol.* Bul-A-8246 (1972).
55. C. R. Blumenstein, *Text. Ind.* 130, 63-69 (July 1966).

56. J. C. Lark, to Standard Oil Co., Indiana. U.S.P. 4,035.531 (1977).
57. J. C. Kibler and R. G. Lappin, to Eastman Kodak Co. British Patent 1,237.671 (1968).
58. R. N. Vachon, to Eastman Kodak Co. British Patent 1,470.873 (1974).
59. J. Fritz, to Rhone-Pragil, France. U.S.P. 3,978.262 (1976).
60. Research Disclosure (anonymous) #15856; (October 1978).
61. H. R. Kuemmerer, to Deering Milliken Corp. U.S.P. 3,061.470 (1962).
62. A. F. Harris, to Monsanto Chem. Co. U.S.P. 2,848.357 (1958).
63. J. C. Prangle, to Dan River Inc. U.S.P. 3,981,836 (1976).
64. J. C. Prangle, to Dan River Inc. U.S.P. 3,983.271 (1976).
65. W. S. Perkins, *Am. Textiles 5*, 20 (May 1976).
66. C. Klyszejko, Solvent Sizing, Proceedings of III Internat. Sizing Symp. Manchester, England (1977).
67. K. C. Schram, to Colloids Inc. U.S.P. 3,728.299 (1973).
68. J. Sano, to Kanebo Ltd. Japan, U.S.P. 3,984.594 (1976).
69. J. E. Pritchard, to Phillips Co. U.S.P. 2,978.362 (1957).
70. J. W. Case, to I.C.I. Ind. U.S.P. 3,476.504 (1969).
71. G. P. Beaumont to Dow Chem. Co. U.S.P. 4,076.629 (1978).
72. J. Zawadzki and T. Janusz, Dry sizing; Proceedings of III Internat. Sizing Symp., Manchester, England (1977).
73. K. Ramaszeder, *Melliand Text. Ber.* (Eng) 8(8), 671-3 (1979).
74. Burlington Ind., Greensboro, N.C. British Patent 1,448.483 (1976).
75. Air Reduction Co., British Patent 863.229 (1961).
76. H. Wechsler, to Borden Chem. Co. U.S.P. 2,966.480 (1960).
77. Eastman Kodak Co. British Patent 1,448.794 (1976).
78. Eastman Kodak Co. British Patent 1,463.106 (1977).
79. G. E. Smedberg, to duPont de Nemours. U.S.P. 3,914.489 (1975).
80. Burlington Ind. Inc., U.S. Patent Application 147,335 (May 6, 1980).
81. T. Rasbottom, *Warp Sizing Mechanisms*, Columbine Press, Manchester, England (1964).
82. J. B. Smith, *Technology of Warp Sizing*, Columbine Press, Manchester England (1964).
83. A. D. Cotney, *The Modern Slasher*, West Point Mach. Co. Bull. (1976).
84. A. V. Douglas, *Tex. Manuf. 90*, 140-3 (1964).
85. R. B. Pressley, *Tex. World 112*, No. 6, 38-52 (1962).
86. Colman Mach. Co., Rockford, IL, Bull.: *Slashing Systems* (1975).
87. R. A. Schutz and P. E. Exbrayat, to Inst. Text. de France. U.S.P. 3,791.132 (1974).

88. D. V. Nesbitt, *Tex. Bull. 90* 46-48 (1964).
89. S. Kuroda, to Kawamoto Ind. U.S.P. 4,025.993 (1977).
90. Tsudakoma Ltd.: *Warping Sizer*, Comp. Bull. 1-30 (1977).
91. R. B. Pressley, *Tex. World 112* No. 9, 52-60 (1962).
92. To Platt Saco Lowell Ltd., British Patent 1,502.604 (1978).
93. West Point Mach. Co. Bull., *High Pressure Sizing* (1979).
94. P. E. Exbrayat and R. A. Schutz, *High Pressure Squeezing.* Proceedings of III Internat. Sizing Symp., Manchester, England (1977).
95. A. F. Kienzl, *Hell. Text. Ber.* (Eng) *6*, 681-684 (August 1977).
96. N. J. Faasen, and H. Van Lingen, *Hell. Text. Ber.* (Eng) *6*, 780-784 (Sept. 1977).
97. S. M. Suchecki, *Textile Industries 143*, 54-61 (February 1979).
98. R. P. Walker, W. S. Perkins, and J. Yadon, *Tex. World 130*, 55-62 (March 1980).
99. J. J. Vincent and K. L. Gandhi, *Tex. Inst. and Industry 12*, 17 (January 1974).
100. Kanebo Ltd., *Amer. Dyestuff Rep. 68*(3), 69 (1979).
101. A. Kannen, *Solvent Sizing.* Proceedings of III Internat. Sizing Symp., Manchester, England (1977).
102. R. E. Ritter, *Tex. Chem. and Color, 1*(10) 22-23 (1969).
103. *ICI Solvent Scouring.* ICI Techn. Service Note TS/IS/4.
104. S. Kuroda, to Kawamoto Ind., Nagoya, Japan. U.S.P. 3,466.717 (1969).
105. W. F. Illman, to Burlington Ind., Greensboro, N.C. U.S.P. 3,990.132 (1976).
106. W. F. Illman, to Burlington Ind., Greensboro, N.C. U.S.P. 3,862.475 (1975).
107. D. J. Shields and J. M. Hawkins, to Eastman Kodak Co. U.S.P. 3,546.008 (1969).
108. A. D. Cotney, Proceedings of the World Conference on Warp Sizing, Clemson, S.C. (May 1982).
109. C. Von Brunn, *Melliand Tex. Ber.* (Eng.) 837-38 (November 1975).
110. F. C. B. Milne, Sizing, dyeing, resin treating in a single pass. Proceedings of III Internat. Sizing Symp., Manchester, England (1977).
111. Milnerized system, British Patent 1,456.213 (1975).
112. J. Torras, Combined Sizing. Proceedings of III Internat. Sizing Symp., Manchester, England (1977).
113. M. A. Hanson and R. D. Gilbert: *Text. Chem. Col. 6*(12) 263-64 (1974).
114. E. Schubert, *Amer. Dyestuff Rep. 39*(6) 181-9 (1950).
115. K. S. Campbell, *Tex. World 109* 73-77 (1959).

116. J. K. Shah and M. C. Sadhu, *Colourage, 23*, 15-20 (September 1976).
117. V. A. Shenai: *Tex. Dyer & Printer, 6*, 30-37 (September 1972).
118. K. J. Schmora and M. Lewin: *J. Polymer Sci. Part A. 1* 2601 (1963).
119. G. C. Amin and S. D. Wadekar: *Ind. J. of Tex. Res. 3*, 20-23 (March 1977).
120. M. Lewin, *Bromine and Its Compounds* (E. J. Joldy, Ed.) E. Benn, Ltd., London, p. 730 (1966).
121. V. Windbichler to Kalle Ag. (Germany) U.S.P. 2,974.001 (1961).
122. L. A. Sitver, B. K. Easton and R. E. Yelin, *Tex. Chem. & Color* 5(5), 79-84 (1973).
123. R. E. Yelin and R. F. Villiers, *Peroxygen Chemicals.* Proceedings of AATCC Conf., Atlanta (1970).
124. M. Lewin and A. Ettinger, *Cell. Chem. & Techn. 3*, 9-20 (1969).
125. L. A. Sitver and R. E. Yelin, to FMC Corp. U.S.P. 3,765.834 (1973).
126. L. A. Sitver and R. E. Yelin, to FMC Corp. U.S.P. 3,740.188 (1973).
127. M. H. Rowe, *Tex. Chem. & Color. 10* 22-28 (1978).
128. L. Kravetz, *Tex. Chem & Color. 5* No. 1, 29 (1973).
129. F. E. Caroprese and J. M. Plutor, to FMC Corp. U.S.P. 3,990.908 (1976).
130. J. E. KATZ, British Patent 1,524,568 (1975).
131. C. T. Greenwood and D. M. W. Anderson, *J. Chem. Soc. 288* (1955).
132. G. N. Richards and D. O'Meara, *Chem. and Ind.*, 40-44 (1958).
133. Kenner Corbett, *J. Chem. Soc.*, p. 2245 (1953).
134. C. R. Williams and D. D. Donnermeyer, *Am. Dyestuff Rep. 57*, 30-37 (June 1968).
135. A. Bose and M. D. Dixit, *Colourage, 25*(8), 23-33 (1978).
136. H. Beerman, *Am. Dyestuff Rep. 61* No. 5 (May 1972).
137. AATCC Intersectional Techn. Paper, *Tex. Chem. & Color 5* No. 11, p. 27-36 (1973).
138. R. E. Kindron and R. R. Yelin, to FMC Corp. U.S.P. 3,796.601 (1973).
139. J. R. Johnson to Burlington Ind. U.S.P. 3,762.897 (1973).
140. R. E. Davidson and M. Sittig, *Water-Soluble Resins.* Reinhold Publ. (1962).
141. A. G. Grunau, Belgian Patent 596.735 (1959).
142. J. A. Moyse, to Hexagon House, Manchester. British Patent 849.368 (1960).
143. E. F. Izard, to E.I. duPont. U.S.P. 2,169.250 (1939).
144. B. H. Kress, to Quaker Chem. U.S.P. 2,968.581 (1961).
145. R. F. Martin, *Process Technology for the Treatment of Textile Finishing Waste Waters*, Vol. II. Clemson Univ., S.C. (1971).

146. C. E. Brian, *Water Pollution Reduction through Recovery of Desize Wastes*. Water Research and Pollution Conf., Univ. of N.C., Chapel Hill (1971).

147. A. T. Walter and G. M. Bryant, to Union Carbide Corp. U.S.P. 3,472.825 (1969).

148. A. T. Walter and G. M. Bryant, to Union Carbide Corp. U.S.P. 3,321.819 (1967).

149. P. Drexler, *Reclamation and Sound Ecology in Sizing*. Proceedings of III Internat. Sizing Symp., Manchester, England, 1977.

150. W. S. Perkins, to Auburn Univ. U.S.P. 4,106.900 (1978).

151. H. Wolf and H. Leitner to BASF, Germany. U.S.P. 4,095.947 (1978).

152. N. C. State Univ. Dept. of Tex. Chem., *Water Pollution Reduction*. E.P.A. Project #12090, Technology Series (January 1972).

153. G. C. Bryan, *Reuse of PVA in Textile Processing*. Presentation at Clemson Univ., S.C. (July 30, 1974).

154. C. A. Brandon, et al. *Tex. Chem & Col.* 5(7), 35-38 (1973).

155. E. G. Kaup, *Chem. Eng.*, pp. 47-55 (April 2, 1973).

156. G. J. Crits, *Ind. Water Eng.* 14(1), 20-24 (1977).

157. R. E. Lacey, *Chem. Eng.* 79, 56-74 (September 4, 1972).

158. J. Roll, *Ind. Water Eng.* 10(3), 16-26 (May 1973).

159. V. Merten, *Desalination by R.O.* MIT press, Cambridge, Mass. (1966).

160. P. Atlas and R. Adams, *Electrostatic Powder Coating of Yarns and Scrims*. Unpublished report, J. P. Stevens and Co. (1977).

2

BLEACHING OF CELLULOSIC AND SYNTHETIC FABRICS

MENACHEM LEWIN / Israel Fiber Institute and Hebrew University, Jerusalem, Israel

1. INTRODUCTION

The art of bleaching has been practiced since the beginning of civilization [1,2]. White fabrics made of wool and of linen were found in archeological excavations, and bleaching operations are mentioned and described in old texts. Primitive people removed much of the soluble soil by washing and masked insoluble soil with dyes. The ancient Greeks and Romans soaked cloth in aqueous alkaline extracts, then laid it in sour milk and trampled it underfoot. The thus neutralized cloth was then spread out on the grass to be whitened by the sunlight and the moisture. Such a process lasted several months, and in order to obtain the desired whiteness it sometimes had to be repeated several times. The use of sulfuric acid (vitriol) instead of the sour milk began in 1750 [2]. The sequence of scour, sour, and bleach continued until the discovery of chlorine by Scheele in 1744 [3], who observed its bleaching effect on plant materials. The actual use of chlorine began in the 1790s when Berthollet produced a bleaching solution by dissolving chlorine in potassium hydroxide. This process was used commercially by Javel in France and the product was marketed as Eau de Javel [1,3]. At the same time, lime and soda ash were used for the scour. Labaraque introduced the use of sodium hydroxide instead of the more expensive potassium hydroxide in the beginning of the nineteenth century. These developments as well as the development in 1798, by Tenant, of the solid bleaching powder laid the foundation of the bleaching technology.

The term "bleaching" encompasses a series of operations designed to produce a clean, white produce. Such operations include today three distinct process stages: desizing, scouring, and whitening by applying oxidizing or reducing chemicals, in order to decolorize and remove the coloring matter from the fabric. These stages are followed by a washing operation. The desizing and scouring processes are being dealt with in separate chapters of this book (i.e., desizing in Chap. 1 of Volume 1, Part B, and scouring in Chap. 3 of Volume 1, Part A). This chapter does not deal with wool bleaching which is described in Chap. 3 of this volume.

The bleaching operation involves solubilizing and removing a large amount—sometimes up to 40% of the fabric—of diversified impurities and extraneous materials. These impurities include natural fats and waxes, pectins and hemicellulose, protein materials, and mineral ingredients, and in addition, a series of materials applied to fiber, yarn, and fabric at the various production stages, including oils, waxes, antistatic agents, and lubricants, as well as sizes composed of starch, polyvinyl alcohol, acrylics, or carboxymethyl cellulose. All these materials remain in the fabric after weaving or knitting operations; moreover, accidental stains of various origins may also be found on the fabrics and have to be removed.

The bleaching process can be carried out at several stages of the fiber-to-fabric production sequence. It is sometimes done on loose fibers, on yarns in hank or package form, on the fabric after singeing and desizing, and before or after mercerization. In each of these cases the bleaching substrate will be different in physical form, in the composition of the impurities, in its response to the action of the bleaching chemicals (i.e., wettability and penetrability of the fiber assembly), and the fibers themselves will be different. Consequently, the composition of the bleaching solutions, the parameters of the process (time, temperature, concentrations, etc.), as well as the equipment used, will be different.

Further complications are introduced by the variability in the structure and properties of the substrate fibers. Such variability is inherent in natural fibers and depends on the variety as well as on the growth conditions of the particular crop or on the blends of fiber bales used.

Synthetic fibers and blends of natural and synthetic fibers require specific changes in the bleaching technology depending on the chemical nature of the fibers, on the composition of the blend, and on the destination of the fabric, and the bleaching processes have to be adapted to them and sometimes specially designed for specific fabrics (Table 2.1).

The bleaching processes aim at achieving several general goals:

1. A high and uniform absorptivity of the fabric for water, dyestuffs, and finishing agents. This is achieved by uniform removal of the hydrophobic impurities of the fiber. It is important for natural fibers and their blends with synthetics.

2. A sufficiently high and uniform degree of whiteness in order to ensure the purity of the bright dye shades.

3. The fabric should not be damaged, and the degree of polymerization should remain high.

4. The whiteness of the fabric should remain stable upon storage.

For the achievement of these goals four major bleaching systems have emerged utilizing hypochlorite, chlorite, hydrogen peroxide, and sulfur derivatives, respectively. These systems are applied in a variety of concentrations, temperatures, times, with a large number of additives (surface-active agents, dispersants, emulsifiers, stabilizers, and activators), and a considerable number of machinery ranges depending on the kind of fabric, its quantities, and bleaching requirements.

Although the bleaching technology appears to be well established and mature, a pronounced trend toward far-reaching changes is being felt in recent years in that field as well as in the general field of wet chemical processing. This trend was triggered by several developments [4-13]: (a) an increase in the share of synthetic fibers in the overall production of textiles; (b) higher percentages of polyester fibers in the cotton-P.E.T. blends; (c) the explosive increase in energy costs; and, (d) the awakening awareness of the necessity for the conservation of the environment. The traditional three-stage, desize-scour-bleach sequence is presently being reviewed, and experiments are being conducted on two-stage and one-stage processes in order to obtain energy and water savings and to decrease effluent and pollution. At the same time the bleaching chemistry is being reviewed, and more sophisticated multipurpose and effective chemical additives are emerging. Significant improvements in machinery performance and the development of new and more efficient production ranges are taking place. Considerable interest is also being shown in improved and expanded process control, and the way appears to be open for the introduction of computer technology into the bleaching system.

Table 2.1 Applicability of Bleaching Processes to Fiber Types[a]

Type of fiber	Alkali	Sodium chorite	Sodium hypochlorite	H_2O_2	Reducing agents
Cellulose	++	++	++	++	++
Animal	—	—	—	++	++
Polyamide	+	++	—	(+)	++
Polyester	(+)	++	+	+	+
Polyacrylonitrile	—	++	+	(+)	+

[a]++ = Process highly suitable.
+ = Process applicable without fiber damage but without achieving a whitening effect.
(+)= Process applicable only with special precautions.
— = Process unsuitable.
Source: Ref. 12.

2. OXIDATION OF COTTON WITH HYPOCHLORITE: FUNDAMENTALS OF BLEACHING

2.1 The Composition of the Oxidizing Solutions

Bleaching with hypochlorite began when Tennant [14,15] manufactured bleaching powder by passing chlorine into lime water at temperatures below 40°C. The bleaching powder appears to be a mixture of $3Ca(OCl)_2 \cdot 2Ca(OH)_2$ with $CaCl_2 \cdot Ca(OH)_2 \cdot 2H_2O$, although its exact composition is difficult to determine due to the coexistence, between 25°C and 40°C, of several solid phases containing varying ratios of $Ca(OCl)_2 \cdot Ca(OH)_2$ and hydrate water. The bleaching powder contains about 35% available chlorine, that is, free chlorine obtained by acid treatment. The bleaching powder is reasonably stable, and one of its first great advantages was its ease of transportation from place to place. This enabled bleaching to be carried out in the textile mill without the need for a close-by, elaborate chlorine-generating plant. Bleaching powder was almost exclusively used for over a century throughout the world, and only after World War I was it replaced by sodium hypochlorite.

Bleaching powder is only partly soluble in water, and $Ca(OH)_2$ remains suspended in the solution and slowly settles out. Upon standing, the calcium hypochlorite solutions absorb CO_2, and insoluble $CaCO_3$ is precipitated with liberation of HOCl.

$$Ca(OCl)_2 + H_2O + CO_2 \rightarrow CaCO_3 + 2HOCl \tag{1}$$

In the presence of sulfuric acid, $CaSO_4$ is precipitated. Addition of sodium carbonate in order to increase the pH of the bleaching solution also precipitates $CaCO_3$. These precipitates accumulate in the mill and are also sometimes deposited on the fabrics imparting to them a harsh hand.

No such precipitates are obtained when using sodium hypochlorite, which can be prepared in the mill by passing gaseous chlorine through solutions of sodium hydroxide:

$$2NaOH + Cl_2 \rightarrow NaOCl + NaCl + H_2O \tag{2}$$

The commercial NaClO solutions contain 14-16% available chlorine. The hypochlorite solutions decompose on standing, particularly when the excess of alkali present is too small and the pH drops due to absorption of CO_2 from the atmosphere. The hypochlorite is decomposed to chlorate:

$$3 \text{ HOCl} \rightarrow ClO_3^- + 3H^+ + 2Cl^- \tag{3}$$

This decomposition proceeds in stages [16-18],

$$HOCl + HOCl \rightarrow HClO_2 + Cl^- + H^+ \qquad \text{slow} \tag{4}$$

$$HClO_2 + HOCl \rightarrow HClO_3 + Cl^- + H^+ \qquad \text{fast} \qquad (5)$$

and its rate increases with decrease in pH, reaching its maximum at pH 7 [19,20]. Since chlorate does not act as an oxidant under the condition applied in bleaching, its formation constitutes a loss of bleaching agent.

Hypochlorous acid decomposes into hydrochloric acid and oxygen. A similar decomposition is also assumed for ClO^-:

$$2HOCl \rightarrow 2HCl + O_2 \qquad\qquad (6)$$

$$2ClO^- \rightarrow 2Cl^- + O_2 \qquad\qquad (6a)$$

Reactions (6) and (6a) are believed to be responsible for the strong oxidizing action of hypochlorite.

The composition of hypochlorite solutions at various pH values has been extensively studied. It is governed by the following two equilibria:

$$Cl_2 + H_2O \xrightleftharpoons{K_h} HOCl + H^+ + Cl^- \qquad (7)$$

$$HOCl \xrightleftharpoons{K_d} H^+ + ClO^- \qquad\qquad (8)$$

The corresponding values of the equilibrium constants at 25°C are $K_h = 4.5 \times 10^{-4}$ and $K_d = 3.2 \times 10^{-8}$ [21,22]. The composition of the system in the pH range 5-10 and 0-5 is shown in Figs. 2.1 and 2.2 [23-26]. It is evident that depending on the pH, different oxidizing varieties are operative in the system.

From the equilibria of Eqs. (7) and (8) the composition of the solution can be calculated by solving the appropriate equations not only at different pH values and temperatures (if the value of the equilibrium constants are known at these temperatures), but also at the various initial concentrations of the oxidizing agent, as can be seen in Fig. 2.3. At a concentration of 10 milliequivalents (meq) per liter, 10% of Cl_2 and 90% of HOCl were calculated, and only at 90 meq/liter do the two varieties appear in equal amounts. A similar picture is obtained when the chloride concentration in the system is increased. This rather pronounced change in the composition will have a far-reaching influence on the kinetics of oxidation and bleaching with hypochlorite and chlorite solutions, as well as on the degradation and nature of the functional groups obtained on oxidation of cellulose.

Hypochlorous acid solutions also contain Cl_2O, which is the anhydride of HOCl, according to the following equilibrium:

$$2HOCl \text{ (aq)} \xrightleftharpoons{K_x} Cl_2O \text{ (aq)} + H_2O \quad (1)$$

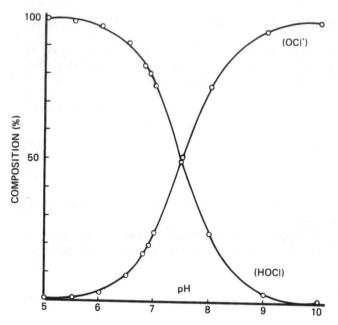

Figure 2.1 Composition of hypochlorite solutions over pH range 5-10 at 25°C. (From Refs. 24 and 26.)

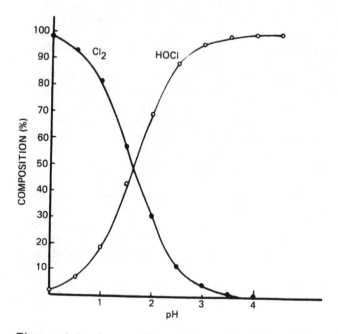

Figure 2.2 Composition of hypochlorite solutions over pH range 0-5 at 25°C. (From Ref. 26.)

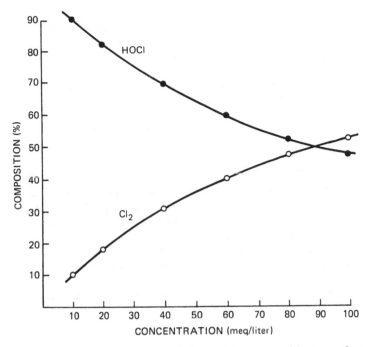

Figure 2.3 Composition of hypochlorous acid at various concentrations, at pH 2 and 25°C. (From Ref. 26.)

$$K_x = \frac{[Cl_2O \ (aq)]}{[HOCl]^2} \tag{9}$$

The value of K_x is 1/115 at 20°C and 1/282 at 0°C [27,28]. It is to be expected that Cl_2O might participate in the oxidation activity of hypochlorous acid.

2.2 The Oxidation of Cellulose: Rate of Reaction

Cotton is readily oxidized by most of the conventional oxidizing agents to various forms of oxycellulose. The oxidation products obtained were reviewed by Meller [29,30]. The oxidation of cellulose is a particularly complex reaction. The Heterogeneous nature of the reaction, the impurities (both carbohydrate and proteinic accompanying the cotton), the many ways in which the monomeric anhydroglucose units may be oxidized, and the fine structural features (e.g., the presence of the crystalline and less ordered regions) render investigations of the oxidation of cotton difficult. Most of the studies reported in the literature on these reactions were predominantly empirical and qualitative, the stress being on the practical aspects, especially since the methods used for the determination of the products of reaction, that is, the functional groups on the cellulose, were not always adequate. The purification of the cotton substrates for the studies also varied, as did the origins of the samples which were not always clearly characterized.

The oxidation of cotton by hypochlorites has been studied by a number of investigators [31-47]. General reviews have been written by McBurney [41] and Holst [48].

The oxidation of cellulose takes place according to the gross equation:

$$HOCl + Cell \rightarrow Cell \cdot O + HCl \qquad (10)$$

in which the weak acid HOCl gives off oxygen, and the strong acid HCl is formed and the pH is decreased. The amount of alkali needed to keep the pH constant corresponds quantitatively to the reacted hypochlorite [23,24,26]. This relationship was found to be valid in the alkaline pH range down to pH 5 (see Fig. 2.4), and it was used for monitoring the reaction rate automatically, without having to withdraw samples for analysis.

This relationship breaks down at the acidic pH values below 4. At pH 4 the ratio of alkali to reacted HOCl is 1.3, at pH 3 it is 2,

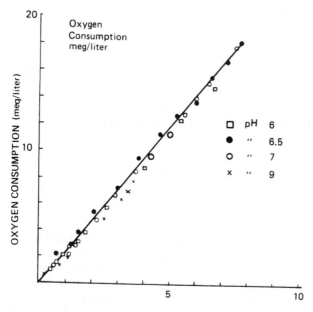

Figure 2.4 NaOH added during oxidation vs. oxygen consumption. (Refs. 24 and 26.)

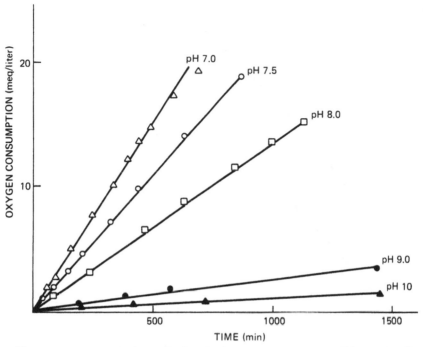

Figure 2.5 Typical oxidation plots: hypochlorite oxidation of purified cotton at 27°C; 10 g/l cotton. (From Refs. 24 and 26.)

and at pH 2 the ratio approaches infinity. At this pH range the
reactions taking place are as follows:

$$2\text{Cell} \cdot \text{H} + \text{Cl}_2 \rightarrow 2\text{Cell} + 2\text{HCl} \tag{11}$$

$$2\text{HCl} + 2\text{HOCl} \leftrightarrow 2\text{Cl}_2 + 2\text{H}_2\text{O} \quad \text{fast} \tag{7}$$

$$2\text{Cell} \cdot \text{H} + 2\text{HOCl} \rightarrow 2\text{Cell} + \text{Cl}_2 + 2\text{H}_2\text{O} \tag{12}$$

Thus upon oxidation with Cl_2 the pH of the oxidizing solution remains
unchanged, and with the increase in the contribution of the Cl_2 to
the overall oxidation, the alkali:HOCl ratio increases.

The plots of consumed oxidant vs. time upon oxidation of purified cotton with hypochlorite were found by several investigators to
be linear [23,24,33,39] and to be fitted best by the following equation:

$$C_t = C_0 - Kt \tag{13}$$

Typical examples of such straight line plots can be seen in Fig. 2.5.
Such plots are highly reproducible if the initial concentration, pH,

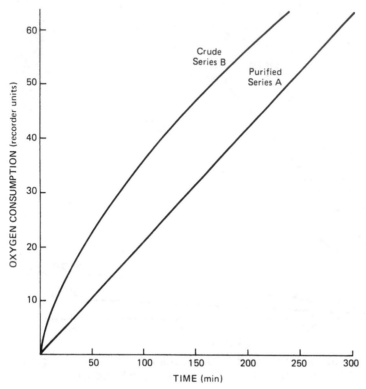

Figure 2.6 Oxidation plots of purified and crude cotton. (From Refs. 24 and 26.)

and temperature are strictly maintained. The slopes of these plots represent the rate constants of the oxidation and are usually expressed in milliequivalents of oxidant per 100 g cotton per min (meg/liter min).

In the case of crude cotton which was not boiled in alkali but only extracted in ethanol and ether, the plots are not linear and behave according to the equation

$$C_t = C_0 - Kt^{1/2} \tag{14}$$

The nonlinearity of these plots (Fig. 2.6) has been attributed to the higher percentage of impurities (3.2% as compared to 0.3 in the purified cotton) which are oxidized at a higher rate than the cellulose itself. When the impurities are oxidized, that is, when the main bleaching action is completed, the rate slows down to that of the cellulose itself.

The most important factor in the kinetics of the oxidation of cotton with hypochlorite is the pH. As seen in Fig. 2.7, the reaction rate reaches a maximum value at pH 7 and then decreases to very low values at pH 9 and 10. This behavior has been recorded by a number of investigators for cotton [32,33,48], for methylglucoside [49], as a model for cellulose, for alginic acid [19], and for the disproportionation of hypochlorite to chlorate and chloride [20]. Since it is unlikely that the cotton and the other substrates underwent a specific variation at the pH range of 6-8, which would explain the strong variations in their oxidation rate, it must be assumed that the variation of the rate with the pH is caused by the changes occurring in the composition of the hypochlorite itself. At this range the concentration of Cl_2 is very small and need not be taken into account when attempting to find a mechanism for the kinetics of the reaction. On the other hand, any such mechanism should consider the relative, rapidly changing with pH, concentrations of both of the other species, that is, the undissociated acid HOCl and the ion OCl^-. An assumption that only HOCl is the oxidizing species may work well in the alkaline pH range [50], in which the variations in the concentration of the OCl^- are small compared with the total concentration of this species. This assumption is not operative at pH values lower than 7.5. If HOCl were the only oxidizing moiety, then the maximum rate of oxidation should occur at pH 4.7, where this moiety has its maximum value (Fig. 2.1).

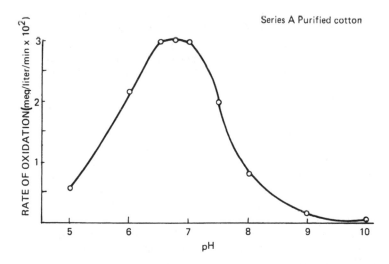

Figure 2.7 Rate of oxidation of cotton vs. pH. Initial concentration of hypochlorite, 37 meq/l; temp., 27°C, 10 g/l cotton. (From Refs. 24 and 26.)

The above considerations suggest that the reactive species is neither of these but is formed out of an interaction or combination of both. Kaufmann [51] suggested that the oxidizing species is the complex (HOCl) (OCl$^-$), and that the rate of oxidation should be given by the equation:

$$\frac{dC}{dt} = KC_{HOCl}C_{OCl^-} \tag{15}$$

which has a maximum at pH 7.43, at which the concentrations of the two varieties are equal, and not at pH 7, at which the actual maximum is found (Fig. 2.7).

Holst [48] suggested that free radicals play an active role in the oxidation with hypochlorite, as evidenced by the strong acceleration in rate in the presence of heavy metal ions and leuco-vat dyestuffs [32,40,52,53], which are known to catalyze the formation of free radicals. It was suggested that a free radical chain is set up as follows:

$$HOCl + OCl^- \rightarrow \cdot ClO + Cl^- + \cdot OH$$

$$\cdot OH + OCl^- \rightarrow \cdot ClO + OH^-$$

$$\cdot ClO + ClO^- + OH^- \rightarrow 2Cl^- + O_2 + \cdot OH$$

According to this scheme the radical $\cdot$OH perpetuates the chain. This mechanism, however, fails to account for the experimental results.

Of the two functions having close maximum values, $[HOCl]^3$ $[OCl^-]$ (maximum at pH 7) and $[HOCl]^2 [OCl^-]^{1/2}$ (maximum at pH 6.9), the second function fits the results very closely (see Fig. 2.8) [24,26]. It seems, therefore, that the variation in the rate of oxidation with pH in the pH range 5-10 may be expressed by Eq. 16:

$$\frac{-dC}{dt} = kC^2_{HOCl}C^{1/2}_{OCl^-} \tag{16}$$

The general shape of this function, (ϕ), is shown in Fig. 2.9.

Another interesting feature of the oxidation of cotton with hypochlorite is the dependence of the rate on the initial concentration of the hypochlorite. It can be seen from Fig. 2.10 that the results lie on three different curves, one for each pH. The influence of the initial concentration is strongest for pH 7, where the ratio of HOCl to ClO$^-$ is the highest, and decreases with increase in pH. When these results are replotted against the function ϕ, all points fall on one straight line (see Fig. 2.11). Because the appearance of $[ClO^-]^{1/2}$ in the function ϕ indicates a radical mechanism, a mechanism was suggested in which the oxidation proceeds by a chain reaction involving transient $\cdot$OH radicals; the main reactive species being HOCl$\cdot$ClO, possibly in the form of Cl_2OOH, a

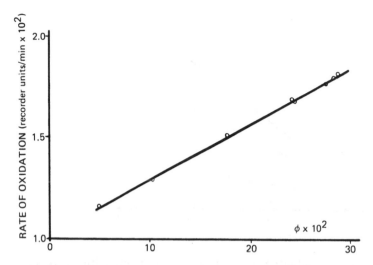

Figure 2.8 Rates of oxidation vs. the function $\phi = [HOCl]^2[ClO^-]^{1/2}$ over the pH range 6.5-8. Rate recorded automatically. Initial conc. of NaOCl, 42 meq/l; temp., 27°C. (From Refs. 24 and 26.)

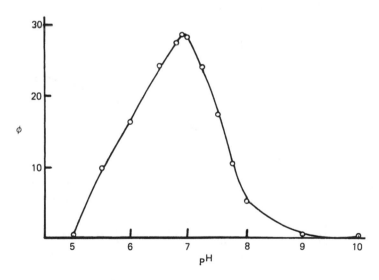

Figure 2.9 Behavior of the function ϕ in the pH range 5-10. (From Refs. 24 and 26.)

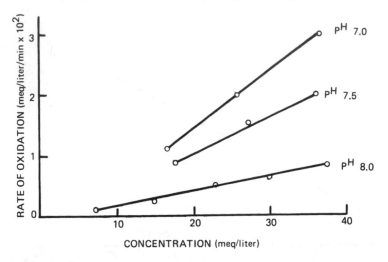

Figure 2.10 Rate of oxidation of cotton with hypochlorite at several initial concentrations. Temp., 27°C; conc. of cotton, 10 g/l. (From Refs. 24 and 26.)

hydroperoxide of Cl_2O, the anhydride of the hypochlorous acid. The hydroperoxide will appear in very low concentrations at equilibrium, which conforms with the general slow rate of oxidation at ambient temperatures.

The scheme of the reactions is as follows [24]:

$$HOCl + HOCl \xleftrightarrow{K_x} Cl_2O + H_2O \qquad (9)$$

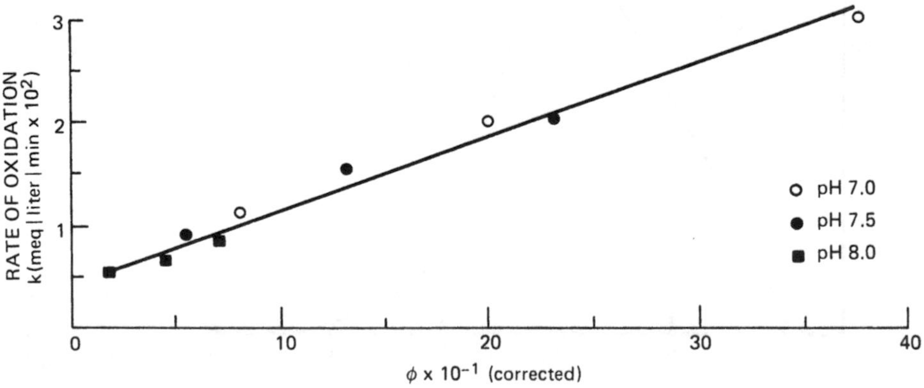

Figure 2.11 Rate of oxidation vs. ϕ at several initial concentrations and pH values corrected for [Cl$^-$]. (From Refs. 24 and 26.)

$$Cl_2O + \cdot OH \xrightarrow{k'} Cl_2OOH \tag{17}$$

$$Cl_2OOH + Cell \xrightarrow{k_{ox}} Oxidation\ products \tag{18}$$

The rate of decrease in oxidant concentration is given by:

$$- \frac{dC}{dt} = k_{ox}[Cl_2OOH] \tag{19}$$

$$= k_{ox}\ k'\ K_x\ [HOCl]^2\ [\cdot OH] = k\ [HOCl]^2\ [\cdot OH]$$

where

$$k = k_{ox}\ k'\ K_x.$$

The radical $\cdot OH$ is assumed to be formed by several equilibria:

$$HOCl \xrightarrow{K_1} \cdot OH + \cdot Cl \tag{20}$$

$$HOCl + \cdot Cl \xrightarrow{K_2} \cdot OH + Cl_2 \tag{21}$$

$$Cl_2 + H_2O \xrightarrow{K_h} HOCl + H^+ + Cl^- \tag{7}$$

Upon eliminating $[Cl_2]$, $[\cdot Cl]$, and $[\cdot OH]$ from these equations and by the use of Eq. (8), one obtains the following expression for the rate of loss of oxidant:

$$- \frac{dC}{dt} = k''\ [HOCl]^2\ [ClO]^{1/2}/[Cl^-]^{1/2} \tag{22}$$

where

$$k'' = \left[\frac{K_1 K_2 K_h}{K_d} \right]^{1/2} \tag{23}$$

This equation contains, in addition to Eq. (16), the value $[Cl^-]^{1/2}$ which was found at small concentrations to decrease the rate of the reaction [1,24,26].

2.3 Functional Groups and Degree of Degradation

When hypochlorite solutions attack cellulose, they not only produce chain scissions (i.e., degradation of the polymer), but also introduce new functional groups (i.e., aldehydes, ketones, and carboxyls). The oxidation is generally regarded as nonspecific, that is no position

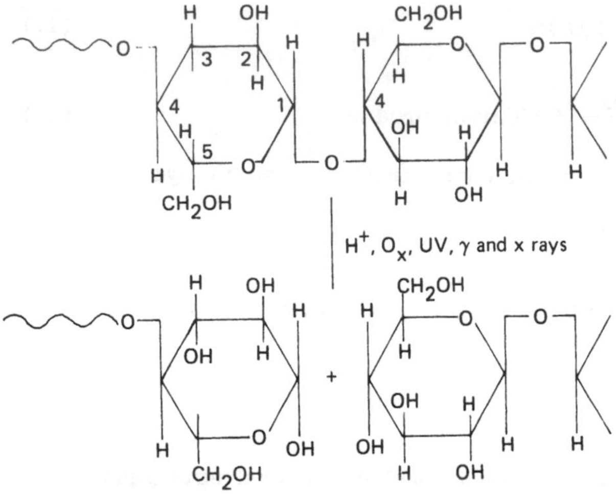

Figure 2.12 Hydrolysis of cellulose.

in the anhydroglucose unit is preferentially attacked. The possible points of attack may include carbons 6, 2, 3, 5, as well as the glucosidic bond, for example, carbons 1 and 4. (See Fig. 2.12). In contrast to this, specific oxidants attack the glucose unit at one point only, forming predominantly one type of functional group; for example, periodate attacks the C_2-C_3 hydroxyls, severing the bond between them and forming a dialdehyde (Fig. 2.13, Structure 6), and N_2O_4 attacks mostly the C_6 hydroxyls, forming uronic carboxyl groups (Structure 5). While the specific oxidants penetrate and swell the crystalline regions of the fibers, the nonspecific oxidants continue to attack the primary oxidation products and produce a highly heterogeneous series of substances [41].

Evidence for this statement was brought forward by Samuelson and co-workers [39,54-57] by hydrolyzing hypochlorite-oxidized cellulose and determining chromatographically the products obtained in the hydrolyzates. Significant quantities of gluconic and cellobiuronic acids were found, confirming oxidation of C_6 hydroxyls to the carboxyl stage [39,54]. However, at the same time the presence of erythronic and glyoxylic acids showed that the C_2-C_3 bond was also severed by hypochlorite oxidation, and C_2 and C_3 carboxyls were formed. In another study gluconic, glucuronic, and cellobiuronic acids were detected in hydrolyzates of cellulose oxidized with chlorine [56].

Water-soluble products formed during oxidation of cellulose with hypochlorite were found to contain oligomeric sugars composed of

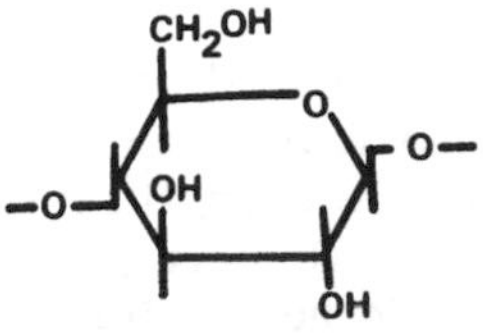

Figure 2.13 Several possible oxidation products of cellulose.

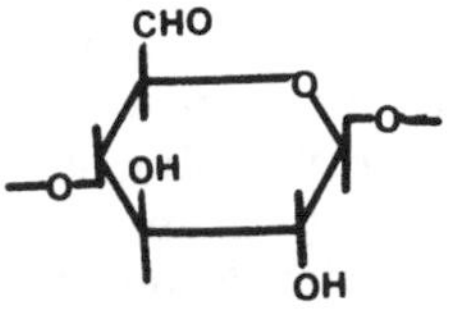

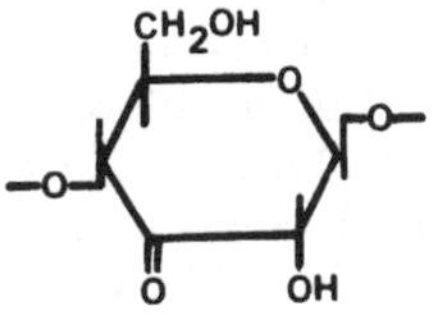

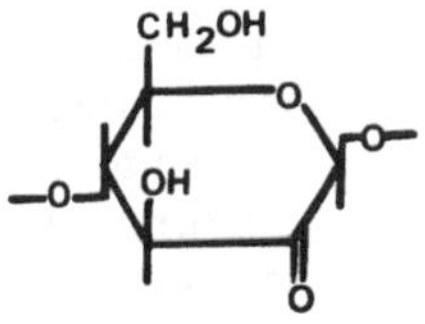

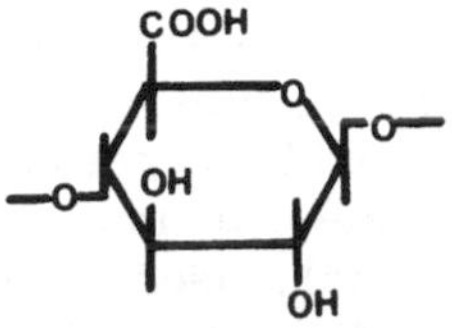

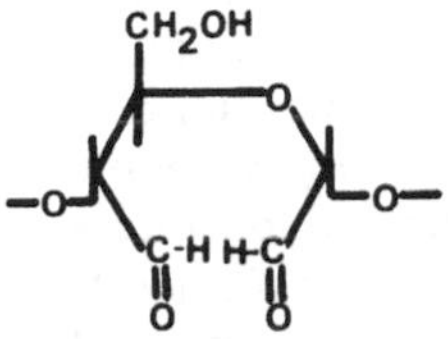

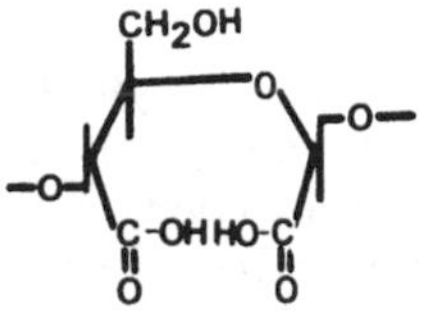

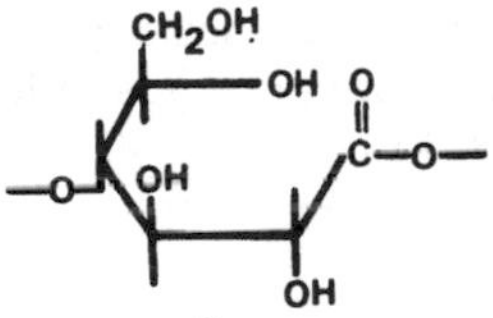

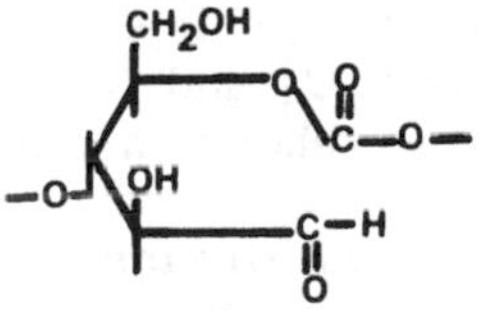

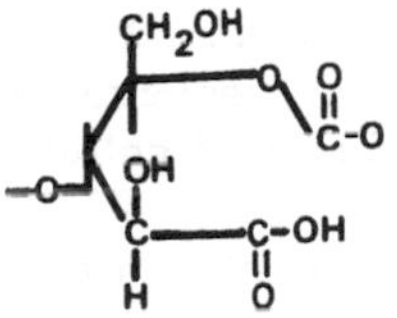

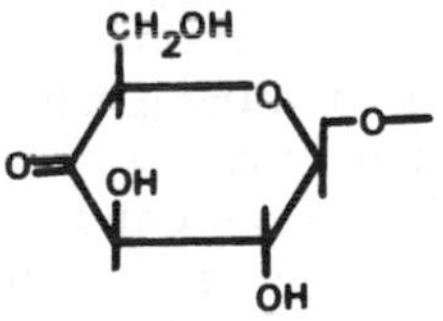

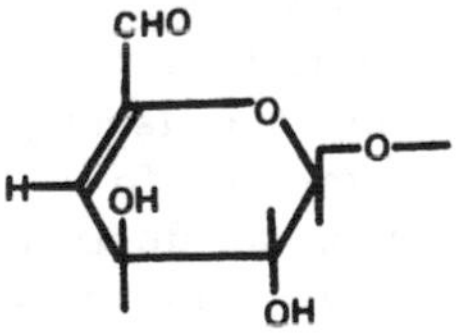

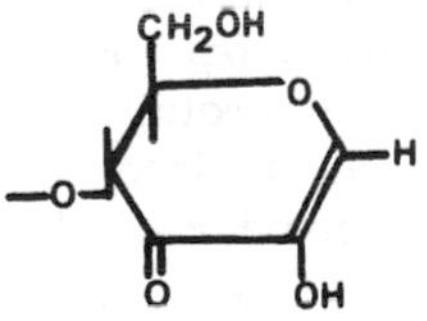

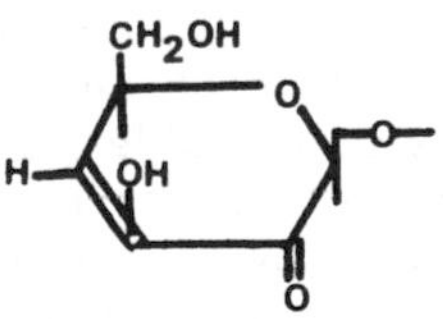

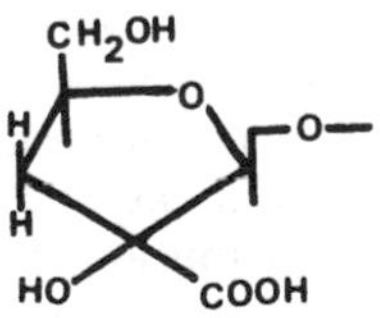

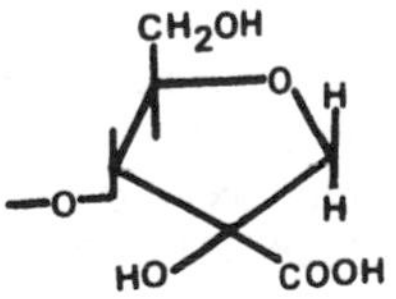

2-7 glucose moieties and a small number of glucose oligomers, with arabinose and erythrose end groups. Glucose was the most abundant monosaccharide, but arabinose and erythrose were also found. Aldobiuronic through aldooctanoic acids, containing glucose and erythronic, arabinonic, and gluconic acid end groups, were found. Glyceric, glycolic, and formic acids were the major monocarboxylic acids, but gluconic, arabinonic, and erythronic acids were also present. Oxalic acid was the most abundant dicarboxylic acid. The presence of these products in solution might explain the loss in weight of about 1% upon oxidation of purified cotton with hypochlorite at a level of 130 meq/100 g cotton in the pH range 5-10, as well as the relatively low stoichiometric yield of about 40% of functional groups (e.g., aldehyde, ketone, and carboxyl groups) found on the oxidized cotton [25,26].

A number of oxidation products were detected upon oxidation of methyl-β-D-glucopyranoside (β-methylglucoside), used as a model compound. With alkaline hypochlorite, C_6 aldehyde, and C_2 and C_3 ketone groups were found. D-glucose and D-gluconic acid were formed in oxidations carried out in the pH range 2-10, indicating cleavage of acetal linkages under conditions not conducive to normal acid hydrolysis. The formation of D-arabinose at the same pH range indicates cleavage between C_1 and C_2 [49]. Several possible oxidation products of cellulose are shown in Fig. 2.13.

Determination of Functional Groups

From the point of view of bleaching, it is of considerable interest to estimate and identify the functional groups remaining on the cellulose as distinct from those on moieties solubilized during the oxidation or during subsequent alkaline treatment. The functional groups, even in small amounts (as they are usually formed during mild oxidation) typical of bleaching, determine to a great extent the character and the behavior of the cotton products upon subsequent processing, that is, dyeing, finishing, laundering, storage, etc. The amount of functional groups formed depends on the severity of the oxidation, that is, the concentration of the oxidant, pH, temperature, concentration of the cotton, time, and the presence of catalysts. Simultaneously with the oxidation, a degradation (e.g., chain cleavages) take place, and the degree of polymerization of the cellulose decreases.

The characterization of the oxidized celluloses has been widely investigated and comprises determinations of the aldehyde, ketone, and carboxyl groups as well as the degree of polymerization [31,32, 42,44,58-66]. The determination of the functional groups requires considerable caution [60], since usually their concentrations are low. In addition the values obtained frequently depend on the accessibility of the cellulose and its state of swelling, which limits the penetration of the reagents and their access to the groups to be determined. Furthermore, the impurities invariably present in the cellulose frequently interfere with several analytical determinations, especially those of the carboxyl groups, which may be rendered inaccurate due to the presence of residual metallic bases in the fibers [58,60]. This is particularly true for methods based on acidimetric titrations of samples treated with salt solutions, followed by estimation of the liberated anions. (For example, the cellulose is soaked in a solution of sodium chloride or calcium acetate, and the liberated hydrochloric or acetic acid is estimated.) All these methods require a pretreatment with dilute mineral acid which subsequently has to be thoroughly removed [58]. The only method for carboxyl determination which does not require such a prewash is the method based on the direct estimation of the methylene blue cations absorbed on the carboxyl groups of the cellulose, from a solution of the hydrochloride buffered at pH 8.4 with veronal [34,63].

One of the first methods used for the determination of the carbonyl content of cellulose was the copper number. This method, which is still being used to a certain extent, is qualitative, since it "does not give a precise measure of the aldehyde groups and may not include the keto groups" [1,61]. It is based on the ability of the carboxyl groups to reduce the copper solutions. The most reliable method for the determination of the aldehyde groups is based on their oxidation with chlorite solutions in the presence of acetic acid [25, 49,60,63]. Sodium chlorite does not react with ketone groups and is thus clearly superior to alkaline hypoiodite [67], which does not differentiate between ketone and aldehyde groups. The carboxyl groups obtained by the oxidation of the aldehyde groups are determined by the methylene blue absorption method after deducting the carboxyl groups present in the cellulose before the chlorite oxidation.

The total quantity of carbonyl groups (aldehydes and ketones) is determined by the amount of sodium cyanide required to transform them to cyanohydrin [59,60,64]. The ketone groups are determined by deducting from this value of the total carbonyls the content of the aldehyde groups as found by the chlorite oxidation [25,60]. For celluloses containing only aldehyde and carboxyl groups, but no ketone groups, the sodium cyanide method yields the same result for the carbonyl content as the chlorous acid oxidation method. This was found to be the case with cotton oxidized by hypochlorite at pH 10 [25] and by hypobromite at pH 12 [26]. For celluloses with virtually no aldehyde groups but containing ketone groups, carbonyls were found only by the sodium cyanide method [69,70]. The use of sodium cyanide was first proposed for starch by Ellington and Purves [64]; the determination of carbonyl groups was based on their conversion to the cyanohydrin groups, hydrolysis of the latter, and determination of the amounts of the ammonia evolved. The method was subsequently modified and simplified [59,60] and involves treatment of the polysaccharide with a measured excess of sodium cyanide

at pH 9.5 at room temperature and the subsequent determination of
the unconsumed cyanide by titration with silver nitrate [25,59,60].
It was established that the alkaline degradation under these conditions is negligible and does not interfere with the results.

The sum of aldehyde and ketone groups can also be determined
by their reduction with sodium borohydride to hydroxyl groups [49a].

Determination of the Degree of Polymerization

The degree of polymerization (DP) of cotton is usually determined
viscometrically. The most prevalent method is the fluidity (reciprocal viscosity) of 0.5% of cotton in a cuprammonium hydroxide solution (Cuam) under strictly defined conditions [42,71]. This method
is still widely used for the control of the degradation in bleaching
processes. The value of the fluidity of native cotton fibers is about
2 and decreases on bleaching. Fluidities greater than 5 indicate
significant degradation, and values of about 3 rhes (reciprocal poises)
are usually aimed at. Fluidities are sometimes determined also in
cupriethylene diamine solutions (Cuen) [71]. The Cuam fluidity
values can be converted to DP values by the use of conversion tables
and a formula developed by Battista [72].

In many cases, as will be seen later, the oxidized celluloses are
sensitive to alkali, and chain scissions occur in the Cuam and Cuen
solutions which render the DP values low and inaccurate [34]. Accurate DP values are obtained for all celluloses when using the nitrate method, in which the cellulose is nitrated in a solution of nitric
and phosphoric acids and phosphorus pentoxide and dissolved in
butyl acetate [65]. The intrinsic viscosity in this solution is determined, and the DP is computed using the equation [66]:

$$[\eta] = 0.0141 \, (DP)^{0.696} \tag{24}$$

Effect of pH of Hypochlorite Solutions on Functional
Groups of the Oxidized Cotton

The pH of the oxidizing hypochlorite solutions was found to have a
profound effect on the functional groups (see Fig. 2.14). Over the
pH range 2-7 the reducing functional groups predominate, as evidenced by the values of the copper number which reach a maximum
at pH 4-5. This maximum coincides with a maximum in the concentration of the hypochlorous acid (see Figs. 2.1 and 2.2). Over the
pH range 8-12, high methylene blue absorptions are obtained, indicating the preferential formation of carboxyl groups on the cellulose.
The celluloses with the high copper numbers are termed "reducing
oxycelluloses" while those formed under alkaline conditions are termed "acidic oxycelluloses." The acidic oxycelluloses are sensitive to
acid. When the carboxyl groups are in the free acid form, the

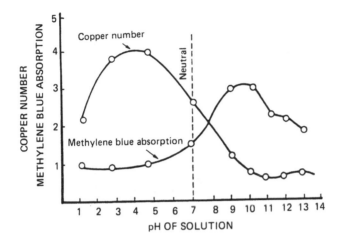

Figure 2.14 Effect of the pH of hypochlorite solutions on the copper number and methylene blue absorption of oxidized cellulose. (From Ref. 31.)

hydrogen ions formed from them in the presence of water bring about hydrolysis of the adjacent glucosidic linkages [72-74], especially before or during drying, or upon storage. This hydrolysis brings about the formation of new chain ends (e.g., of new carbonyl groups). In addition, the carboxyl groups render the adjacent glucoside linkages more susceptible to hydrolysis in strong acid [49].

Quantitative data on the functional groups obtained upon mild oxidation with about 130 meq/100 g cotton in the pH range 5-10 are given in Table 2.2 [25]. It is seen from the table that over the whole range of pH the functional groups account for approximately 40% of the oxygen consumed in the oxidation, suggesting that the mechanism of oxidation is basically the same for all pH values. This is strengthened by the fact that the yields of the cotton obtained after the oxidation amounted to 99% for all pH values. It is believed that the 60% of the oxidant not accounted for by the functional groups may be at least partly explained by the formation of water-soluble products, which were shown to be oxidized further upon filtering of the cotton fibers during oxidation runs. These water-soluble products, which form 1% of the original cotton, are short-chain materials (see above) and are likely to consume relatively large quantities of oxidant. Another part of the oxidant, unaccounted for, might have been consumed on oxidation of the low-molecular-weight fractions formed due to chain cleavage during the oxidation and remaining in the cellulose phase. This fraction remains in the fiber upon

Table 2.2 Color Intensity and Functional Groups for Cotton Oxidized with Hypochlorite at Different pH values

pH	$10^3 \times D/Ox$	$10^2 \times \dfrac{CHO/Ox}{mmol}$ meq Ox	$10^2 \times \dfrac{CO/Ox}{mmol}$ meq OX	$10^2 \times \dfrac{COOH/Ox}{mmol}$ meq OX	Yield functional groups $\dfrac{meq \times 100}{meq\ Ox}$
2	4.75	9.1	6.5	2.9	42.8
3	2.85	6.9	8.1	1.6	36.4
4	2.37	5.7	9.8	1.4	36.6
5	3.20	7.15	9.6	2.2	42.2
6	3.47	7.1	8.85	2.4	41.4
7	3.06	6.9	7.4	2.7	39.4
8	2.28	6.15	5.4	4.5	41.0
9	1.22	3.9	1.3	7.0	38.4
10	0.60	0.26	0.0	10.3	41.8
11.5	0.85	—	—	—	—

Source: Refs. 26 and 86.

filtration but is washed out in the thorough washing operation pre-
ceeding the functional group determinations [25,73]. The short
chains in this fraction will contain a relatively higher proportion of
functional groups since they are formed from scission of chains in
the less ordered regions (LOR) of the cotton and are susceptible to
oxidation to a higher degree than the average polymeric cotton chains
which are also included in the inaccessible crystalline regions. It
has been calculated that for a degree of degradation of 0.04 and an
oxygen consumption of 130 meq/100 g cotton, 1% of the dissolved
polymer contains all chains of a DP up to 37, amounting to 15% of
the total number of chains [75]. Such effects of fractionation seem
to be generally accepted for oxidation of cotton [73,76].

It is seen from Fig. 2.15 that the carboxyl group content increases steadily with the increase in pH in the range 5-10 [25]. The
sum of aldehyde and carboxyl groups per 100 g cotton is constant
over the whole pH range, which indicates that the same site is oxidized
either to aldehyde or to carboxyl depending on the pH.

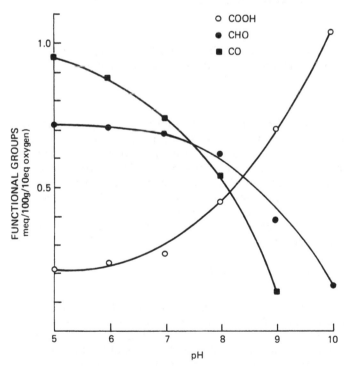

Figure 2.15 Functional groups formed by oxidizing cotton with hypochlorite over the pH range 5-10. (From Refs. 25 and 26.)

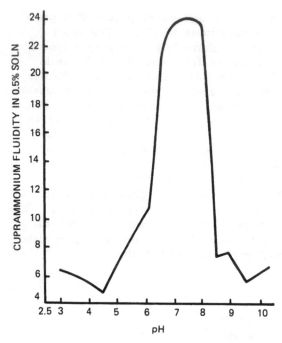

Figure 2.16 Degradation of cotton with self-buffered hypochlorite solutions (3 g available chlorine/liter) after 5 hr of treatment. (From Ref. 32.)

The relative amounts of carbonyl and aldehyde groups found at any pH value are indicative of the rates of their formation. The ketone content is seen in Fig. 2.15 to rise rapidly with the decrease in pH, and at pH 5 the number of ketone groups is higher than that of the aldehyde groups, while at pH 10 no ketone groups are found. It is evident that the number of sites oxidized in the alkaline pH range is smaller than in the acidic pH range [25].

The presence of ketone groups in hypochlorite-oxidized cellulose was also reported by Kaverzneva [77] and by Pinte and Rochas [78].

Effect of pH of Hypochlorite Solutions on the Degree of Polymerization

It was early recognized that bleaching or oxidizing cotton in the neutral pH range brings about a far-reaching degradation. (see Fig. 1.16) The maximum rate of degradation occurs at pH 7 similarly to the maximum in the overall rate of oxidation by hypochlorite (see Fig. 2.7). Since bleaching is carried out within fixed times, and since within the fixed time span the rate of the oxidation and of the

degradation (e.g., decrease in DP or increase in fluidity) is highest
in the pH range 6-8, this range was termed the "dangerous zone,"
to be avoided in practical bleaching.

The change in pH does not influence the mechanism of degradation, as a straight-line relationship is obtained between the degree
of degradation and the consumed oxygen [25]. Similarly, a straight
line relationship is obtained between the intrinsic viscosity and the
consumed oxygen (see Fig. 2.17). The linearity of this plot extends
only to a degree of polymerization of about 350, corresponding to a
consumption of 200 meq of oxygen per 100 g cotton, when the devia-
tions from linearity due to the crystallinity of cotton occur. This
linearity of the plot over the whole pH range 5-10 indicates a random
degradation mechanism. This is supported by the ratio of the weight
average to the number average (by osmotic pressure) degrees of
polymerization being close to 2, as seen in Table 2.3. In random
degradation, by analogy to random condensation polymerization, the
number of bonds broken per unit time is a constant provided the
total number of bonds present is large compared with the number of
bonds broken, that is, as long as the DP of the degraded cellulose
is high.

From the values of the DP of the oxidized cotton and of the
original cotton, the degree of degradation (α) can be calculated

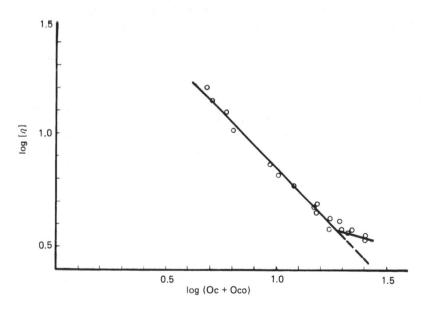

Figure 2.17 Intrinsic viscosity (nitrate method) of oxidized cotton
vs. consumption of hypochlorite in the pH range 5-10. (From Refs.
25 and 26.)

Table 2.3 Ratio of Molecular Weights in Oxidized Cellulose

Oxygen consumed, meq/100 g	pH	(DP)w	(DP)n	Ratio
177	8	350	185	1.90
71	8	630	366	1.72
29	8	1200	599	2.03

Source: Refs. 24 and 26.

according to the formula of Montrol and Simha [25,104]. The degree of degradation is defined as the ratio of the number of scissions S per base mol of cotton divided by the initial number average degree of polymerization: $\alpha = S/N_0$. Using the value of S and knowing the number of functional groups formed by oxidation per base mol of cotton, the number of functional groups formed per scission or per macromolecular chain of cellulose can be computed (see Table 2.4). In a similar way the number of consumed oxygen atoms related to one scission is calculated. For the oxidation of cellulose in the pH range 5–10, 26 oxygen atoms are consumed per each scission. The higher this number, the less is the efficiency of the degradation.

It can be deduced from Table 2.4 that the oxidative attack takes place on a number of sites on the macromolecular chain at random.

Table 2.4 Functional Groups Formed per Scission upon Oxidation of Cotton with Hypochlorite over the pH Range 5-10

pH	Groups per scission		
	−COOH	−CHO	−CO
5	1.1	3.6	4.8
6	1.2	3.5	4.4
7	1.3	3.4	3.6
8	2.0	3.0	2.6
9	3.4	1.9	0.6
10	5.0	0.8	0.0

Source: Refs. 25 and 26.

The number of these sites decreases with the increase in pH. They constitute sensitive spots on the chain and may bring about chain cleavages in alkali or during storage or aging [25]. This random oxidation and degradation along the chains brings about the relatively large number of short-chain fragments found in the hypochlorite oxidizing solutions [39,54-57].

2.4 Alkali Sensitivity of Oxidized Celluloses: The "Peeling" Reaction

According to the β-alkoxyl-carbonyl elimination mechansim [79], any strongly negative group in a position β to an ether, where the α-carbon atom is carrying a hydrogen, will render the ether bond sensitive to alkalis.

$$R_1O-\overset{|}{\underset{|}{C}}{}^\beta-\overset{|}{\underset{|}{C}}{}^\alpha-R_2 + OH^- \rightarrow [R_1O-\overset{|}{\underset{|}{C}}-\overset{|}{\underset{|}{C}}{}^--R_2] + H_2O \qquad (25)$$

$$\rightarrow R_1O^- + \overset{|}{C}=\overset{|}{\underset{|}{C}}-R_2$$

The hydrogen atom on the α carbon atom adjacent to a ketone or aldehyde group will become acidic enough to be removed by a base. The removal of the hydrogen atom will be followed by an elimination of the alkoxyl group from the β carbon atom so that an unsaturated product will be formed along with the cleavage of the etheric bond [30,79].

Hypochlorite-oxidized celluloses containing aldehyde and ketone groups will therefore be susceptible to chain cleavage, as can be seen in Eq. (26) for a C_6 aldehyde containing cellulose and in Eqs. (27) and (28) for C_2 and C_3 ketones containing celluloses, respectively. In each case a conjugated double bond is formed. In the case of the ketone groups a rearrangement can take place to a stable metasaccharinic acid end group [102,103].

$$CH_2OH \quad \xrightarrow{OH^-} \quad CH_2OH \quad \xrightarrow{H + R'-OH} \quad CH_2OH$$

3   **13**   **16**

4   **14**   **15**

In the case of hydrolyzed cellulose (hydrocellulose) the electronegative group is the aldehyde group of the reducing end of the chain and is in the position gamma to the glycoxy group which is the rest of the chain. The first step in this case is the formation of an enediol which is converted to the ketose, being beta to the glycoxyl group and enabling the elimination of the chain, according to the Lobry-de-Bryun-Van Eckenstam transformation [80]:

$$
\begin{array}{ccc}
\text{H}-\text{C}=\text{O} & & \text{H}-\overset{\displaystyle |}{\underset{\displaystyle |}{\text{C}}}-\text{OH} \\
| & \rightarrow & \| \\
\text{H}-\text{C}-\text{OH} & & \text{C}=\text{O} \\
| & & |
\end{array}
\tag{29}
$$

In the case of the C_6 aldehyde [Eq. (26)], a cleavage of the chain will be produced at the glucosidic bond at C_4. This will form a new reducing chain end which will undergo the Lobry-de-Bryun-Van Eckenstam transformation [Eq. (29)]. The ketone group formed at C_2 will produce an elimination of another glucose unit similar to hydrocellulose. Thus a stepwise depolymerization (known as the "peeling" reaction) will set in which will continue until the whole length of the chain located in the less ordered regions (LOR) of the polymer will be depolymerized, or until a "stopping" reaction sets in [Eq. (27),(28)] in which metasaccharinic acid end groups are formed. Strong evidence has been brought forward to the effect that the "stopping" reaction occurs at the borders of the crystalline regions due to steric hindrance to the penetration of the hydroxyl ions into the crystalline region and due to the lack of freedom of movement of the monomers involved [30,80-84,87].

In the case of Eq. (27), the chain scission occurs on C_4, forming a new chain with a reducing end, which means that the depolymerization reaction will propagate along the chain and the hot alkali solubility will be high. With C_3 ketone group, the chain scission will occur at C_1, producing a nonreducing chain end and a diketo derivative which may be transformed to a saccharinic acid (Eq. (28)]. Thus, although the chain will be cleaved, the depolymerization reaction will not continue, and there will consequently be no hot alkali solubility of such a cellulose.

Aldehyde groups on C_2 and C_3, as obtained by periodate oxidation (6), will produce a rupture of the pyranose ring upon alkaline treatment with a resulting decrease in the DP, but no peeling-off reaction will subsequently occur:

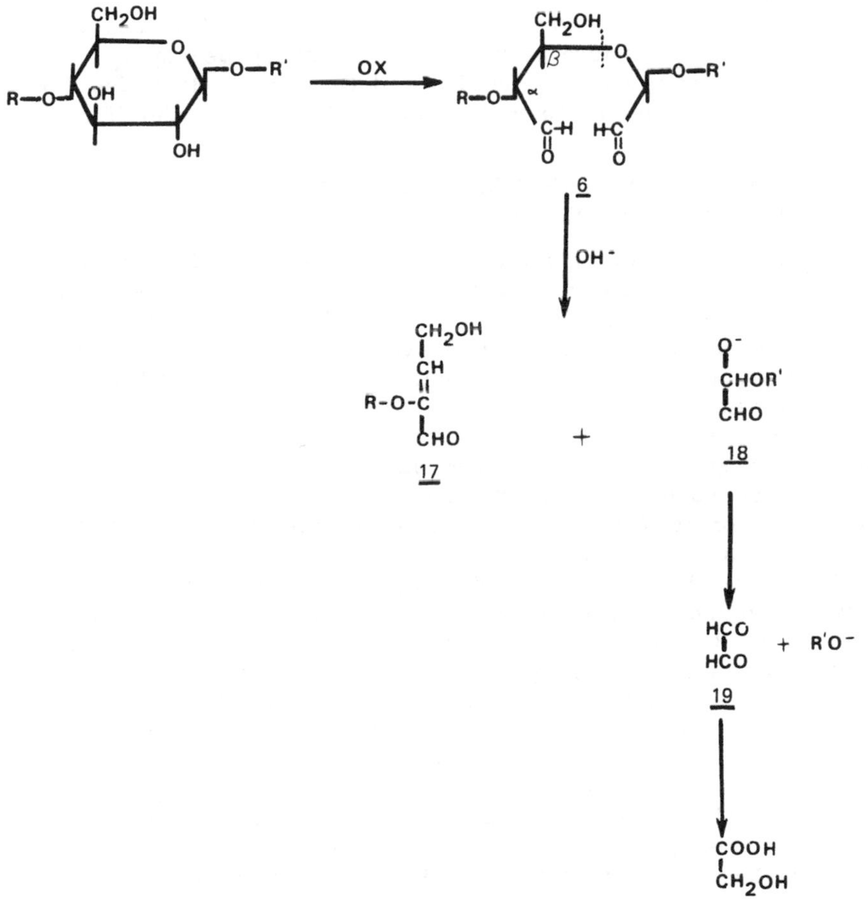

Of the resulting 2 fragments, one will be a hemiacetal of glycol-aldehyde (18) which may split and leave a nonreducing chain end. The other will be a derivative of erythrose (17) and relatively stable [30].

An oxidative split between C_1 and C_2, producing an aldehyde group on C_2 (Structure $\underline{9}$, Fig. 2.13), will bring about a cleavage at C_4, similar to a C_2 ketone group.

The carbonyl groups formed by oxidation and hydrolysis of cellulose may thus be classed in two groups:

1. *Active carbonyls* which in alkaline solutions bring about chain scissions and progressive depolymerization of the chain, resulting in a high hot alkali solubility. To this group belong C_1 and C_6 as well as C_2 aldehydes from C_1-C_2 scissions and C_2 ketones. These carbonyls constitute "peeling centers" on the chains (Fig. 2.12).
2. *Nonactive carbonyls* which may bring about chain scissions but do not induce peeling. To this group belong C_3 aldehydes from a C_3-C_4 scission and C_3 and C_2 aldehydes from a C_2-C_3 scission, as well as C_3 and C_4 ketones [80] (Fig. 2.12).

2.5 The Yellowing of Modified Celluloses in Alkaline Solutions and Color Reversion on Aging

It has long been known that when hypochlorite-oxidized cotton is placed in aqueous alkali, the solution acquires a yellow-brownish or even a dark brown-black color especially on boiling. A similar discoloration is obtained in the alkaline extraction stage of the pulp bleaching processes and upon steeping pulp sheets in alkali in the viscose process.

The yellowing phenomenon was recently systematically investigated [80,85-95] by Lewin and co-workers and was found to be significantly related to the functional groups formed upon oxidation and to the behavior of bleached and oxidized cottons upon aging and their performance in use.

The color obtained upon boiling oxidized or hydrolyzed cotton in alkali is found to be stable for several hours of standing and to strictly obey Beer's Law (Fig. 2.18). The color intensity "D" of the alkaline extracts increases with time of extraction. The increase is steep at first and then continues at a diminishing rate up to a limiting value. This behavior of D closely resembles solubilization curves obtained for a variety of oxycelluloses in alkali reported by Meller [97]. D increases linearly with the increase in the extent of oxidation or with the time of acid hydrolysis (see Fig. 2.19). The linearity of this relationship was found to hold up to an oxidation level of at least 140 meq of NaClO/100 g cotton at the whole pH range of 5-11.5.

It is of interest to note that a similar straight-line relationship between reversion in color on aging, as measured by the pc (post color) method [98], and oxygen consumptions up to 75 meq/100 g

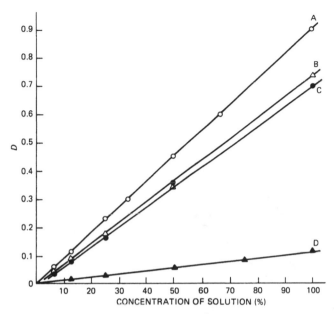

Figure 2.18 Absorbance of yellow extracts vs. concentration: obeyance of Beer's Law. Extracts prepared by 1 hr boiling under reflux of a suspension of oxidized cotton in an alkaline solution, cooling, and measuring light absorbance in a Hilger Spekker type absorption meter using No. 1 filter (λ_{max} = 460). A, B, C, NaClO oxidation; D, NaBrO oxidation, A, D, extraction with 5% NaHCO$_3$; B, extraction with 5% Na$_2$CO$_3$; C, extraction with 0.1 NaOH. (From Refs. 26 and 86).

of cellulose was obtained for periodate-oxidized linters by several investigators [99,100]. The post color number (pc) is defined by the equation:

$$pc = 100[(\frac{k}{s}) \text{ after aging} - (\frac{k}{s}) \text{ before aging}] \qquad (31)$$

where k/s is the ratio of the coefficient of absorption to the scattering coefficient and is related to the reflectivity of an "infinite" pile of sheets as follows:

$$\frac{k}{s} = \frac{[1 - Rx]^2}{2 Rx} \qquad (32)$$

where k = coefficient of absorption, s = coefficient of scattering, and Rx = reflectivity of an "infinite" pile of sheets, that is, a sufficient number of sheets so that adding more does not alter the measurement significantly.

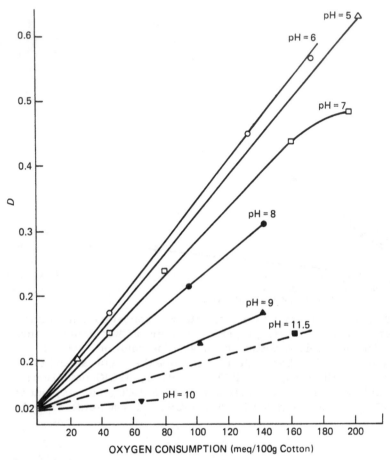

Figure 2.19 Absorbance vs. oxygen consumed at various pH values; NaClO oxidation. Extraction with 5% NaHCO$_3$ for 1 hr. (From Refs. 26 and 86.)

The ratio k/s was shown [98] to be linearly related to the amount of dye added to a cellulose pulp and was therefore believed to be a relative measure of the amount of colored materials present. The pc values are obtained from measurements of reflectance of cellulose sheets before and after overnight aging at 120°C against a MgCO$_3$ standard. The relationship between pc numbers and D values for a number of cotton samples oxidized to various extents by hypobromite at pH 10 is shown in Fig. 2.20. D increases with pc, and a linear relationship is obtained between D and 1/pc. No such correlation was found for hydrolyzed celluloses [85]. This is not sur-

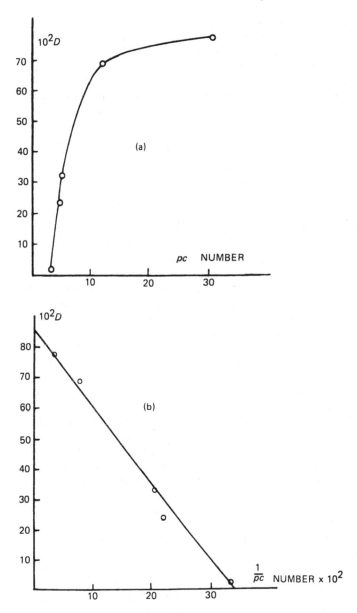

Figure 2.20 Relationship between reversion in color (pc) and yellowing (D). (a): D vs. pc; (b): D vs. 1/pc. (From Refs. 26 and 86).

prising, since while D was found to increase linearly with hydrolysis time, pc values were found by a number of investigators not to change even for strongly hydrolyzed samples [99-101]. The hot bicarbonate extraction of oxidized cotton was found to decrease considerably the pc number. Thus pc values of 22.2 and 13.4 were found before and after 1 hr extraction at the boil with 5% $NaHCO_3$ (D = 0.72), respectively. It appears that the alkaline extraction, while removing from the cotton a part of the groups responsible for the color reversion (presumably the groups contained in the short chains which depolymerized fully and dissolved in the alkali), leaves intact the greater part of these groups which reside on longer chains not fully depolymerized. The alkali extraction cannot, therefore, serve as a treatment preventing the color reversion of bleached cottons.

As seen in Fig. 2.19 and Table 2.2, D is highly dependent on the pH of the oxidation. No correlation is found between D and carboxyl and ketone groups, while unequivocal evidence is found for the correlation of D with the aldehyde group content in Table 2.2. A strong similarity exists between plots of D vs. pH and of aldehyde groups vs. pH [26,86] constructed from Table 2.2. These variations of D with pH are in remarkable agreement with results obtained on reversion of color on aging versus pH [106-109]. For over-bleached pulp with NaClO a maximum in color reversion at pH 6-8 and a minimum at a pH above 9.5 was obtained [106]. Maxima of the same order of magnitude were found for samples oxidized at pH 2.5 and 6 and a minimum at pH 4 [107]. A maximum was also found at pH 11.5 [108,109].

Further proof that the aldehyde groups are responsible for the yellowing is summarized in Table 2.5. D decreases sharply upon oxidation with sodium chlorite and reduction with sodium borohydride [26,86]. Qualitatively similar results were obtained on such post-treatments of oxidized celluloses for reversion of color determined by the pc method [107,108,110-112].

Highly significant linear correlations are found to exist between D and aldehyde group content for a series of cottons modified by hypochlorite and hypobromite oxidation over the whole pH range and by acid hydrolysis [26,80,86].

Straight line plots of D versus aldehyde groups for three kinds of modified cotton are presented in Fig. 2.21. The slopes of the straight line plots, however, are different, depending on the nature of the aldehydes formed in each particular case of oxidation or hydrolysis. The contribution of the C_1 aldehydes of the hydrocellulose to the yellowing can be clearly identified, calculated, and compared to the contribution of the aldehydes at other positions on the chain by using plots similar to those in Fig. 2.21. Such a calculation is not possible with the pc method, since no color reversion on aging was observed on hydrolyzed cellulose.

Table 2.5 Influence of chlorite and Borohydride Treatments on Yellowing[a]

pH	Oxidant	D	D, Chlorite	D, Borohydride
5	Hypochlorite	0.631	0.071	0.046
6	Hypochlorite	0.561	0.053	0.060
9	Hypochlorite	0.176		0.035
10	Hypochlorite	0.039	0.028	0.022
8	Hypobromite	0.772[b]		0.046[c]

[a]Extractions made with 5% $NaHCO_3$.
[b]Solubility in 5% $NaHCO_3$ during extraction 22.4%.
[c]Solubility in 5% $NaHCO_3$ during extraction 1.3%.
Source: Refs. 26 and 86.

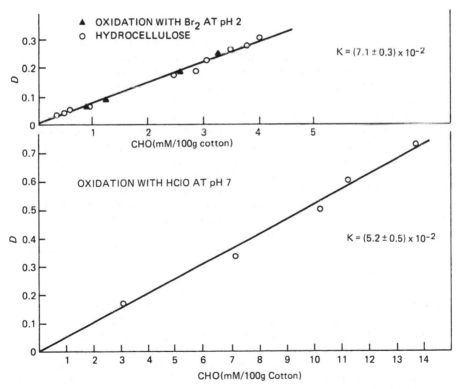

Figure 2.21 D vs. aldehyde groups for cotton oxidized with hypochlorite at pH 7, with Br_2 at pH 2 and for hydrocellulose. (From Refs. 26 and 86.)

Table 2.6 Activity of Aldehyde Groups in Several Mildly Oxidized Celluloses[a]

Treatment	$\dfrac{D}{CHO} \times 10^2$	$\dfrac{(\frac{1}{DP})}{D} \times 10^4$	Reference
NaClO, pH 7	5.2	67	25,86,80
Cl_2, pH 2	5.4	52	80
Br_2, pH 2	7.1	53	80,86,87
Br_2, pH 3-7	5.8	106-66	80
HCl, 5 N, 25°C	7.1	220	80,86

[a]D in extinction units; boiling 1 hr in 5% $NaHCO_3$ [81,87]. −CHO in millimols per 100 g cotton; chlorite oxidation and methylene blue absorption [25,68]. DP by nitration [25].
Source: Ref. 80.

All the C_1 aldehydes are "active" and bring about the peeling reaction under the standard conditions of the alkaline extraction used in the determination of D, characterized by a value of the slope of 7.1×10^{-2} (see Table 2.6). If the value of the slope for an oxidized cellulose is smaller than for hydrocellulose, it means that a part of the aldehydes is not active in peeling. It is seen from Table 2.6 that oxidation of cotton with Br_2 at pH 2 produces the same slope as an acid hydrolysis, which indicates that all aldehyde groups produced in this oxidation are "active." Oxidation with hypochlorite at pH 7 with a slope of 5.2×10^{-2}, on the other hand, produces only 73.2 active aldehyde groups (5.2/7.1) of all aldehyde groups formed. Thus, 27% of the aldehyde groups formed upon hypochlorite oxidation are inactive and would appear to be located on C_2 and C_3 formed in scission of C_2-C_3 or on C_3 following a C_3-C_4 scission.

The plots in Fig. 2.21 can be easily utilized for the determination of the aldehyde group content of oxidized cellulose. Such a method is rapid and simple [80]. The plots of the reciprocal of the weight average degree of the polymerization (DP) vs. D were found to be straight lines for hydrolyzed as well as for the oxidized celluloses (see Fig. 2.22 and Table 2.6). The smallest slope is obtained for hydrocellulose in which all aldehydes are not only active aldehydes but also formed directly from scissions. While in the case of hydrocellulose one aldehyde is formed per scission, in the case of oxidized cellulose the number varies from 3.6 at pH 5, to 1 at pH 10. When a plot like that in Fig. 2.22 is constructed for an oxidation or bleaching with a given oxidizing agent at a given pH, it is possible to use it in order to obtain the DP of an unknown sample from its D value.

In view of the known dependence of the tensile strength on 1/DP (see Fig. 2.23) of cotton fibers and yarns [105,106], the D values will at the same time give an indication of the stregnth of cotton samples.

2.6 Functional Groups and Stability of Modified Cotton on Aging and Storage

Anhydridizdtion

There is a pronounced and significant parallelism between the D and pc values of oxidized cotton. Both values change similarly with the increase in consumed oxygen, with the aldehyde content, with the

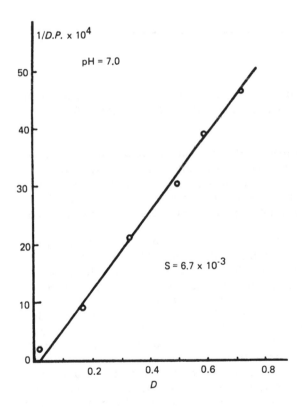

Figure 2.22   1/DP vs. D for cotton oxidized with hypochlorite at pH 7.   (From Ref. 80.)

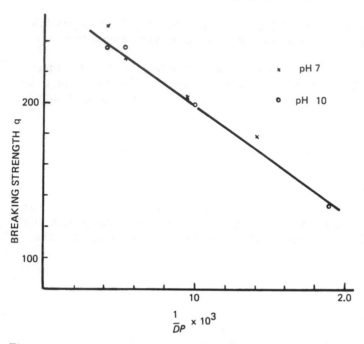

Figure 2.23 Breaking strength vs. 1/DP for unpurified cotton yarns oxidized with hypochlorite. DP by nitration. (From Ref. 26.)

increase in hot alkali solubility, with the change in pH of the oxidizing medium, and upon posttreatment with chlorite and borohydride solutions. The pc drops upon alkaline extraction. Increasing the pH of the water, in which the sheets were prepared for testing by the pc method, from 6 to 11, greatly increases the reversion in color [113].

This parallelism, however, breaks down when considering hydrocellulose and carboxyl-containing celluloses. The fact that C_1 aldehydes produced on hydrolysis do not cause reversion in color was explained by assuming the formation during aging of anhydride linkages between C_1 aldehydes and C_2 and C_6 hydroxyls [87] forming α-[(1 → 2)-anhydroglucosan] and β- or levoglucosan [(1 → 6)-anhydroglucosan]. Such links have been postulated as first stages in the pyrolysis of cellulose [114]. Meller attributed to the anhydridization the decrease in 1% hot alkali solubility of hydrocelluloses containing such lingages [119]. Heating hydrolyzed cotton to temperatures of 105–140°C for 12–72 hr, as practiced in the literature for the pc method, may be sufficient to produce the (1 → 6) anhydride linkages. Once formed, these linkages will remain stable under the aging conditions and will not contribute to the production of the yellow color

substances, thus explaining the lack of reversion of color in hydrocelluloses or in other oxidized celluloses containing C_1 aldehydes.
In celluloses oxidized with hypochlorite, other aldehyde groups are
also formed which do not yield the stable $(1 \rightarrow 6)$ linkages and will
therefore contribute to the reversion.

The presence of water during aging would appear to inhibit the
formation of the $(1 \rightarrow 6)$ linkages. On the other hand these linkages
might hydrolyze under prolonged storage at high relative humidities.

The above considerations are capable of explaining many inex-
plicable results reported in the literature. The lack of reversion of
hydrocellulose [115-117], the acceleration of the rate of reversion on
aging and its extent upon increasing the moisture content of the oven
during aging [113,116,118-120], the post-oven reversion of aged samples upon storage, and the fact that the post-oven reversion did not
occur on samples aged at high humidities [113] all seem to be due
to the anhydridization and stabilization of the terminal aldehydes [87].

Anhydridization does not occur on storage at ordinary temperatures and relative humidities. Thus, the yellowing due to terminal
C_1 aldehyde groups in bleached celluloses will happen in practice and
will be determined by measuring absorbance in bicarbonate extracts,
while it will not be determined in the pc aging test [26,87].

Autohydrolysis by Carboxyl Groups

The rate of the autohydrolysis of carboxyl containing celluloses caused by hydrogen ions derived from the carboxyl groups [73,74] increases with temperature and moisture content of the samples and
can be almost stopped at 0°C or on complete dryness of the samples.
In the pc aging test the cellulose is entered wet into the aging oven
and the temperature is raised to at least 100°C. These are eminently
suitable conditions for the autohydrolysis. A higher moisture content of the oven increases the effect.

The autohydrolysis occurs at the glucosidic linkage adjacent to
the carboxyl group and brings about a chain scission and the formation of a new terminal C_1 aldehyde on the same monomeric unit in
which the carboxyl group is located. If the carboxyl group is located on C_6, the $(1 \rightarrow 6)$ anhydride linkage will not be formed on
storage, and the terminal C_1 aldehyde will contribute to reversion.
This might explain [26,87] the observed influence of the carboxyl
groups on the reversion, on aging, and the synergistic effect of
carboxyls and carbonyls as well as the lack of dependence of this
effect on humidity during aging described in the literature [113,121].

The autohydrolytic effect does not apply to the D determination.
The carboxyls are neutralized in the alkaline solution. If, however,
the carboxyl-containing cottons are stored for prolonged periods,
the autohydrolysis will produce new aldehydes which will increase
the yellowing.

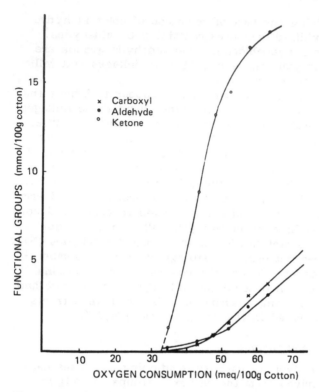

Figure 2.24 Oxidation of purified cotton with Br_2 at pH 2 at 27°C. Functional groups vs. oxygen consumption. (From Ref. 26.)

The effect of the carboxyl groups on brightness reversion is believed to be small and is estimated at about 10% of the effect caused by the equivalent amount of carboxyl groups [120,121].

No evaluation was found in the literature of the effect of the carboxyl groups on the actual yellowing of cellulose on storage.

Effect of Ketone Groups: Ketocellulose

No direct evidence is found in the literature on the contributions of ketone groups to alkali solubility or to yellowing. Data usually pertain to celluloses containing both aldehydic and ketonic carbonyls. The view that ketone groups cause brightness reversion was based on measurements of pc of chlorite-treated NaClO oxycelluloses [113]. This chlorite-postoxidized cellulose had a high carboxyl group content, and pronounced autohydrolysis might have occurred.

Although, as seen above, C_2 ketones should produce chain cleavages as well as the peeling reaction and, therefore, high brightness reversion, and although high D values should have been obtained in the few cases when almost pure ketocellulose was prepared, no such effects were noticed. One such ketocellulose is obtained upon oxidation of cotton with Br_2 at pH 2 [26,80,86] (see Fig. 2.24). In spite of the high ketone group content, the yellowing values obtained are very low and correspond only to aldehyde groups formed in small amounts during the oxidation. A similar ketocellulose is obtained upon alkaline oxidation of cotton with hydrogen peroxide [69,80] (see below, Sec. 5.1). Evidently, the ketone group must have been produced on C_3 and possibly to a small extent on C_4.

The existence of a stable C_3 keto derivative has been, however, questioned, and it was postulated [122-124] that ketocelluloses 20 and 21 (see Fig. 2.25) will both be readily converted, via the respective carbanions 22 and 23, into a common tautomeric 2,3-enediolate anion, 24, in dilute alkali. Chain cleavage may now occur by two competitive routes. Elimination of O_1 from 24 will leave a chain fragment terminated by 26 with a second fragment with a nonreducing chain end [see Eq. (28)], and no peeling will be initiated. Alternatively, O_4 will be cleaved from 24 according to Eq. (27), and a new reducing moiety 25 will be formed which will initiate peeling. The route 24-26 is preferred since no peeling or yellowing takes place in ketocelluloses. The preference of this route was explained [93] on the basis of calculated electronic distribution in pyranoses [115], according to which O_1, having a higher electronegative charge than O_4, will be a better leaving group, bringing about the formation of 26 rather than 25.

2.7 The Yellowing Chromophore

The Structure of the Chromophore

A yellow discoloration similar to that of the bicarbonate extract of cellulose is obtained on other reducing polysaccharides such as amylose, and on mono- and oligosaccharides, such as glucose, α-methylglucoside, D-xylose, D-galactose, D-fructose, D-mannose, L-rhamnose, D-arabinose, D-maltose, D-cellobiose, cellotriose, lactose, sodium glucuronate, and D-galacturonic acid [80,86]. Nonreducing sugars, such as melizitose, raffinose, gluconic and absorbic acids, and sucrose remained unchanged upon the bicarbonate extraction.

During the peeling reaction monomeric anhydroglucose units are progressively detached from the cellulose. If it is assumed that these units degrade in the boiling bicarbonate solutions in a similar way to glucose yielding quantitatively similar colored materials, it is possible to calculate from the optical densities of the alkaline cellulose extracts, D, and of the alkaline glucose solution boiled under similar conditions, D_{gl}, the weight loss during hot alkaline extraction of modified celluloses. The quantitative correlation seen in Fig. 2.26,

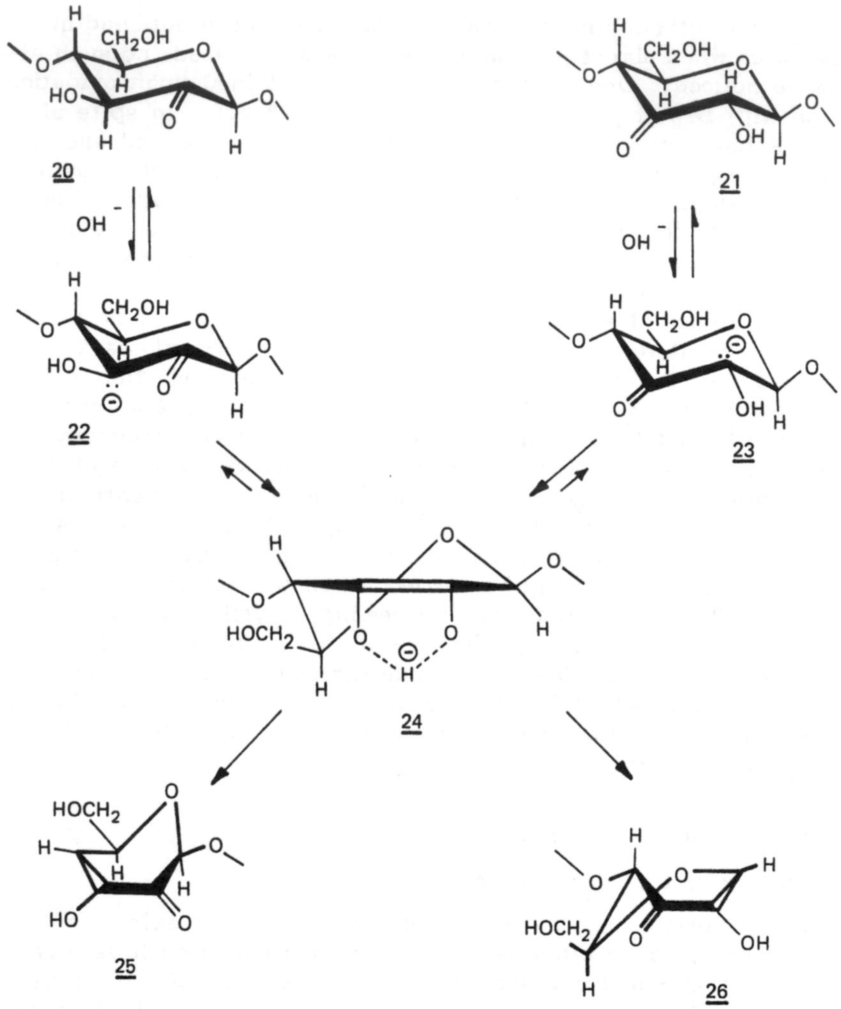

Figure 2.25 Mechanism of the peeling reaction of ketocellulose. (From Ref. 93.)

typical for a wide range of modifications of cellulose, establishes the validity of this assumption [80].

Furthermore, this quantitative correlation indicates that the colored substances are highly significant products of the β-alkoxyl elimination (peeling) process. While during this process mainly saccharinate molecules, which do not absorb in the visible and UV range, are liberated [126], a chromophore responsible for the yellowing of the solution is formed as a minor product, having a part of its spectrum in the range of the visible wavelengths [80].

The ultraviolet spectrum of the bicarbonate extract of hydrocellulose (Fig. 2.27) reveals a maximum at $\lambda = 290$ nm. Upon acidification a blue shift occurs to 256 nm with an accompanying hyperchromic effect that could be fully reversed by adding alkali. The spectra at the various pH values pass through an isosbestic point at 270 nm, proving that the chromophore is a single material and not a mixture, and that it is basically stable at the whole pH range. An apparent pK value of the dissociation, responsible for the shift, of 8.5, was determined. Dialysis through a cellophane membrane removed all of the UV activity, indicating that the chromophore is of low molecular weight [80,92].

The UV spectrum of a boiled alkaline solution of cellobiose showed two absorption peaks at 288 and 260 nm which on mild acidification merged into a single peak at 260 nm, which below pH 3 shifted to 250 nm. D-Glucose upon a similar treatment revealed a peak at

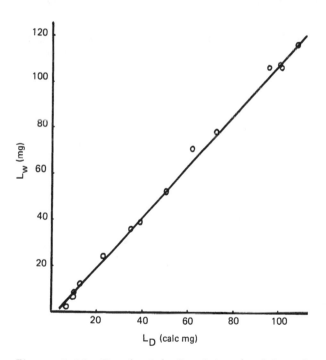

Figure 2.26 Gravimetrically determined loss in weight (L_w) on bicarbonate extraction vs. loss in weight calculated from the D values on the basis of D of glucose, L_D. Pima cotton, hydrolyzed with 5 N HCl at 25°C. (From Ref. 80.)

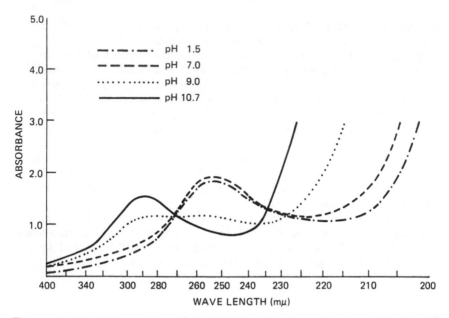

Figure 2.27 UV spectra of 0.6 M bicarbonate extracts of hydrocellulose at several pH values, measured against water in the reference
beam. (From Ref. 80.)

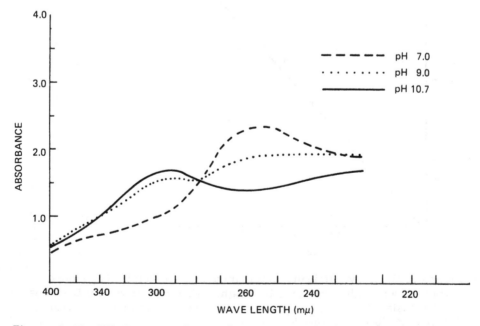

Figure 2.28 UV Spectra of cellobiose at several pH values, measured
against a glucose solution in the reference beam. Both solutions,
4 mmol/l, boiled for 1 hr in 0.6 M $NaHCO_3$. (From Ref. 80.)

Table 2.7 Apparent Molar Absorption Coefficients (ε') of Carbohydrate Alkaline Extracts

Substrate	pH of chromophore solution	λ max (nm)	ε'
Hydrocellulose	1.5	252	945
	7.0	253	1010
	10.5	290	895
Cellobiose-D-glucose[a]	1.5	247	1025
	7.0	259	1050
	10.5	294	820
D-Glucose	1.5	245	1000
	7.0	260	1180
	10.5	261	1050

[a]Difference spectrum with equimolar solutions of alkali-treated D-glucose in the reference beam and alkali-treated cellobiose in the sample beam.
Source: Ref. 92.

260 nm which shifted without an isosbestic point to 245 only below pH 3 [80,88,127]. When the spectra of the boiled cellobiose were taken with an equimolar solution of a similarly treated D-glucose in the reference beam (see Fig. 2.28), they fully resembled those of the hydrocellulose (Fig. 2.27) and exhibited a similar pH behavior, isosbestic point, pK, and maximum wavelength. The apparent molar absorption coefficients calculated for cellulose by dividing the absorbance at the peak by the weight loss, and calculated for the sugar solutions by dividing the absorbance by the initial weight of the sugar (see Table 2.7) were also found to be very similar, thus corroborating further the identity of the chromophore obtained from cellulose and cellobiose.

The "depolymerization" of cellobiose takes place similarly to that of cellulose, producing a molecule of D-glucose and a molecule of 4-deoxy-D-glycero-2,3-hexodiulose (Fig. 2.13, 12) which is also obtained from a reducing chain end of cellulose [see Eqs. (29) and (27)]. The chromophores obtained from each of the above two fragments are therefore identical with those formed separately from D-glucose and cellulose.

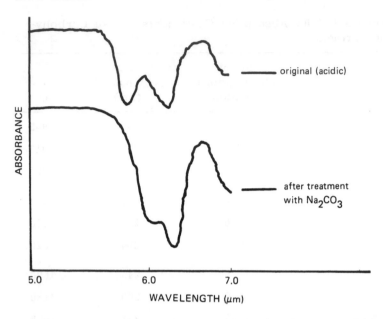

Figure 2.29 IR bands of chromophore in dialyzate of bicarbonate extract of hydrolyzed cellulose. (From Ref. 80.)

The infrared spectrum of the chromophore-containing dialyzate
(Fig. 2.29) showed two strong bands at 1730 and 1600 cm^{-1}. Upon
treatment with sodium carbonate the band at 1730 disappeared and
the band at 1600 deepened. Addition of strong hydrochloric acid

Table 2.8 Structure of the Chromophore

IR absorption (cm^{-1})	Structure[a]	Literature values (cm^{-1})
1730	27	1740-1710
1600	28	1630-1540
1560	29	1560-1550
1410	29	1390-1320

[a]

$$27, R^1-\underset{\underset{R^2}{|}}{\overset{\overset{O}{\|}}{C}}-CH-\overset{\overset{O}{\|}}{C}-R^3 \ ; \ 28, R^1-\overset{\overset{O---H-O}{\|}}{C}-\underset{\underset{R^2}{|}}{C}=C-R^3 \ ; \ 29, R^1-\overset{\overset{O---M---O}{\|}}{C}=\underset{\underset{R^2}{|}}{C}=C-R^3$$

27 is in acid; 28, in alkali; and 29, metal chelate structure.
Source: Refs. 80 and 92.

restored the band at 1730 cm^{-1}. The band in the acidic solution is ascribed to an aliphatic β-diketone, whereas the 1600 band is ascribed to the tautomeric β-hydroxy-α,β-enone (Structures 27 and 28 in Table 2.8). Several fractions of the chromophore-containing hydrolyzate exhibited bands assignable to the metal chelate of a β-diketone (27 in Table 2.8). Addition of ferric chloride to the chromophore solution produced a deep yellow color typical for Structure 29. The lack of color upon addition of 2,4-dinitrophenylhydrazine is consistent with the known formation of colorless pyrazoles from β-diketones [80,92].

Influence of Alkaline Earth Cations on Chromophore Formation

It is assumed that the chromophore is derived from the compound (30) formed according to the Eq. (33):

(33)

There appears to be a competition between the formation of the isosaccharinic acid 31 which is colorless and does not absorb in the UV and of the chromophore from the α-diketone moiety (30). The conversion of 30 to 31 was found to be accelerated in the presence of divalent cations [128]. This acceleration appears to be responsible for the lower amounts of chromophore formed in the presence of alkali earth cations as compared to alkali metal cations (see Fig. 2.30) [94]. The effect of the different cations on the amount of chromophore formed, as deduced from the slopes of the plots of the absorbance against the amount of cellulose dissolved (i.e., from the values of ε', varies with the nature of the cation and its valency as follows:

$$Na^+ = K^+ \geqslant Li^+ \geqslant Ba^{2+} > Sr^{2+} \geqslant Ca^{2+}$$

The most effective cation is Ca^{2+}. The earth alkaline cations do not decompose the chromophore once it has been formed but prevent its

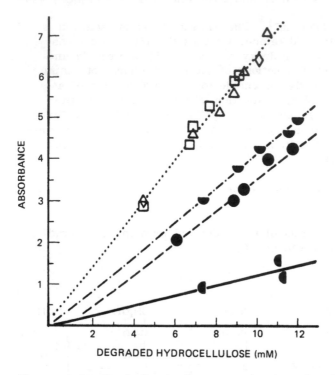

Figure 2.30 Variation in the concentration of yellowing chromogen (absorption coefficient at λ_{max} = 288-292, pH 10.4) as a function of loss in weight [$C(C_6H_{10}O_5)$, mM] of hydrocellulose (250 mg) during boiling in lithium ($\triangle\cdots\triangle$), sodium ($\square\cdots\square$), potassium ($\triangle\cdots\triangle$), calcium ($\bullet\!\!-\!\!\bullet$), strontium ($\bullet\text{---}\bullet$), barium ($\ominus-\cdot-\cdot-\ominus$) hydroxide solutions (25 ml) of initial alkalinity $C(OH^-)$ = 0.02 M for 6 hr. (From Ref. 94.)

formation by a specific effect. A minimum concentration of the divalent cations is needed for this effect, and an increase in their concentration above 0.1 M will not enhance it [94].

The alkaline earth cations will thus decrease the discoloration of liquors during alkaline processing of celluloses and starches. It is, however, also possible that the brightness of the cellulose itself will be improved by the divalent cations. When the depolymerization by the alkali in the peeling reaction proceeds to the end in a particular chain, that is, when the stopping reaction sets in at the border of the crystalline region, it is believed that a metasaccharinic acid (MSA) unit is being formed from a diketone variety similar to Structure 30. It is possible, however, that also in this instance instead of a small part of the saccharinic acid, a chromophore will be formed. This chromophore will remain attached to the fiber and will not dissolve. The divalent cations which are known to accelerate the benzylic acid rearrangement to form MSA may interfere in this case as well and decrease the amount of chromophore formed and thus enhance the brightness of the cellulose.

2.8 Oxidation of Cotton at Low pH Values

As indicated above [Sec. 2.2, Eq. (12)], the oxidation with hypochlorite below pH 4 is carried out by a mixture of Cl_2 and HOCl, the composition of which changes with the initial concentration of the oxidants and the pH.

Here, the dependence of the rate of the oxidation upon the ϕ function postulated above for the pH range 5-10 is valid only to a limited extent. The value of the ϕ function at pH 4 is 1.08×10^{-6} as compared to 3.4×10^{-6} at pH 5-10, while the rate constant at pH 4 is about 60% of that of pH 5. The value of the ϕ function at pH 3 is only 3.0×10^{-7}. It has to be assumed that an additional component of the solution becomes operative at pH 4, and its contribution increases as the pH decreases [26].

The oxidation of cotton with Cl_2 over the pH range 4-2 is accelerated by diffuse light [26]. Its energy of activation is low. The rate of oxidation reaches a minimum at pH 3 and increases at pH 2. An induction period of about 10-15 hr which precedes the oxidation is observed (see Fig. 2.31). The induction period is preceded by a *reversible sorption* of Cl_2 on the cotton. The extent of the sorption which is completed after about 90 min decreases with increase in pH and temperature. The sorption proceeds according to the Langmuir isotherm [26].

The functional groups formed upon chlorine oxidation are similar to those of hypochlorite oxidation at pH 5-10 (Fig. 2.32). The first 25 meq per 100 g cotton were consumed on the sorption and only further consumed chlorine-produced functional groups. The oxidation with Cl_2 is slower than with hypochlorite; however, the properties of the cellulose are similar.

Little is known about the mechanism of this oxidation. The kinetics of the oxidation suggest a free radical mechanism [26]. Such a mechanism was suggested together with an ionic process for pulp bleaching. The latter seems to predominate in lignin reactions, while the cellulose reactions are considered to be free radical processes. This would explain the very significant reduction of carbohydrate degradation upon additions of radical scavengers [129].

2.9 Accelerated Oxidation of Cotton with Hypochlorite in the Presence of Dyestuffs and Metallic Hydroxides

Leuco vat dyestuffs accelerate the oxidation of cotton with hypochlorite [130-132]. The physical and chemical properties of the fibers

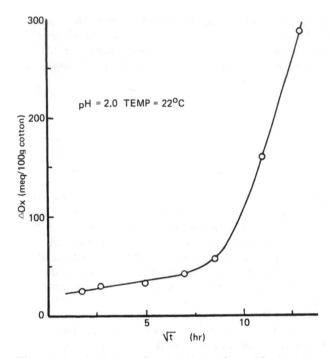

Figure 2.31 Rate of oxidation of purified cotton with Cl_2 at pH 2. (From Ref. 26.)

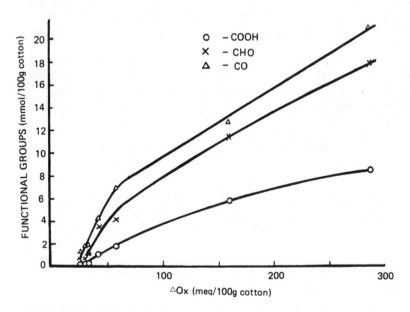

Figure 2.32 Oxidation of purified cotton with Cl_2 at pH 2: functional groups vs. consumed chlorine. (From Ref. 26.)

indicate that the mechanism of the accelerated oxidation does not depend on the nature of the dye and on physico-chemical modification of the fine structure of the cotton. The amount of oxygen consumed under identical conditions is said to be governed by the amount of dye present. The main reaction appears to be the oxidation of the C_6 hydroxyls to aldehyde groups and carboxyls. The presence of reactive dyes in the reacted form decreased the rate of the oxidation under acidic and neutral conditions. Maximum degradation occurred at pH 7 [135].

The oxidation of cellulose by hypochlorite is accelerated by the presence of metallic deposits on the cotton [134-140]. The effect is shown by cobalt, iron, chromium, nickel, copper, and manganese hydroxides in their lower state of oxidation.

Although the ketone group content was not directly determined in these studies, one may detect their presence from the difference in the copper number values obtained after chlorous acid oxidation and after borohydride reduction of the oxidized samples. The amounts of carboxyl groups formed and the copper number values increase with the increase in the amount of hypochlorite consumed in the oxidation and with concentration of metal salt deposited. The alkali aolubility values indicate that considerable amounts of active aldehydes were formed. The dependence of the rate on the pH is in most cases similar to the uncatalyzed oxidation, that is, the maximum rate is obtained at pH 7-7.5, except in the case of iron, where the maximum rate is observed at pH 6.1.

2.10 Basic Chemical Requirements Desired in a Bleaching Process

The data and discussion presented above (Sec. 2) enable us to draw a number of conclusions regarding the chemical requirements of a bleaching process.

1. The oxidation of the impurities and the coloring materials on the surface and in the lumen of the fibers has to be performed at a rate which is much faster than the rate of the oxidation of the cellulose. The parameters of the bleaching process have to be chosen so as to increase the difference between these two rates and to ensure the slowest possible rate of the cellulose oxidation. Such parameters are the pH, the concentration of the bleaching agent, the time and the mode of circulation of the bleaching liquor, the liquor ratio, the temperature, and the presence of additives such as surfactants, activators, and catalysts.

2. The oxidation of the cellulose, if it could not be avoided, should proceed in such a way as to produce the lowest possible yield of functional groups, calculated on the consumed oxygen. This would be possible if the oxidation could be directed and confined to the chain ends. In this case the oxidation would consume several

monomeric units at each chain end and produce soluble products, while the main cellulose chain would not be attacked.

3. Another requirement is a high consumption of oxygen, that is, a high number of oxygen atoms per chain scission. This would ensure a low degree of degradation and prevention of damage to the fiber.

4. Of particular importance is a low content of aldehyde and especially "active" aldehyde groups on the bleached fibers, since this would minimize chain scission and dissolution on alkaline laundering and upon mercerization. It would express itself in a low value of the slope of yellowing versus aldehyde groups and would result in little yellowing on laundering. Furthermore, it would decrease the yellowing of the fabric upon storage.

5. Another important requirement is a low carboxyl group content. This would increase the stability of the cellulose on storage, that is, it would prevent the autohydrolysis reaction and the formation of new active aldehyde groups with consequent yellowing and weakening of the fiber.

6. It should be noncorrosive and not increase the pollution potential of the mill effluents.

3. BLEACHING WITH HYPOCHLORITE

3.1 Technological Considerations

The impurities, mostly located on the surface and in the lumen, are more reactive in the bleaching process than the cellulose and are oxidized first and at a faster rate than the cellulose. Conditions are usually sought for the possible complete elimination of the oxidation of the cellulose, such as are believed to exist in chlorite bleaching. In the case of hypochlorite, however, some oxidation of the cellulose takes place simultaneously, albeit at a much slower rate than that of the oxidation of the impurities.

The kinetic curves of bleaching usually consist of an initial steeper slope, which corresponds to the oxidation of the impurities followed by a flattening of the curve due to the oxidation of the cellulose (Fig. 2.6). The variation of the rate of the oxidation of the impurities with pH is generally similar to that of the cellulose component. It increases as the pH decreases from 11 to 7. The degraded impurities probably continue to be oxidized in solution, thus affecting the rate of oxidation even after their removal from the fibers.

In the first stage of oxidation of the impurities, using up to 12 meq/100 g cotton (about 0.45 g active chlorine), the physical properties, including textile strength, elongation at break, energy at break, as well as the DP, are changed very slowly, while at the higher consumptions the deterioration is rapid (see Fig. 2.33). The

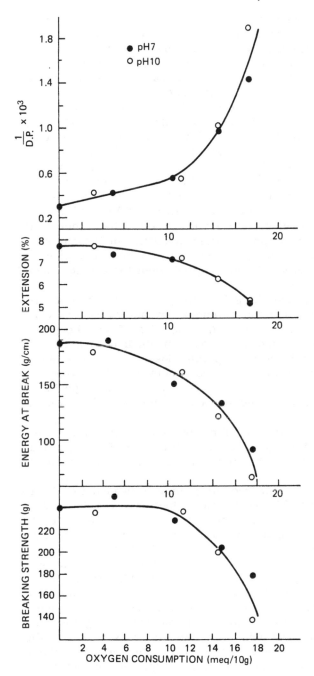

Figure 2.33 Bleaching of unpurified cotton yarns: properties of yarns vs. oxygen consumption (From Ref. 26.)

effect of the consumed oxygen on the changes in the physical properties and the DP measured is about the same at both pH values of
7 and 10, thus lending further proof to the view that the attack on
the cotton fibers is a function of the oxygen consumption and not
directly of the pH. The changes in the physical properties are seen
to be directly related to the DP (see Fig. 2.23).

The bleaching with hypochlorite cannot be carried out in one
stage. First, the fabric has to be desized. On desized fabric about
1.8% owf (on the weight of the fabric) of active chlorine is needed
for complete bleaching. If after the desizing an alkaline boil is applied, the amount needed is 0.5-1% owf, depending on the purity of
the fibers and the state of the fabric. This amount usually suffices
also for the complete removal of the seed husks. It is desirable to
apply to the fabric only these required quantities of hypochlorite,
since in this case the danger of fabric damage would be rather slight.
It is, however, necessary to apply higher quantities of bleaching
agent in order to speed up the operation, which would otherwise last
for 6-8 hr [141]. The process is interrupted when the correct amount
of bleaching agent has been consumed.

The initial pH of the bleaching is usually 11.5-12. It is maintained during the process between 11 and 9, by adding sodium carbonate as buffer, since according to Eq. (10) (Fig. 2.4), the pH
decreases during bleaching, and the amount of alkali needed to maintain the pH at a constant level corresponds closely to the reacted
NaClO. Maintaining the pH above 9 is especially important when
bleaching dyed goods. The stability of dyestuffs to the hypochlorite
bleaching baths has been tabulated for a large number of dyestuffs
[141].

After the bleaching stage the fabric is rinsed with water, de-
chlorinated with a reducing agent (e.g., sulfite, bisulfite, hydrosulfite, or thiosulfate), rinsed with water, treated with acid (souring),
and finally rinsed.

It has been stated that the reducing stage is needed for the removal of the residual hypochlorite solution which may remain entrapped in the interstices of the yarns and fabric as well as in order to
remove the chloroamines, which may split hydrochloric acid and cause
hydrolysis of the cellulose. It is also believed [141] that the reducing stage stabilizes the whiteness. It is evident that this treatment
is also capable of removing the reversibly sorbed chlorine on the
cellulose. It may be assumed that the pH of the solution drops low
enough during the rinsing operation to produce elementary chlorine
which might be rapidly sorbed. This might happen during the souring operation with the mineral acid when the pH drops to 2-3 in the
instance where no antichlor stage was applied. The reducing stage
is not needed if a peroxide stage follows the hypochlorite bleach,
since the peroxide acts as a reducing agent of the chlorine residues.

The souring stage is practiced as the last stage, when no further alkaline stages are being applied. It is designed to remove the residual alkali from the cloth, but in addition it is known that the whiteness of the fabric increases by 1-2%. This may be due to the shift in the spectrum of the yellowing chromophore in neutral and acidic pH values (see Fig. 2.27). Chromophoric diketone group-containing units remain attached to the chain ends of the cellulose where the peeling reaction in alkali was interrupted by the rinsing. Some dissolved chromophores, peeled off the cellulose, may also remain occluded on the fiber. It is clearly seen that the tail of the spectrum at 400 nm is significantly lowered at the lower pH values (Fig. 2.27). When the fabric is treated again with alkali, the color due to these chromophores will return as a result of the reversible shift of the chromophore with pH. The whiteness of the fabric will decrease. This "indicator" effect has been observed in practice.

3.2 Hypochlorite Bleaching Processes

The processes for bleaching based on hypochlorite vary with the kind of the fabric, the amounts to be bleached, and the machines to be used. They can be carried out in several systems: kier, jig, package, winch, pad-roll, and other impregnation processes.

A typical formulation for a *kier* bleaching consists of 2-4 g/liter active chlorine, 0.1-0.2% wetting agent, NaOH to pH 11.5, and 2-4 g/liter Na_2CO_3. The liquor ratio is 1:3 up to 1:7, and the time of bleaching is 3-4 hr at 20°C. The fabric is then passed through a number of baths: cold water; 3-5 g/liter sodium pyrosulfite (65% SO_2) or 8-12 g/liter sodium sulfite (24% SO_2); cold water; a formic or acetic acid souring bath; cold water [141].

The formulation of the bleaching solution for the *jigger* is similar to that of the kier, but the bleaching time is only 1-2 hr, and the recommended temperature is 20-25°C. Similar formulations are used for cotton-polyester.

For the *package* bleaching with a liquor ratio of 1:10, a formulation is recommended which contains 1-2 g/liter active chlorine, 0.1-0.2% wetting agent, NaOH to pH 11, and 1-2 g/liter Na_2CO_3. The time of treatment is 1 hr at 20-25°C. The same formulation is used for *winch* bleaching with a liquor ratio of 1:20 to 1:40.

Hypochlorite bleaching can be carried out by partially or fully continuous processes. To carry out the process usually a padder is needed while no additional machinery is essential. The impregnated fabric is simply stored for the required length of time at room temperature.

For the *pad-batch* process a solution of 2-4 g/liter active chlorine, 0.5-1% wetting agent, NaOH to pH 11.5, and 2-4 g/liter Na_2CO_3 is used with a pickup of 100%. For cotton-polyester a pickup of 70%

is taken. The temperature of impregnation and stowing is 20°C and the stow time is 2 hr.

For the *J-box and conveyor processes* the impregnating solution contains 2-5 g/liter of active chlorine, 0.5-1.0% wetting agent, NaOH to pH 11.5, and 2-4 g/liter Na_2CO_3 at a pickup of 100-120% and a temperature of impregnation and stowing of 20°C and a stowing time of 1-2 hr.

In the *fully continuous open-width process* the short treatment times of 10-15 min are compensated for by higher concentrations of the impregnating solutions: 10-15 g/liter active chlorine, 5-8% of soda, and a liquor ratio of 100% at 20°C.

The hypochlorite bleaching process suffers from a number of disadvantages:

1. There is a danger of damaging the fabric due to accidental lowering of pH.
2. The process is slow. It is carried out at low temperatures, and it is therefore difficult to integrate into a rapid continuous operation.
3. There is a danger of yellowing upon storage.
4. The relatively high salt loads in the process effluents are undesirable for ecological reasons.

On the other hand, the process is regarded as the chapest bleaching procedure and can be applied with low-cost, simple chemicals on small as well as on large batches of fabric in a variety of machines.

3.3 Accelerated Bleaching

The slow rate of bleaching, the excess of active chlorine needed in the bleach liquors, and the need for strict control of the chlorine concentration of the liquor and of the fluidity of the fabric during bleaching triggered a series of attempts to accelerate the hypochlorite processes.

Bleaching at Elevated Temperatures [142,143]

The stability of hypochlorite solutions at higher temperatures depends on the pH, and at pH values above 10 appears to be satisfactory: upon boiling for 3 hr 5% decomposed, while at pH 9 a loss of 15% and at pH 7 a loss of 60% were obtained.

It was also found that the temperature does not have any specific effect on the fluidity and functional groups of the oxidized cotton. In experiments performed at pH 11 at 20°C and 60°C in which the same level of consumed oxygen was reached after 30 and 16 min, respectively, the carboxyl group contents, the copper numbers, and the fluidities were virtually the same for the same level of oxygen consumed [142] (Fig. 2.34). It is also evident that for the same

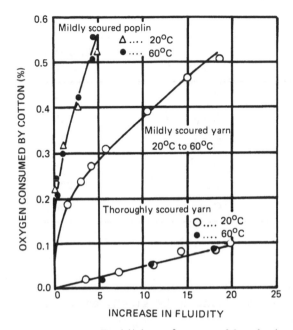

Figure 2.34 Fluidities of cotton bleached with hypochlorite at 20°C and 60°C at various oxygen consumptions. (From Ref. 142.)

fluidity value, the fabric containing more impurities consumes more oxygen, indicating that the impurities are preferentially attacked at both temperatures.

It was recommended to bleach at 70°C for 1 hr in a solution of 0.9-1.6% chlorine owf at a pH range of 8.6-12.8 and with the addition of 10-30 g/liter of NaOH [1,142].

Influence of UV Irradiation

Ultraviolet irradiation was found to accelerate bleaching with NaClO. The whiteness increased with the time of irradiation, with little influence of the concentration of active chlorine in the solution. At the same time, however, a strong degradation was observed: in the nonirradiated fabric bleached at 25°C, a whiteness of about 83% with a DP of 1700 was noted, while in the irradiated sample a brightness of 81% with a DP of 400 was obtained. Stowing the fabric after a short UV irradiation of 1 min produced an accelerated bleaching effect but slightly lower whiteness and DP. It was concluded that UV irradiation is not suitable for NaClO bleaching [144].

Table 2.9 Lifetime of Singlet Oxygen in Various Solvents

Solvent	Lifetime, μsec	Solvent	Lifetime, μsec
H_2O	2	C_6H_6	24
CH_3OH	7	CH_3COCH_3	26
50% D_2O + 50% CH_3OH	11	CH_3CN	30
		$CHCl_3$	60
C_2H_5OH	12	CS_2	200
C_6H_{12}	17	CCl_4	700

Source: Ref. 146.

Mixtures of NaClO and H_2O_2

The activity of NaClO bleaching was found to be greatly enhanced by the addition of H_2O_2 [1]:

$$H_2O_2 + ClO^- \rightarrow HO_2^{\cdot} + Cl^- + \cdot OH \tag{34}$$

In this way new ·OH radicals are introduced which may react with Cl_2O and produce a higher concentration of the hydroperoxide Cl_2OOH, which appears to be the active species in hypochlorite bleaching. In addition the $HO_2^{\cdot}$ and the HO_2^- are effective in the alkaline bleaching liquors. The ·OH may also terminate the chains by interacting with $HO_2^{\cdot}$:

$$HO_2^{\cdot} + \cdot OH \rightarrow H_2O + O_2 \tag{35}$$

It was recently suggested [144] that NaClO interacts with H_2O_2 at 100°C or upon UV irradiation producing singlet oxygen $^1O_2$, which is a very powerful oxidizing agent, albeit its lifetime is 2 microseconds (μsec) in aqueous medium. The lifetime of singlet oxygen varies with the solvent in which it is produced (see Table 2.9), and in CCl_4 lifetimes of about 700 μsec have been reported [145,146].

When fabrics impregnated with H_2O_2 are thermally treated at 100°C, high brightness values are obtained, similar to those obtained in stowing at 25°C. The DP values are, however, lower. UV irradiation accelerates the bleaching, but the DP and the whiteness decrease. With mixtures of NaClO and H_2O_2, obtained by first impregnating the fabric with NaClO and subsequently with H_2O_2, bleaching times of 2 min can be obtained upon IR or UV irradiation at 180°C. When the fabric is UV irradiated during 5 sec and subsequently IR heated for 2 min at 100°C, a whiteness of 80% GE and a DP of 1700

are obtained, as compared to 78.5% whiteness and a DP of 1900 with
only IR heating at 180°C. It was found [144] that by preimpreg-
nating the fabric with an activator (Hexahalofen), which is said to
increase the lifetime of the singlet oxygen, preferably in the presence of a stabilizer (believed to enhance the selective oxidation of
unsaturated bonds), the bleaching results are much improved. At
a ratio of NaClO: H_2O_2 of 1:1, whitenesses of 82-83% were obtained
within 2 min of IR heating at 100°C of freshly impregnated fabric.
DP values of 2000-2100 were obtained as compared to a DP of 2400 of
the original fabric.

The Hypochlorite-Hypobromite System

Bromide is quantitatively converted to hypobromite by oxidation with
hypochlorite over the whole pH range 7-14 [147].

$$HOCl + Br^- \rightarrow HOBr + Cl^- \tag{36}$$

The reaction rate is linearly dependent on the hydrogen ion concentration. At pH values below 9 the oxidation of the bromide is very
rapid. If at pH 7-9, hypochlorite is present in excess of the stoichiometric quantity, other reactions set in, which consist of the decomposition of the resulting hypochlorite-hypobromite, yielding chlorate, bromate, and chloride, as well as bromite and chlorite as intermediates (Fig. 2.35) [17,18]. Hypobromite is a catalyst for the decomposition of hypochlorite to chlorate. The rate of the decomposition of the mixture increases with the concentration of the components
and with the decrease in the pH. The half-time value of a 2.1×10^{-2}
mol per liter solution of NaClO containing 0.7×10^{-2} mol per liter of
NaBrO at pH 9.1 is 15 hr as compared to 4000 hr for a solution of
NaClO alone. At pH 7 the corresponding values are 10 min and
22 hr.

The composition of hypobromite solutions is governed by the
following three equilibria:

$$[H^+] \, [BrO^-] \leftrightarrow K_{d(Br)} \, [HBrO] \tag{37}$$

$$[H^+] \, [Br^-] \, [HBrO^-] \leftrightarrow K_{h(Br)} \, [Br_2] \tag{38}$$

$$[Br^-] \, [Br_2] \leftrightarrow K_{c(Br)} \, [Br_3^-] \tag{39}$$

These are composed, depending on the pH and on the bromide concentration, of four oxidizing species [149]:

$$Ox = [BrO^-] + [HBrO] + [Br_2] + [Br_3^-] \tag{40}$$

Using the equilibrium constants $K_{c(Br)} = 0.063$ [150], $K_{h(Br)} = 5.8 \times 10^{-9}$ [151,152], and $K_{d(Br)} = 2 \times 10^{-9}$ [153,154], the composition can
be calculated (Fig. 2.36) for various pH values and Br^- concentrations.

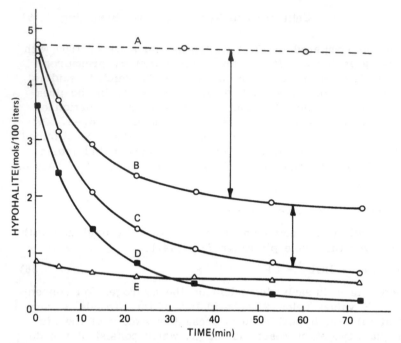

Figure 2.35 Decomposition of hypochlorite (A) and of a hypochlorite-hypobromite mixture (B-E) at pH 7. Change in concentration of bromate plus hypobromite plus hypochlorite (B), of hypochlorite plus hypobromite (C), of hypochlorite (D), and of hypobromite (E), with time. (From Ref. 18.)

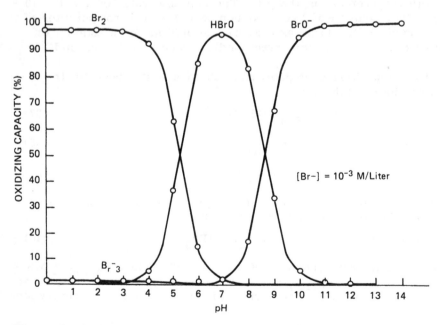

Figure 2.36 Composition of hypobromite solutions at pH 0-14 and $[Br^-] = 10^{-3}$ mol/liter. (From Ref. 149.)

Table 2.10 Redox Potentials of Chlorine and Bromine Compounds

	Volts
$Cl_2 + 2e^- \leftrightarrow 2\ Cl^-$	1.358
$HClO + H^+ + 2e^- \leftrightarrow Cl^- + H_2O$	1.49
$ClO^- + H_2O + 2e^- \leftrightarrow Cl^- + 2OH^-$	0.90
$ClO_2^- + 2H_2O + 4e^- \leftrightarrow Cl^- + 4\ OH^-$	0.76
$ClO_2 + e^- \leftrightarrow ClO_2^-$	0.954
$Br + 2e^- \leftrightarrow 2\ Br^-$	1.087
$HBrO + H^+ + 2e^- \leftrightarrow Br^- + H_2O$	1.33
$BrO^- + H_2O + 2e^- \leftrightarrow Br^- + 2\ OH^-$ (if NaOH)	0.70

Source: Ref. 155.

It is to be expected that the differences in composition between hypochlorite (Figs. 2.1, 2.2) and hypobromite will bring about differences in their oxidizing and bleaching activity. While the redox potentials of Br_2, HBrO, and BrO^- are lower than those of the corresponding chlorine compounds [155] (see Table 2.10), the rates of oxidation by the bromine compounds are much more rapid. On the activity of chlorite and chlorine dioxide it was stated [156,157] that the reason for their not attacking the cellulose is their low oxidation potential. Since the oxidation potential of hypobromite is close to that of chlorite, both should oxidize with comparable rates. It has already been pointed out by Giertz [158] that the redox potential cannot be the decisive factor in determining the reaction rates, and other factors, such as steric hindrances and specific reaction mechanisms, (i.e., activation energies) are of greater importance. It has indeed been found that while the activation energy of the oxidation of cellulose with NaClO is in the range of 11-18 Kcal, with NaBrO values of 7.1-8.4 were recorded throughout the whole pH range [26].

The oxidation by NaBrO appears to be governed by a different mechanism than that by NaClO. A zero-order rate of oxidation was found for NaClO (Fig. 2.5), whereas in the case of NaBrO the consumed oxygen is clearly related to the square root of time [Eq. (14)] throughout the pH range 2-13 (Fig. 2.37). This together with the low activation energy indicates a diffusion mechanism for the NaBrO oxidation.

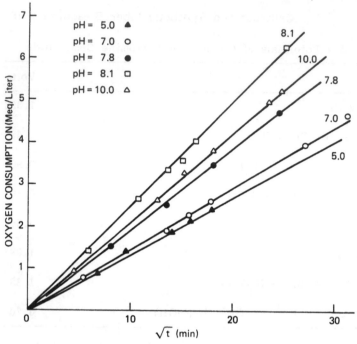

Figure 2.37 Rates of oxidation of purified cotton with NaBrO over the pH range 5-10. (From Ref. 26.)

Table 2.11 Halftime Values of Oxidation of Cellulose with NaClO and NaBrO in the pH Range 5-10[a]

pH	$t_{1/2}$ NaBrO	$t_{1/2}$ NaClO	$\dfrac{t_{1/2} \text{NaClO}}{t_{1/2} \text{NaBrO}}$
10	625	21,000	33.2
9	725	10,000	13.8
8	556	2,050	3.7
7	1420	667	0.47
6	1630	908	0.56
5	1845	3,630	1.57

[a]27°C; 10 g/liter of purified Deltapine cotton. Initial oxidant conc: 0.04 mol/liter.
Source: Ref. 159.

Table 2.12 Functional Groups and Optical Density of Alkaline Extract for Cotton Oxidized with Hypobromite at Different pH Values

pH	10^3 ×D/Ox	10^2 ×CHO/Ox mmol/meq Ox	10^2 ×CO/Ox mmol/meq Ox	10^2 ×COOH/Ox mmol/meq Ox	Yield of functional groups meq/meq Ox × 100
2	0.06	0.15	13.7	0.014	28.0
3	0.40	0.69	11.8	0.13	26.0
4	0.57	0.74	9.9	0.63	23.8
5	1.60	2.9	9.1	1.5	30.0
6	2.63	4.2	7.9	5.5	46.2
7	4.57	4.6	6.9	7.8	54.2
8	5.90	5.8		8.9	60.2
8.3	5.83	5.4	3.1	10.8	
8.7	3.47	4.1			
9	2.28	3.8	2.1	11.3	57.0
10	1.59	3.1	1.0	14.0	64.2
11	0.83	2.7	0.0	12.7	56.2
12	0.58	0.9	0.0	9.7	40.6

Source: Refs. 26 and 87.

The rate of oxidation of cotton by NaBrO decreases with increase in pH except at pH 8 where a maximum appears, resembling the maximum at pH 7 for NaClO (Fig. 2.37). The rate of the NaBrO oxidation is much higher than that of NaClO, but the ratio of the rates depends on the pH (Table 2.11). Significant differences in the yields of the functional groups (cf. Tables 2.12 and 2.2) formed per milliequivalent of consumed oxygen between the two oxidants were noted. With NaClO about 40% and with NaBrO up to 60% were accounted for in the alkaline pH range. This higher yield expresses itself in a higher carboxyl content, while the aldehyde content is significantly lower in the NaBrO oxidation. A steep increase in the amount of ketone groups along with a decrease in aldehyde and carboxyl groups is observed.

Hypobromite was found to brominate lignin and coloring matter in cottonseed husks [17,148], in pulp [17], and in flax fibers [160]

at pH 9-12. For cottonseed husks 2.45 mmol and for flax 1.8 mmol hypobromite were consumed in the bromination (substitution) reaction. It is known [158] that substitution with hypochlorite occurs only at pH values below 7. The bleaching effect of hypobromite thus appears to be also qualitatively different from that of hypochlorite.

The addition of relatively small quantities of Br^- to the NaClO bleaching solution greatly accelerates the process (Fig. 2.38). A small amount of bromide corresponding to 5% of the active chlorine present decreases the bleaching time by 50% and facilitates particularly the removal of the cottonseed husks. The rapid brominating effect and the accelerated oxidation accelerate the decrease of the pH during the bleaching operation (Fig. 2.39). Fewer aldehyde groups are formed, but the degradation does not seem to be inhibited. The number of oxygen atoms per scission, which is 26 for the oxidation with NaClO, is only about 16 for hypobromite in the alkaline range. The application of the bromide therefore needs caution.

The hypochlorite-hypobromite system operates according to the following gross reactions:

$$NaClO + S \xrightarrow{k_{11}} S.O + NaCl \tag{41}$$

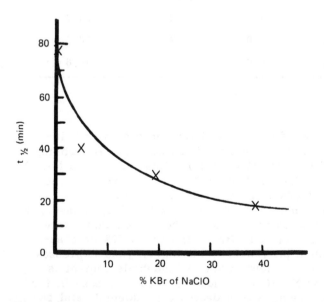

Figure 2.38 Accelerating effect of bromide on NaClO bleaching of cotton seed husks. Half-time values vs. % KBr of NaClO, calculated as active chlorine. (From Ref. 47.)

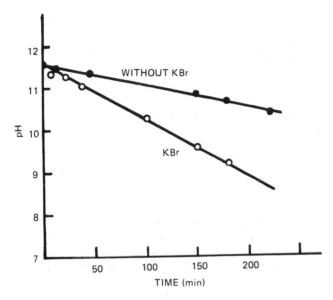

Figure 2.39 pH vs. time of bleaching of cotton seed husks with NaClO and NaClO + NaBrO. (From Ref. 47.)

$$NaClO + Br^- \xrightarrow{k_{22}} NaBrO + Cl^- \qquad (36)$$

$$NaBrO + S \xrightarrow{k_{33'}} S.O + NaBr \qquad (42)$$

$$NaBrO + S \xrightarrow{k_{33''}} S.Br + NaOH \qquad (43)$$

When $k_{33'} + k_{33''} \gg k_{22}$ there will be no accumulation of NaBrO in the solution since any NaBrO produced will be immediately reduced to bromide. The bleaching will therefore be performed with a very low concentration of hypobromite. This is the case in the strongly alkaline pH range above pH 11. At lower pH values the concentration of NaBrO reaches its maximum value and remains constant so long as hypochlorite is present in the solution and thereafter it decreases.

When bleaching cotton with NaClO at a liquor ratio of 1:20, at room temperature with 0.9-1.0 g/liter active chlorine and an initial pH of 9.0, the bleaching time is 80-90 min. When bromide is used, the initial concentration of NaClO can be reduced to 0.7-0.8 g/liter active chlorine since a more complete exhaustion of the hypobromite is obtained. The bromide, 0.25-0.30 g/liter, is added 20 min after the beginning of the bleaching operation, at a pH of 7.5, and the bleaching is continued for an additional 25 min down to a concentration of 0.1 g/liter active chlorine. It was stated [161] that the resulting bleached cotton fabric had a lower copper number and fluidity than cotton bleached to the same degree of brightness with hypochlorite alone. Due to the ease with which the cottonseed husks were bleached, kier boiling time could be reduced [159,161].

4. BLEACHING AND OXIDATION OF COTTON WITH SODIUM CHLORITE

The use of sodium chlorite as a bleaching agent began in 1939, and despite its higher cost compared to H_2O_2 and $NaClO$ it attracted wide interest due to several established advantages [117,148,163-167,170, 171]:

1. Sodium chlorite is an effective bleaching agent. It produces high-whiteness fabrics and at the same time appears to attack only the impurities of the cotton without visibly degrading the cellulose when applied at the range of conditions used in bleaching processes.
2. Traces of metal ions do not catalyze the decomposition of sodium chlorite or chlorous acid, and do not bring about a degradation of cellulose.
3. Sodium chlorite can be used advantageously for the bleaching of other cellulosic and noncellulosic fibers and their blends. It is particularly useful for rayons and flax for which mild bleaching agents and longer bleaching times are needed.
4. A precooking or scouring stage is not essential, and since the cellulose is not degraded, conditions can be adapted to permit an even bleaching of cottons heavily contaminated with seed husks.
5. The weight loss is lower, and due to the incomplete removal of the waxes, the hand of the chlorite-bleached fabric is softer as compared to $NaClO$ and H_2O_2 bleaches.
6. Due to the acid conditions of the bleaching, only a slight swelling of the fabric is obtained.
7. The bleached fabrics have low residual alkali, which facilitates removal of chemical residues. Relatively small amounts of water are needed for rinsing and removing the chlorite bleach liquor.

4.1 Chemical Reactions of Chlorite and Chlorine Dioxide

Chlorine dioxide is obtained from chlorates by reaction with chlorides in a strongly acidic solution [Eq. (48)]. Chlorous acid formed in Eq. (5) reacts with excess of chlorate yielding ClO_2. A number of additional reactions take place at the same time, their relative rates depending on the conditions [17,186-189:

$$HClO_2 + HClO \xrightleftharpoons[\text{low pH}]{\text{high pH}} HClO_3 + HCl \tag{5}$$

$$HClO_3 + HClO_2 \leftrightarrow 2ClO_2 + H_2O \tag{44}$$

$$HClO + HCl \leftrightarrow Cl_2 + H_2O \tag{7}$$

$$5HClO_2 \leftrightarrow 3HClO_3 + Cl_2 + H_2O \tag{45}$$

$$2HClO_2 \leftrightarrow HClO_3 + HClO \tag{46}$$

$$HClO_2 + HCl \leftrightarrow 2HClO \tag{4}$$

$$Cl_2 + 2HClO_2 \leftrightarrow 2ClO_2 + 2HCl \tag{47}$$

$$2HClO_3 + 2HCl \leftrightarrow ClO_2 + Cl_2 + 2H_2O \tag{48}$$

$$HClO_3 + 5HCl \leftrightarrow 3Cl_2 + 3H_2O \tag{49}$$

At low acidity the reverse reaction of Eq. (5) takes place [177,178], and ClO_2, Cl_2, ClO_3^-, and Cl^- are formed in various proportions. With the decrease in pH the rate of formation of chlorate will rapidly increase [16-18]. Reaction (7) is a rapid equilibrium and thus chlorine will be present, albeit in low concentrations, in the bleaching solutions at the acidic pH range. Its concentration will depend on the concentration of the reactants, on the pH, and on the presence of substrates capable of reacting with it [189]. As bleaching proceeds, more chlorite and chlorine dioxide will be reduced to chloride and the concentration of the latter will rise. This will increase the concentration of Cl_2 via reactions of Eqs. (4), (7), (47)-(49).

Several of these reactions have been shown to be sensitive to the way the reagents are mixed [189], and different products are obtained according to the conditions. Thus, when acid is added slowly to an aqueous chlorite solution, mainly ClO_2 is produced. If, however, the same amount of acid is added rapidly or when the chlorite is added to the excess acid, mainly Cl_2 is formed [189].

The concentration of Cl_2 in the hot chlorite bleaching baths will not be high, since it will be very rapidly consumed by the impurities of the cellulose and by the cellulose itself. However, the amounts formed during a bleaching run may be significant and may produce functional groups typical of chlorine and HClO at the given pH values.

In order to suppress the formation of Cl_2 and HClO during the commercial production of chlorites, a reducing agent, such as sulfur

Table 2.13 Effect of pH on $HClO_2$ Concentration and Rate of Bleaching[a]

pH	$HClO_2$ (millimol/liter)	Approx. time for bleaching at 100°C (hr)
2.6	30	1/4
3.5	4.4	1/2
4.5	0.4	1 1/2
5.5	0.04	5
7.0	0.0014	30
8.5	0.00004	100

[a]Concentration of $NaClO_2$, 0.2 M.
Source: Ref. 167.

dioxide, methyl alcohol, formaldehyde, or oxalic acid, is added which transforms the HClO to chloride.

Sodium chlorite is produced by oxidizing ClO_2 with sodium peroxide:

$$ClO_2 + Na_2O_2 \rightarrow 2NaClO_2 + O_2 \qquad (50)$$

It is stored in the form of a powder containing 1% sodium hydroxide [190,191].

The activity of sodium chlorite depends strongly on the pH of its solutions. With increase in pH, the chlorous acid ($HClO_2$) is neutralized, its concentration decreases, and simultaneously the time needed for bleaching incrreases (see Table 2.13). The composition of sodium chlorite solutions over the pH range 1-10 is shown in Fig. 2.40. The dissociation constant of chlorous acid [1,164,165,170] is 1.1×10^{-2} (pK$\sim$2):

$$HClO_2 \leftrightarrow H^+ + ClO_2^- \qquad (51)$$

$$K_{HClO_2} = \frac{[H^+][ClO_2]^-}{[HClO_2]}$$

It is therefore a medium-strength acid and hydrolyzes in water only to a limited extent:

$$NaClO_2 + H_2O \leftrightarrow NaOH + HClO_2 \qquad (52)$$

In order to decrease the pH and to obtain the necessary concentration of chlorous acid, acidic or acid-releasing substances (i.e., activators) are needed. According to several authors, chlorous acid decomposes in aqueous solutions to chlorine dioxide, chlorate, chloride, and oxygen [117,166,170] as follows:

$$5ClO_2^- + 2H^+ \rightarrow 4ClO_2 + Cl^- + 2OH^- \tag{53}$$

$$3ClO_2^- \rightarrow 2ClO_3^- + Cl^- \tag{54}$$

$$ClO_2^- \rightarrow Cl^- + 2O \tag{55}$$

All three reactions depend on the pH, as can be seen in Fig. 2.41, in which the composition of chlorous acid solutions after standing in a stream of nitrogen at 95°C for 1 hr is plotted. Above pH 5 the solution decomposes very slowly, and it becomes virtually stable above pH 7. Below pH 5 the decomposition yields mainly ClO_2 and chlorate. The amounts of oxygen formed by Eq. (55) are small and change very little with pH. At 95°C they amount to only 3% of the decomposed chlorous acid [170]. The contribution of each of the reactions of Eqs. (53), (54), and (55) to the decomposition of sodium chlorite at various pH values at 60°C is shown in Fig. 2.42 in the presence and in the absence of cotton. It is evident that the presence

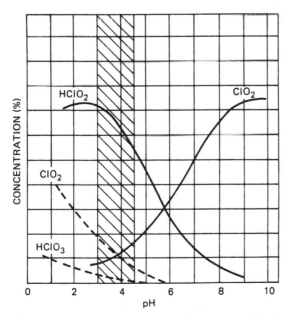

Figure 2.40 Composition of sodium chlorite solutions over the pH range 1-10. (From Ref. 141.)

of the cotton influences only the reaction of Eq. (55): that is, less oxygen is formed in the presence than in the absence of cotton. A part of the oxygen could have been used for the bleaching or oxidation of the cotton.

The decomposition or hydrolysis of ClO_2 appears to proceed in a manner different from that of the chlorite solutions (see Fig. 2.43). While chlorate formation increases with increase in pH in ClO_2 solutions, the opposite is seen for chlorite. The reactions responsible for this decomposition are Eqs. (44) and (48). Chlorine dioxide is used extensively in the bleaching of pulp. The pH range at which it is used is similar to the pH of chlorite bleaching. According to

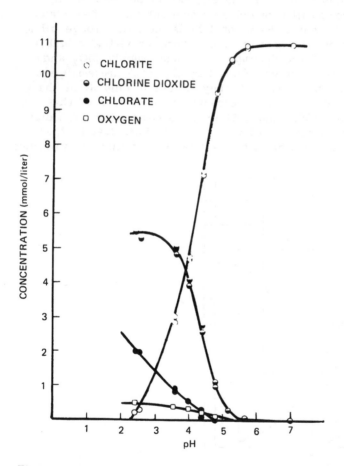

Figure 2.41 Decomposition of 1 g/l $NaClO_2$ solution as a function of pH. Quantities of products formed in solution after 1 hr at 95°C. (From Ref. 166.)

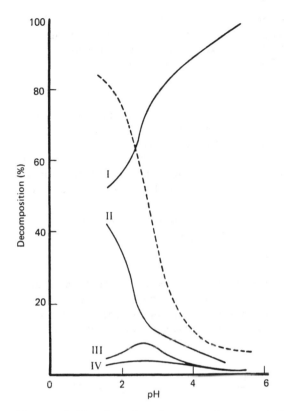

Figure 2.42 Calculated percentage contribution of different reactions in the breakdown of $NaClO_2$ in solutions of different pH values. --- Total decomposition after 60 min at 60°C. (I) Contribution of Reaction (53) to total decomposition, measured both for the blank and in the presence of cotton. (II) Contribution of Reaction (54), both for blank and in presence of cotton. (III) Contribution of Reaction (55), blank only. (IV) Contribution of Reaction (55) in the presence of cotton. (From Ref. 166.)

Rapson four reactions of ClO_2 are operative during the bleaching of pulp [197,198]:

1. $2ClO_2 + 2OH^- \rightarrow ClO_2^- + ClO_3^- + H_2O$ (56)

This reaction occurs in the aqueous solution without the participation of the bleaching substrate (S), and its rate increases with pH (see Fig. 2.43).

2. $ClO_2 + S \rightarrow ClO_2^- + S \cdot Ox$ (57)

In this reaction the impurities of the cellulose are bleached and the cellulose is degraded; thus, the degree of polymerization decreases and the hot alkali solubility increases, indicating the formation of active aldehydes. The rate of this reaction increases with pH.

$$3. \quad ClO_2 + S \rightarrow Cl^- + S.Ox \tag{58}$$

In this reaction the impurities are bleached but the cellulose is not degraded and its rate may not depend on the pH. The evidence that the reactions of Eqs. (57) and (58) occur is provided, according to Rapson [197,198], by the finding that at pH 7, when the rate of the ClO_2^- reaction with the substrate is very slow, all the chlorite formed should have remained in the solution [i.e., 80% based on Reaction (57) alone], and virtually no chloride ions should have been found. Experimentally, however, only 30% of the ClO_2 is converted to ClO_2^-, and large amounts of chloride ions are found.

$$4. \quad ClO_2^- + S \rightarrow S.Ox + Cl^- \tag{59}$$

This reaction occurs simultaneously with (57) and (58), and unlike Reactions (56) and (57), its rate decreases with the pH.

Evolution of ClO_2 in the chlorite solutions is generally believed to be detrimental to the textile bleaching process, although it is

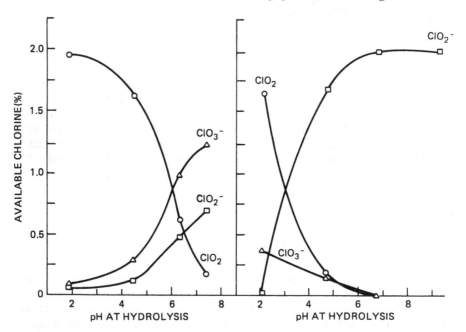

Figure 2.43 Hydrolysis of aqueous solutions of ClO_2 (left) and $NaClO_2$ (right) after 3 hr at 70°C. (From Ref. 169.)

unclear whether ClO_2 is essential for the bleaching effect. Agster believes that ClO_2 does not act directly as a bleaching agent [192], since a bleaching effect is obtained only in the presence of water, when it is converted to chlorite and chlorate according to Eq. (44). The bleaching action is due to the free acid $HClO_2$, as is evident from Table 2.13 and Fig. 2.40 [192]. According to Peters [1] the chlorite ions must be at least partly responsible for the bleaching effect since the results in Table 2.13 show "that the rate at which the bleaching at the higher pH values occurs is apparently faster than that which would be anticipated from the quantity of chlorous acid present."

Nevertheless ClO_2 exists in the bleaching solutions in the acidic pH range and causes corrosion of the equipment as well as toxic fumes, of objectionable odor, especially when its concentration reaches 0.02 g/liter at 80°C. The evolution of ClO_2, similarly to the formation of chlorate, constitutes a loss of bleaching agent.

Addition of sodium chloride to chlorine dioxide as well as to sodium chlorite solutions increases the efficiency of bleaching by suppressing the formation of chlorate [201,202]. At the same time the brightness also increases slightly. The presence of chloride appears to interfere with the mechanism of chlorate formation, possibly by decreasing the concentration of HOCl and forming Cl_2 by Eq. (7). The presence of Cl_2 together with chlorine dioxide was found to increase sinergistically the brightness and the color stability of pulps, thus explaining the increase in brightness upon addition of chloride [202]. The maximum synergistic effect of Cl_2 occurs when it is present at about 5% of the available chlorine.

According to Rapson [198-202], in acid solution, ClO_2, $HClO_2$, and HClO, in mobile equilibrium with elemental chlorine, are all involved in the reaction whether ClO_2 or ClO_2^- is used as the oxidizing agent.

4.2 Activation of Chlorite Solutions

The pH of the chlorite solutions decreases during the bleaching process even without the addition of acid. The $HClO_2$ is reduced to HCl which is a stronger acid. In addition it has been stated that a part of the NaOH originally present in the $NaClO_2$ solutions is being sorbed on the cellulose at room temperature and even in the temperature range of 100-140°C [193]. The sorption occurs also on the impurities. Nevertheless, additional acidification is needed in many cases, especially with long liquor processes and when scoured cloth is used.

The starting pH in the impregnation bleaching processes is usually 5-7, and the acid is released slowly through the action of activators. Considerable attention has been given to the activation, and many chemicals have been studied for this purpose (see Table 2.14). The activators should preferably have several functions in

Table 2.14 Several Activators for Bleaching with Sodium Chlorite

Activator	Temperature (°C)	Bleaching method	Time (hr)	Whiteness (%)	Ref.
$CH_2(OH)(SO_3Na)$ + Na_2SO_3 + Na_2CO_3	20-30	Pad-roll	12-16	75	173
Citric acid, sodium citrate, or triethylcitrate	90	Bath	2	—	174
$NH_4H_2PO_4$	100	Pad-steam	2	—	176
$(NH_4)H_2PO_4$ + $ClCH_2COOH$	95-98	Bath	2	80-84	177
$NaBrO_2$	85	Pad-steam	2		178
$CuSo_4$	98	Pad-steam	0.5		179
$H_3PO_4 + H_2C_2O_4$	95	Bath	1.3	80	180
Na malonate	90	Bath	2		
HCHO	90	Bath	1.5		164,182
H_2O_2	95	Bath	1.5	86	183

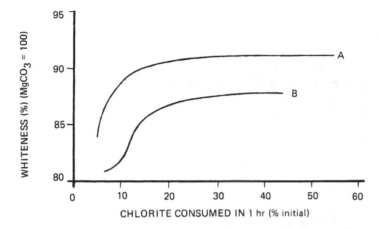

Figure 2.44 Effect of NaH_2PO_4 on the whiteness obtained upon bleaching with 1.75 g/l $NaClO_2$. Curve B, without phosphate. Curve A, with 2 g/l of NaH_2PO_4. Temperature, 80°C; pH, 4-4.5. (From Ref. 191.)

addition to the gradual acidification of the bleaching solution: (a) to prevent or decrease the rate of ClO_2 formation and to minimize corrosion and odor formation; (b) to decrease the formation of chlorate and thus to increase the utilization of the chlorite and improve the economics of the process; (c) to increase the rate of bleaching at a higher pH range possibly without ClO_2 formation; (d) to decrease the temperature of the process; (e) to increase the whiteness obtained through better utilization of the chlorite.

One such activator is sodium dihydrogen phosphate. It acts as a buffer, and in concentrations of 1-5 g/liter, depending on the liquor ratio, keeps the pH of the bleaching solution in the range of 4-4.5. As can be seen in Fig. 2.44, the addition of 1 g/liter of NaH_2PO_4 brings about a significant increase in whiteness [191].

It can be seen from Fig. 2.45 that 10 activators applied in pad-roll bleaching of a huck toweling fabric gave higher reflectance values than a control chlorite bleach when steaming for 70 min with saturated steam at 100°C was applied [166]. The fluidity values of the bleached fabrics were only slightly higher than those of the unbleached samples. When steaming was done at 130°C for 15 min, no enhancing effect of the activators on the reflectance was noticed while the fluidity increased in all cases. It is of interest to note that 98% of the seed husks were removed by the steaming at 100°C, while only 75% were bleached at higher temperatures and shorter times [166]. Only in the presence of four of the activators listed in Fig. 2.45 was chlorate formed in the solution, while when using NH_4NO_3, $(NH_4)_2SO_4$, $Zn(NO_3)_2$, and $MgCl_2$,

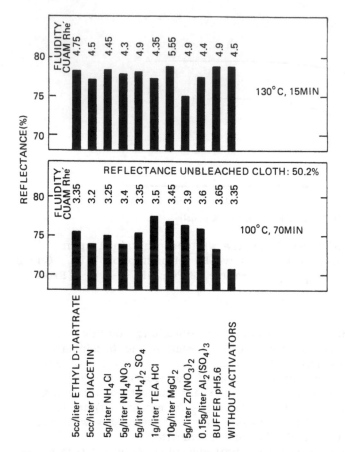

Figure 2.45 Reflectance and fluidity obtained in chlorite bleaching with different activators. Bleaching by cold padding, rolling up of the fabric, and steaming with saturated steam. Chlorite conc., 10 g/l (100% NaClO$_2$); wetting agent, 0.75 g/l; pickup 100%. (From Ref. 166.)

no chlorate was formed [166], indicating a change in the relative rates of the chlorite decomposition reactions of Eqs. (53), (54), and (55) in the presence of some activators. This is not surprising in view of the sizable number of sensitive reactions and equilibria involved in the chlorite bleaching system and considering the different modes of activity of the various activators.

Many of the activators are hydrolyzable compounds (cf. diethyl tartrate and ethyl lactate [184]) which slowly generate acids; ammonium salts resemble esters in yielding free acids upon boiling in aqueous solutions, sodium chloroacetate liberates hydrochloric and glycolic acids during steaming. When sodium bromite was applied as an activator, bleaching of a cotton poplin containing 12% starch sizing with $NaClO_2$ could be carried out in the pH range 9-11 [178]. Improved whiteness, complete desizing, better hydrophilic porperties, reduced bleach consumption, and decreased evolution of ClO_2 were claimed. The fabric was impregnated to a 55% pickup in a bath containing 20 g/liter of 80% $NaClO_2$, 1.4 g/liter active Br as $NaBrO_2$, and 2 g/liter softener; steaming at 85°C was carried out for 2 hr, followed by rinsing in Na_2CO_3 and hot and cold water. It appears that in this case direct interactions between the chlorite and the bromite as well as between the highly reactive products of their reduction (i.e., hypochlorite and hypobromite) during bleaching took place. The interactions between hypochlorous and hypobromous acids in the alkaline pH range were discussed previously. (See Sec. 3.3, discussion on "The Hypochlorite-Hypobromite System.")

The use of formaldehyde as an activator is of particular interest. It permits bleaching to be carried out at room temperature and in the neutral and weakly acidic pH range [141]. It appears to interact with chlorite, forming hypochlorite with consequent acceleration of the bleaching.

Another combined process results when hydrogen peroxide is added to the chlorite bleaching bath. The bleaching is carried out in two stages. First the bath is acidified to pH 4 and brought to 85°C. After 30 min in which the chlorite exerts its effect, the pH is brought up to 11 when the bleaching effect of the H_2O_2 takes place [195]. H_2O_2 can also be added to the chlorite bath in order to suppress the formation of ClO_2. This can be effective at pH values above 3.5. When bleaching is to be done at lower pH values (e.g., for stripping of dyestuffs), hermetically closed equipment has to be used [162].

Sodium chlorite can be activated with ultraviolet light. In the acidic pH range (below pH 4) chlorine dioxide is formed, while in the alkaline pH range the chlorite is converted into hypochlorite [163]. The empirical rate equation for the photochemical decomposition was found to be

$$- \frac{d(ClO_2^-)}{dt} = kI_0[ClO_2^-]^{0.5} \tag{60}$$

where I_0 is the intensity of the incident radiation; Eq. (60) remains valid in acid as well as in alkaline media, but the rate constant is greater in an acid medium. When scoured cotton samples were immersed in a 1% sodium chlorite solution, squeezed to 75% retention, and irradiated for 10 min with a low-pressure quartz mercury discharge lamp at a distance of 8 cm, reflectances of 85% and 83% and DP values of 1500 and 1300 were obtained for acidic (pH 5) and alkaline (pH 9) solutions, respectively.

Table 2.15 Fluidity and Reflectance Factor of Cotton Fabric
Bleached at Different pHs

pH	Chlorite consumed (g/100 g cotton)	Fluidity	Reflectance factor[a]
2.6	1.7	4	0.84
3.5	1.7	3.5	0.87
4.6	1.7	3	0.89
5.5	1.6	3.5	0.91
7.0	1.5	7	0.88
8.0	0.8	11	0.86
8.5	0.9	14	0.82

[a]Measured with an EEL Reflectometer, with the blue filter, relative
to a magnesium oxide standard.
Source: Ref. 167.

In another development, ionizing radiation was applied to acti-
vate chlorite bleaching of a fabric impregnated with 25 g/liter $NaClO_2$
and 1 g/liter of a wetting agent (8:1, ethylene oxide-butylphenol con-
densate) at 25°C, squeezed to 65% wet pickup. A 2-Mrad dose of
gamma rays (^{60}Co) at 9.1 rads/hr was applied. A significant increase in reflectance and a small decrease in breaking strength were
achieved [172].

Prolonged storage of chlorite bleaching solutions is sometimes
desired but difficult to achieve due to their inherent instability. It
has been suggested to add to the chlorite a small amount of an amine
[171]. A solution of 25 g/liter $NaClO_2$ is stabilized for 7 days by the
addition of 0.3 ml/liter of trimethylenediamine adjusted to pH 10.6
with HCl. After 2 hr of steaming in a pad-steam bleaching process
of a cotton drill fabric, the pH dropped to 5.5, the $NaClO_2$ consumption was 68%, and the whiteness obtained was 97% relative to $MgCO_3$.

4.3 Oxidation of Cellulose with $NaClO_2$ and ClO_2

While it is generally accepted that sodium chlorite does not degrade
cellulose when used under optimal bleaching conditions [196], it was
reported that when a large excess of chlorite is used or when the pH
is high, the fabric is strongly degraded. When 3.7 g $NaClO_2$ were
used to bleach and oxidize 100 g of a desized fabric at the pH range
7-6.2, a fluidity of 18 rhes was obtained, while when the same amount
was used at a pH of 4.5-5.4, a fluidity of 5 resulted [164].

Upon increasing the pH of the bleaching solution, the amount of chlorite needed to obtain a given brightness decreases but at the same time the fluidity increases (see Table 2.15) [167]. At the high pH values above 7 no chlorine dioxide is formed. The percentage of chlorate formed also decreases with increase in pH (see Table 2.16). The utilization of sodium chlorite in bleaching of the organic impurities, however, increases with increasing pH and amounts to 70-80% at the preferred bleaching pH of 3.5-4.6 [167].

Little is known about the functional groups formed upon chlorite oxidation of cotton. It has been established that chlorous acid at pH 3 oxidizes aldehyde groups quantitatively to carboxyls, and in the case of C_1 aldehydes, an ester of gluconic acid is formed [63]. No oxidation of the cotton occurred with a 0.2 M solution of $HClO_2$ at pH 3. When, however, a 1 M solution was used [148] for long reaction times exceeding 24 hr, a pronounced degradation occurred, and the copper number as well the carboxyl group content increased sharply. It was suggested that chlorine dioxide or even a complex between chlorous acid and chlorine dioxide was responsible for this oxidation [1,148]. It is, however, also possible that small amounts of hypochlorous acid and chlorine are formed in the system and attack the cellulose (see above, Sec. 2.8).

Oxidation of cellulose by ClO_2 appears to yield similar results to those of $HClO_2$. The fluidity and carbonyl and carboxyl group content increase with the pH of oxidation, as is seen in Fig. 2.46 [148].

The presence of the carbonyl groups in celluloses treated with chlorine dioxide appears to be responsible for the yellowing and hot alkali solubility. Chlorine dioxide does not seem to improve the stability of pulp brightness, that is, it does not affect the carbonyl groups formed by previous oxidation in a manner similar to chlorous acid [112,169,197-199].

According to Samuelson [200] the oxidation of C_1 aldehydes to carboxyls by chlorous acid and ClO_2 is a slow reaction. At the same time, the hydroxyls at C_6 in the cellulose chains are oxidized, this oxidation being the predominant reaction, as evidenced from the relative quantities of gluconic and glucuronic acids found by chromatography in the hydrolyzates of the oxidized celluloses [200].

The fact that significant quantities of erythronic and glyoxylic acids were found in the hydrolyzate indicates that by prolonged oxidation of cellulose by ClO_2 and chlorous acid, the C_2-C_3 bond is cleaved with the possible intermediate formation of C_2 and C_3 aldehydes, which are subsequently oxidized to carboxyls [200]. The general pattern of the oxidation of cellulose with chlorite and ClO_2 appears to be similar to that of the oxidation with Cl_2 and hypochlorite in the pH range 2-10, although the ratio of the products obtained and the relative rates at which they are obtained may be vastly different.

Table 2.16 Efficiency of Utilization of Chlorite at Various pHs

pH of buffered chlorite solution	Percentage of consumed chlorite found as		Percentage of consumed chlorite used in oxidation of organic matter	Chlorite (g/100 g cloth) to produce a reflectance factor of 0.85
	ClO_2	ClO_3^-		
2.6	1.1	34.0	65	1.9
3.5	2.5	23.8	74	1.75
4.6	1.3	18.6	80	1.5
5.5	0.4	12.4	87	1.4
7.0	nil	6.6	93	1.4
8.0	nil	2.1	98	—
8.5	nil	1.3	99	—

Source: Ref. 167.

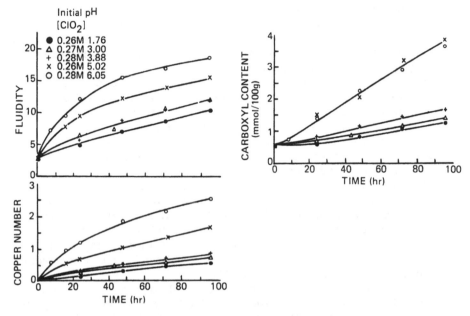

Figure 2.46 Effect of pH on the fluidity, copper number, and carboxyl content of cellulose oxidized by solutions of ClO_2 at 20°C. (From Ref. 148.)

Only scarce information is available on the possible presence of ketone groups in chlorite-oxidized cellulose. The formation of α-hydroxy monoketo groups during the oxidation of cotton with $HClO_2$ was reported [203]. Similar to the other ketocelluloses, no enhanced yellowing or dissolution was found upon alkaline extraction of this ketocellulose [204].

4.4 Chlorite Bleaching Processes

Bleaching of cotton fabrics with chlorite can be carried out in a number of systems: Kier, jig, package, winch, pad-roll, and other impregnation processes.

A basic consideration in the chlorite bleaching practice is the need for a pretreatment and aftertreatment of the fabric. There are a number of pretreatments used in practice which range from an enzyme desizing and a soda ash scour to a 2-3 hr boil in 10-30 g/liter sodium hydroxide in the presence of wetting agents. The severity of the pretreatment used depends on the state of the fabric, on the amount of size and impurities, and on the final use to be made of the fabric.

Inadequate pretreatment can bring about a number of undesirable effects: after-yellowing of the high whiteness obtained; the seeds husks are bleached but not removed from the fabric, residual wax-low hydrophilicity; uneven dyeing and bleaching due to differences in cotton quality and machine performance [141].

In many cases a posttreatment with boiling soap and soda is needed in order to improve the wettability of the bleached fabric, since sodium chlorite removes only a part (up to 50%) of the fats and waxes of the unscoured or only mildly scoured fabrics. Another aftertreatment stage which is sometimes needed is rinsing in a reducing bath containing hydrosulfite or bisulfite in order to remove the residual oxidizing agents.

A typical formulation for the *kier bleaching* consists of 1-2% of 80% $NaClO_2$ owf, 0.25-0.5% NaH_2PO_4 owf, and wetting agent. The pH is adjusted with formic acid to 3.8-4.2. After circulating the liquor for 15-20 min for the saturation of the fabric, the temperature of the liquor is raised to 80-85°C in 1-1.5 hr and kept at this temperature for an additional 2-3 hr. The fabric is then aftertreated with soap and soda ash and washed [191].

In the *package* bleaching, in which the liquor ratio is 5-8:1, the bleaching solution is composed of 2-4 g/liter $NaClO_2$, 0.5-1 g/liter NaH_2PO_4, 1-2 g/liter $NaNO_3$, wetting agent, and formic acid to pH 3.8-4.2. After cold circulation for 10 min, the temperature is raised to 80-85°C in 1-2 hr and maintained for 1-2 hr. When bleaching in the *winch* at liquor ratios 1:20 to 1:40, the bleaching liquor is composed of 1-3 g/liter $NaClO_2$ (80%), 1-3 g/liter $NaNO_3$, 1 g/liter NaH_2PO_4, and wetting agent. Formic acid is added to give a pH of 3.8-4.2, after 5-10 min of running. The temperature is raised to 80°C and running continued for 1.5-2.5 hr [191].

In the impregnation processes the bath is composed of 20 g/liter of $NaClO_2$ (80%), 2-3 g/liter of wetting agent, formic acid to give a pH of 6-6.5, and a given amount of activator. The padding is done at 20-40°C with a pickup of 100-120%. The fabric is then stored at 85-95°C on a roll, J Box, or brattice conveyor for 1-4 hr [141].

It is possible to bleach cotton in the open-width process at room temperature with $NaClO_2$, 2-2.5% owf, together with 0.3% formaldehyde as activator, and 0.05-0.1% soda ash. The impregnation is done at 20-25°C. The pickup should be 85-100%. The cloth is taken off the mangle in batch form. It is covered with polythene sheet and tied onto the batch roller and left overnight [191].

4.5 Corrosion

Corrosion is caused mainly by the ClO_2. Activators and buffers which decrease the formation of ClO_2 and also of elementary chlorine counteract corrosion. Commonly used metals like iron, cast iron, mild steels, and even stainless steels cannot be used for bleaching with chlorite. High (2.5-4.5%) molybdenum-containing stainless steels are resistant to acidified chlorite. So are titanium, glass, and ceramics. Ceramics are used for lining iron vats and for pumps, ducts, and valves. Teflon and polypropylene have also been successfully used.

The addition of 0.1-2 g/liter of $NaNO_3$ has been claimed to reduce significantly the corrosion of conventional stainless steel equipment [194]. A similar effect can be obtained by periodic passification with nitric acid [185].

The corrosive effect depends to a considerable degree on the liquor ratio and mainly on the extent of the area of contact between the solution and the fabric. In jigs and winches, where a large volume of liquor is not directly in contact with the cotton, corrosion and odor formation are high. In the case of impregnation bleaching, the problem is less severe [191]. The gaseous chlorine dioxide is "filtered" or consumed by the impurities of the fabric. Another way is to "dilute" and neutralize the ClO_2 with gaseous ammonia formed from ammonium salts added to the initial alkaline chlorite solution [166]:

$$NH_4^+ + OH^- \rightarrow NH_3 + H_2O \tag{61}$$

It has been found that in all cases where the fabric is in contact with a corrosion point, the cotton becomes severely degraded [166].

5. BLEACHING AND OXIDATION OF COTTON WITH HYDROGEN PEROXIDE

5.1 Introduction

Hydrogen peroxide is the most widely used bleaching agent for textiles, and over 85% of all fabrics are bleached with it. It came into use around 1878 [1]. At that time, and until the 1930s, mostly barium peroxide and later sodium peroxide were used. Hydrogen peroxide was applied only in several mills for bleaching of specialized items such as silk, some wool fabrics, and cotton fabrics containing vat-dyed yarns. This situation changed in the 1930s when the use of H_2O_2 rapidly spread and replaced previously used hypochlorite bleaching processes. The change was triggered by two developments: the ready availability of concentrated and stable H_2O_2 solutions and the intensive and continuing technical service accompanied by research and development programs, which was offered to the textile plants by the H_2O_2 suppliers [2].

Hydrogen peroxide is a weak acid, and its dissociation constant

$$H_2O_2 \leftrightarrow H^+ + HOO^- \tag{62}$$

Table 2.17 Decomposition of Concentrated Solutions of H_2O_2 at Various Temperatures

Temperature (°C)	Time (days)	% Decomposed
20	180	0.2
20	365	0.4
40	180	1.5
40	365	3.0
60	30	1.5
80	7	1.9
100	1	1.3

Source: Ref. 191.

in water is 1.78×10^{-12} [15]. When undissociated it is relatively stable. Hydrogen peroxide is potentially unstable from the point of view of thermodynamics. It decomposes exothermically according to the equation

$$H_2O_2 \rightarrow H_2O + 1/2\ O_2 \tag{63}$$

liberating 22.62 Kcal/mol. The activation energy of this decomposition is, however, high (50 Kcal/mol), and it is not reached in aqueous solution even on boiling [13]. The decomposition is, however, easily catalyzed by heavy metals and easily oxidizable materials. Commercial concentrated H_2O_2 solutions usually contain small amounts of inorganic or organic stabilizers which preserve the stability in the presence of inadvertent contamination. (Table 2.17) [191]. Hydrogen peroxide has a number of important advantages over other bleaching agents:

1. It produces a stable white color, and the bleached fabrics are highly hydrophilic since the fats and waxes are solubilized and removed by the hot alkaline solutions used.
2. Its reaction products are relatively nontoxic and innocuous, and it decomposes to oxygen and water, thus reducing greatly the effluent pollution of the bleaching plants.
3. When applied under optimum conditions, the cellulose and lignin are not degraded during bleaching. Cottonseed husks can be effectively bleached, often without a previous boiling operation.
4. It is compatible with most fibers and can be applied to a wide variety of fabrics under a wide range of bleaching conditions and machinery.

5. The number of operations and stages in the bleaching processes can be reduced, and continuous, one-stage processes can be worked out.

6. Bleaching and dyeing can sometimes be combined in a single operation.

7. Hydrogen peroxide bleaching, when applied under the appropriate conditions, is close to fulfilling the six chemical requirements outlined in Sec. 3.10.

As against these advantages several disadvantages are cited in the literature:

1. Bleaching is slow unless high temperatures are applied. Energy costs are high.

2. Catalytic decomposition of the H_2O_2 can occur along with catalytic degradation of the cellulose due to iron, nickel, copper, cobalt, and lead hydroxides present in the bleaching solution or in the fabric.

3. The above metals and their alloys cannot be used as materials of construction of H_2O_2 bleaching containers. Passivated molybdenum-stainless steel, wood, stonework, and silicate- and cement-coated mild steel or cast iron can be used for bleaching equipment. Coatings with glass and plastic laminates can also be applied.

5.2   Decomposition of H_2O_2

The decomposition of H_2O_2 has been investigated by many workers for several decades [205-214,1,69]. It is a complicated reaction and is influenced by a considerable number of variables, such as temperature, pH, time, quality of water, presence of stabilizers, catalysts, and activators, their concentration, and chemical nature. It is also influenced by heterogeneous reactions on surfaces, on the fabric, and on any colloidal matter which may be present. The reproducibility of the experimental results reported in the literature on H_2O_2 decomposition was in many cases low [205], but nevertheless, a general picture of this reaction has lately emerged.

The rate of H_2O_2 decomposition increases with temperature, and activation energy values of 15 [205] and 19.4 Kcal/mol [208] were obtained by two different investigators for the homogeneous reaction (without fabric) in the alkaline pH range. In the presence of cotton fabric, values ranging from 5.8 to 28.7 Kcal/mol were obtained [208].

The rate increases with increase in pH (Figs. 2.47 and 2.48) and NaOH concentration. It is believed that the reaction proceeds through the decomposition of the HO_2^- ion. Erdey [215] suggested that the decomposition occurs through a reaction between H_2O_2 and HO_2^-:

$$H_2O_2 + HO_2^- \rightarrow H_2O + OH^- + O_2 \tag{64}$$

$$HOO^- \rightarrow OH^- + 1/2\ O_2 \tag{64a}$$

According to Eq. (64) there should be a maximum in the decomposition rate when the pH = pK, that is, when 50% of the H_2O_2 has dissociated. Since the pK at 60°C is 11, the maximum rate should occur at pH 11. Such a maximum has recently been obtained [214] in experiments in which the ratio of OOH⁻ to H_2O_2 was gradually changed from 0:10 to 10:0. At 0.0°C the maximum rate was obtained at a ratio of 0.7:1, whereas it shifted to 0.3:1 at 40°C. This shift indicates, according to Isbell, that a chain reaction occurs simultaneously with the ionic or nonchain reaction, and its contribution to the overall process increases with increase in temperature [214]. The chain reaction may be started by traces of a catalyst. The decomposition of H_2O_2 has been assumed by Haber and Weiss to take place in the presence of catalysts by Eqs. (65) and (66), that is, by the formation of hydroxyl and hydroperoxide radicals [211,212]:

$$H_2O_2 + F_e^{2+} \rightarrow \cdot OH + OH^- + F_3^{3+} \tag{65}$$

$$OOH^- + F_e^{3+} \rightarrow \cdot OOH + F_e^{2+} \tag{66}$$

$$H_2O_2 + OH^- \leftrightarrow OOH^- + H_2O \tag{62}$$

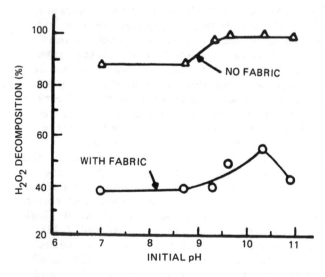

Figure 2.47 H_2O_2 decomposition in absence and presence of fabric as a function of pH in the alkaline range. (From Ref. 209.)

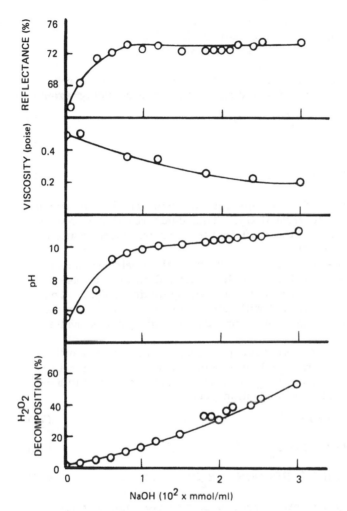

Figure 2.48 Effect of variation of NaOH concentration on bleaching results at 90°C. Reaction time 30 min. Concentrations in solution: NaHPO$_4$, 0.00325; CaCl$_2$, 0.00386; H$_2$O$_2$, 0.148 mmol/ml. (From Ref. 208.)

$$\cdot OOH \rightarrow H^+ + \cdot O_2^- \tag{66a}$$

$$H_2O_2 + OOH^- \leftrightarrow \cdot OH + H_2O + \cdot O_2^- \tag{67}$$

$$\cdot OH + \cdot O_2^- \rightarrow OH^- + O_2 \tag{68}$$

According to Isbell [214] the reaction between H$_2$O$_2$ and the OOH$^-$ ion affords [Eq. (67)] hydroxyl radical, superoxide radical and

and water. Oxygen is evolved from the reaction between the hydroxyl radical and the superoxide radical [Eq. (68)]. In this way one mol of hydrogen peroxide is decomposed by one mol of hydroperoxide anion in a nonchain reaction.

Decomposition can, however, also take place by a chain reaction that begins with the superoxide radical, which is a reducing agent:

$$H_2O_2 + \cdot O_2^- \rightarrow \cdot OH + OH^- + O_2 \tag{69}$$

$$OOH^- + \cdot OH \rightarrow H_2O + \cdot O_2^- \tag{70}$$

$$3H_2O_2 \rightarrow 2\cdot OH + 2H_2O + O_2 \tag{71}$$

Reactions (62), (70), and (71) represent a chain reaction by which O_2 and water are formed from H_2O_2. This chain reaction would appear to be responsible for the shift in the maximum decomposition rate discussed above [214].

Further evidence for the contribution of the chain mechanism was suggested by Isbell [217] from the kinetic data of the base-catalyzed oxidation of ketoses with the hydroperoxide anion. The rate of this oxidation increased with pH up to pH 12, while in the range 12-14 no further increase occurred, indicating a free radical reaction.

The decomposition of hydrogen peroxide in the acidic solution is thought to be a heterolytic bond breakage. The addition of acid reverses the dissociation of H_2O_2 [Eq. (62)] and thereby stabilizes it. On the other hand a protonization occurs:

$$HOOH + H^+ \rightleftharpoons (HOOH_2)^+ \tag{72}$$

$$(HOOH_2)^+ \rightarrow H_2O + 1/2\, O_2 + H^+ \tag{73}$$

$$(HOOH_2)^+ + RCOO^- \rightleftharpoons RCOOOH + H_2O \tag{74}$$

The protonized H_2O_2 can either decompose to water and oxygen [Eq. (73)] or form more or less stable peroxy acids [Eq. (74)] which are suitable for bleaching in acidic medium [206].

The homolytic bond breakage

$$HOOH \rightarrow 2(\cdot OH) \tag{75}$$

would require a high activation energy of 48 Kcal/mol and is therefore not likely to occur.

5.3 Oxidation of Cotton by H_2O_2

The oxidation of the cellulose by H_2O_2 takes place simultaneously with its decomposition. Bleached cotton appears to have a catalytic effect on the decomposition (Fig. 2.49) [69]. The rate of

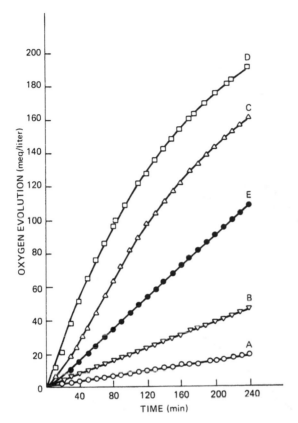

Figure 2.49 Rate of evolution of O_2 from H_2O_2 solution for several cotton concentrations. (1) without cotton; (2) 2.5 g/l; (3) 5 g/l; (4) 10 g/l; (5) 10 g/l of cotton previously extracted with 0.1 N HCl for 6 hr. Initial H_2O_2 concentration, 0.3 N; pH, 9.5; 0.01 mol/l pyrophosphate stabilizer. (From Refs. 69 and 80.)

decomposition yielding gaseous oxygen increases with increasing concentration of cotton from 0.0 to 10 g/liter. The lowest curve in Fig. 2.49 represents the decomposition of H_2O_2 without cotton. The catalytic effect of the purified cotton fibers persisted also after a thorough extraction of heavy metal traces was performed on the cotton with 0.1 N HCl for 6 hr. A similar catalytic effect of purified cotton was also obtained at 50°C in the presence of sodium silicate (5 g/liter) and sodium hydroxide (0.5 g/liter) for high concentrations of 200-500 g/liter bleached cotton [1,221].

The oxidation of the cotton was found to start 15-50 min after the beginning of the evolution of the O_2 (Fig. 2.50). The ratio of

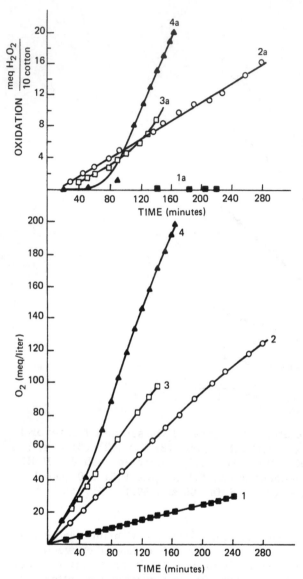

Figure 2.50 Rates of simultaneous decomposition of H_2O_2 and of oxidation of purified cotton by H_2O_2 at several pH values: 1,1a, pH 7.0; 2,2a, pH 9; 3,3a, pH 9.5; 4,4a, pH 9.7. Concentrations as in Fig. 2.49. (From Refs. 69 and 80.)

Table 2.18 Rate of Simultaneous Decomposition of H_2O_2 and Oxidation of Cotton[a]

Cotton, g/liter	pH	K_{dec}	K_{ox}	$\dfrac{K_{dec}}{K_{ox}}$
0.0	9.5	0.08	—	
10	7.0	0.12	0.0	
10	9.0	0.47	0.053	8.9
10	9.5	0.70	0.053	13.2
10	9.7	1.22	0.085	14.4
10 (HCl treated)	9.5	0.45	0.0	

[a]Cotton, 10 g/liter; $Na_4P_2O_7$, 0.01 mol/liter; initial H_2O_2 conc., 0.3 N; temperature, 80°C; time, 4 hr; rate constants in meq/min of O_2.
Source: Refs. 69 and 80.

the H_2O_2 decomposed as O_2 and that consumed on oxidation was found to remain constant throughout the 280 min of experiments [69], indicating the possibility of a single mechanism governing both phenomena. The rates of decomposition and oxidation increased with the pH, but at the higher pH values the decomposition rate increased faster and the ratio of the rates increased (Table 2.18). Very little oxidation occurred at pH 7. It also almost did not occur in the case of the HCl-extracted cotton, indicating that traces of a catalyst are needed for the oxidation reaction, and that the HOO⁻ ion is by itself not very effective in oxidation [69,80].

The main functional groups formed on the cellulose by the H_2O_2 oxidation appear to be ketones (Fig. 2.51), while very few aldehyde and carboxyl groups were found. Hydrogen peroxide at pH 9.5 is thus producing a *ketocellulose*, similarly to the oxidation of cotton with Br_2 at pH 2 (Fig. 2.24). The ketone groups in the H_2O_2-oxidized cotton are "inactive" and do not bring about yellowing [Eqs. (27), (28)] and color reversion (Table 2.19) [69,80]. Hartler and co-workers [222] found that the brightness reversion of kraft pulps, bleached with H_2O_2 in the presence of several stabilizers, did not increase and even decreased slightly with the increase in oxygen consumption, although it was accompanied by an increase in carbonyl content and a decrease in viscosity.

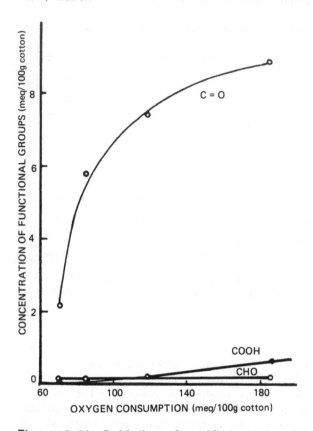

Figure 2.51 Oxidation of purified cotton with H_2O_2: functional groups vs. consumed oxidant concentrations as in Fig. 2.49. (From Refs. 69 and 80.)

It is of interest to note that the solutions in which the oxidations were carried out, although alkaline and at 80°C, remained colorless, showing that no yellowing chromophores were formed or peeled off during oxidation. The low D values obtained (Table 2.19) correspond to the low amount of aldehyde groups determined on the cotton [69] which are formed during the chain scission.

Table 2.20 reveals further unique features of this oxidation. The number of oxygen atoms consumed per chain scission is above 100 as compared to 26 and 18 in the cases of hypochlorite and hypobromite [25,26] oxidations. Furthermore the yield of the functional groups is less than 10% as compared to 40% and 80-90% for hypochlorite and bromine oxidations. Ivanov and co-workers also suspected low yields of functional groups upon oxidation with high concentrations of H_2O_2 when 6.5% of the oxidized cellulose was solubilized [223].

Table 2.19 Functional Groups, Degradation, and Yellowing of Cotton Oxidized by Hydrogen Peroxide[a]

Ox. conc.[c] meq/100 g cotton	Carboxyl (meq/100 g)	Aldehyde	Ketone	Yield (%)	D Yellowing	pc	Oxygen atoms per scission	DP (nitrate)
0.0[b] (blank)	1.79	0.09	0.0		0.008	0.06		
70	0.03	0.11	2.16	3.3	0.010	0.16	189	2270
85	0.05	0.12	5.80	7.1	0.012	0.12	222	2150
118	0.21	0.21	7.42	6.7	0.016	0.16	273	2060
185	0.63	0.19	8.92	5.3	0.015	0.27	128	1160
10[d]	0.08	0.0	0.78	8.6	0.008	0.06		

[a]Consistency, 10.0 g/liter. Temperature, 80°C. Stabilizer, 0.01 M $Na_4P_2O_7$; pH, 9.5-9.7.
[b]Purified cotton, treated with $Na_4P_2O_7$ and without H_2O_2 in the oxidation apparatus at 80° for 100 min.
[c]H_2O_2 consumed only on oxidation.
[d]Cotton extracted with 0.1 N HCl for 6 hr.
[e]All data after deduction of the value of the blank.
Source: Refs. 69 and 80.

Table 2.20 Comparison of Properties of H_2O_2 and NaClO-Oxidized Cellulose

	Ox.	pH	-COOH	-CHO	-CO	Yield %	Ox. sciss.	Groups per scission			Yellowing D units
			(meq/100 g)					-COOH	-CHO	-CO	
NaClO	130	8	2.33	16.0	14	41	26	2	3	2.6	0.8
H_2O_2	118	9.5	0.21	0.21	7.42	6.7	118	0.24	0.49	17.2	0.015

Source: Refs. 69 and 80.

It appears that the attack on the cellulose is not directed at random to all carbons of the anhydroglucose units of the chains. It is directed mainly to the C_3 hydroxyl groups on random monomeric units in the noncrystalline parts of the chains. As many as 8-17 keto groups are formed per polymeric molecule, e.g., per scission. A larger percentage of the peroxide would seem to be consumed in the oxidation of the chain ends, producing a progressive "wet combustion" of the monomers along the chains and a solubilization of the small fragments produced.

5.4 Mechanism of Oxidation and Decomposition

The oxidation of a hydroxyl group in a cellulose chain can occur by an attack of a free radical:

$$\cdot OH + R-\overset{\overset{\textstyle H}{|}}{\underset{\underset{\textstyle R}{|}}{C}}-OH \rightarrow R-\overset{\overset{\textstyle \cdot}{}}{\underset{\underset{\textstyle R}{|}}{C}}-OH + H_2O \tag{76}$$

$$R-\overset{\overset{\textstyle \cdot}{}}{\underset{\underset{\textstyle R}{|}}{C}}-OH + HOOH \rightarrow R-\overset{}{\underset{\underset{\textstyle R}{|}}{C}}=O + H_2O \tag{77}$$

Carbonyl groups can be formed in this way on a number of monomeric units of the chain. The source of the $\cdot OH$ radicals can be at least twofold:

1. It may stem from a reaction between H_2O_2 and OOH^- [Eq. (67)]. This reaction takes place in alkaline H_2O_2 at temperatures of about 40°C. It was found that oxidation of alditols started under these conditions without an iron catalyst [229]. The oxidizing $\cdot OH$ and the reducing $\cdot O_2^-$ may interact as in Eq. (68) and produce O_2 or they may react with other substances of the environment such as polysaccharides. Since in the case of cellulose the oxidation is confined to the less ordered regions of the polymer and occurs on the surface, there might be an adequate local concentration of radicals for an oxidation to take place.

The $\cdot O_2^-$ radical is believed by Isbell to have a catalytic effect similar to the effect of ferrous ion in the Fenton reaction [213]. The $\cdot OOH$ radical is an acid, and under alkaline conditions it exists in the form of the $\cdot O_2^-$ which is not capable of abstracting a tightly bound hydrogen atom, but is particularly effective in cleaving hydroperoxides [Eq. (80)]. This reaction is similar to (69), which produces additional $\cdot OH$ radicals. Thus a supply of $\cdot OH$ radicals will be ensured even without iron ions.

2. The •OH radicals will be formed by iron catalysis. The presence of iron ions will have an effect similar to $\cdot O_2^-$, according to the mechanism of Haber and Weiss [211], [Eqs. (65),(66)]. The effect, however, is according to Isbell, limited "by the low solubility of ferric and ferrous hydroxides" [229].

Isbell postulated recently [224] that hydrogen peroxide in alkaline solutions adds on to aldehyde groups or potential aldehyde groups of carbohydrates, forming a hydroperoxide, which is similar to an aldehyde sulfite or hydrate:

Anomeric pyranoses
and furanoses

Anomeric pyranosyl and
furanosyl hydroperoxides + H_2O

$$CH_2OH \cdot \underset{OH}{\overset{H}{C}} - \underset{OH}{\overset{H}{C}} - \underset{H}{\overset{OH}{C}} - \underset{OH}{\overset{H}{C}} - \underset{O}{\overset{}{C}} - H + H_2O_2 \rightleftharpoons CH_2OH \cdot \underset{OH}{\overset{H}{C}} - \underset{OH}{\overset{H}{C}} - \underset{H}{\overset{OH}{C}} - \underset{OH}{\overset{H}{C}} - \underset{OH}{\overset{OOH}{C}} - H$$

(acyclic adduct)

This adduct is formed relatively rapidly. Its formulation is reversible, and it exists for some time in the solution until it decomposes. It can decompose by two routes depending on the conditions. At low temperatures, preferably in the presence of magnesium hydroxide, oxygen will be produced by Eqs. (79) and (80):

$$\underset{R'}{\overset{}{R}}C{=}O + H_2O_2 \rightarrow \underset{R'}{\overset{OH}{R}}C{-}OOH \tag{78a}$$

$$\underset{R'}{\overset{OH}{R}}COOH + OOH^- \rightarrow \underset{R'}{\overset{OH}{R}}CO\cdot + \cdot O_2^- + H_2O \tag{79}$$

$$\underset{R'}{\overset{OH}{R}}COOH + \cdot O_2^- \rightarrow \underset{R}{\overset{OH}{R}}CO\cdot + O_2 + OH^- \tag{80}$$

$$\underset{R'}{\overset{OH}{R}}CO\cdot + H_2O_2 \rightarrow \cdot OH + \underset{R'}{\overset{}{R}}COOH \tag{81}$$

In this way a high concentration of •OH radicals will be produced, and the sugar will catalyze the H_2O_2 decomposition with evolution of oxygen, or another oxidation reaction will occur, as discussed above [214,228]. If this mechanism operates in the reactions of bleaching and oxidation of cellulose, it might explain the catalytic effect of the cotton on the decomposition. It might also tentatively explain the sorption of H_2O_2 which was observed by several investigators [1,231]. This sorption was found to decrease with increase in pH, presumably by the more rapid decomposition of the hydroperoxide at higher alkalinity. Upon raising the temperature, for example, by steaming the peroxide-sorhed fabric, the fluidity and the whiteness increased.

According to Isbell another mechanism prevails at slightly elevated temperatures. The acyclic hydroperoxide formed in reaction (78) will decompose either by a free radical mechanism or by an ionic mechanism to yield formic acid and the next lower aldose:

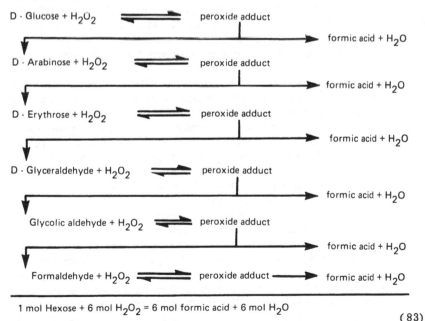

(a) Free-radical mechanism   (b) Ionic mechanism

(82)

The process is then repeated until complete degradation to formic acid is obtained:

D - Glucose + H_2O_2 ⇌ peroxide adduct
→ formic acid + H_2O

D - Arabinose + H_2O_2 ⇌ peroxide adduct
→ formic acid + H_2O

D - Erythrose + H_2O_2 ⇌ peroxide adduct
→ formic acid + H_2O

D - Glyceraldehyde + H_2O_2 ⇌ peroxide adduct
→ formic acid + H_2O

Glycolic aldehyde + H_2O_2 ⇌ peroxide adduct
→ formic acid + H_2O

Formaldehyde + H_2O_2 ⇌ peroxide adduct → formic acid + H_2O

1 mol Hexose + 6 mol H_2O_2 = 6 mol formic acid + 6 mol H_2O

(83)

While the most prevalent and rapid route for this process appears to proceed through the acyclic form, the reaction can also proceed, albeit slower, via the formation of the hydroperoxides of

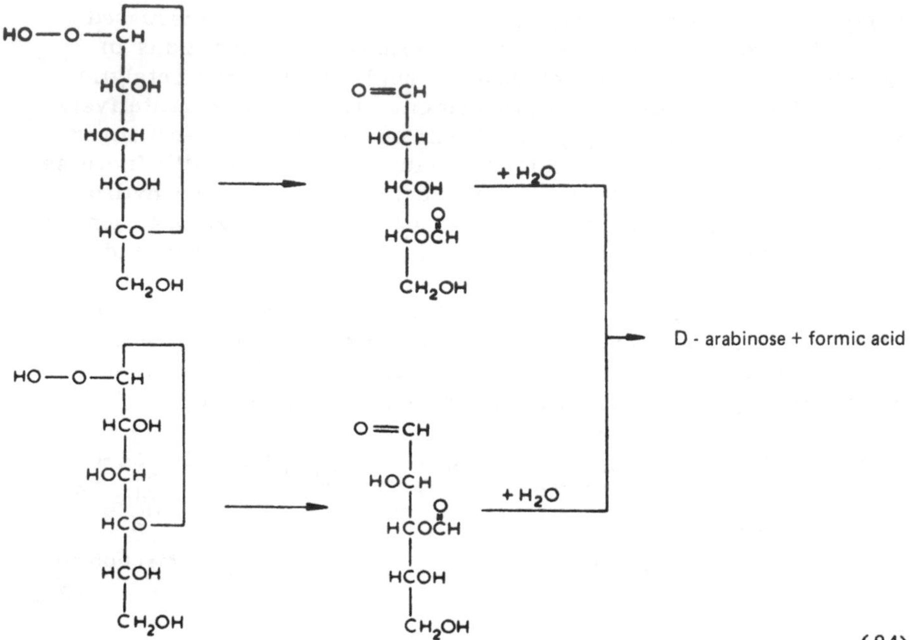

HO—O—CH
HCOH
HOCH
HCOH
HCO⎤
CH₂OH

→

O=CH
HOCH
HCOH
HCOCH (with =O)
CH₂OH

+ H₂O

HO—O—CH
HCOH
HOCH
HCO⎤
HCOH
CH₂OH

→

O=CH
HOCH
HCOCH (with =O)
HCOH
CH₂OH

+ H₂O

→ D - arabinose + formic acid

(84)

the cyclic pyranoses and furanoses. The corresponding formyl esters will be formed and hydrolyzed under the alkaline conditions to
the same products obtained by the acyclic form [224]. The acyclic
reaction appears to be the dominating reaction since the hexoses
which have the lowest aldehydo-form content have the lowest reaction
rate.

The fact that the ferrous ion significantly enhances the oxidation of most of the hexoses [227] indicates that the free radical mechanism is operative, but the ionic mechanism is not excluded. The
same products were formed in the catalyzed and uncatalyzed reactions.
The reaction rate did not increase with pH in the range 12-14 thus
indicating that it is not lose catalyzed supporting further the free
radical mechanism.

Ketoses are oxidized with H_2O_2 in a similar way, but from each
mol of hexulose 4 mol of formic acid and 1 mol of glycolic or glyoxylic acid are formed. The mechanism consists of a nucleophilic addition of a hydroperoxide anion to the carbonyl form of the ketose
followed by oxidative cleavage of the hydroperoxide adduct to glycolic acid and the next lower aldose. This aldose is degraded entirely to formic acid. The cleavage takes place at the C_2-C_3 bond.

A similar reaction sequence occurs for disaccharides and polysaccharides [Eqs. (85a,b)]. The mechanism for the slower stage
(85b) involves a shift of Compound 35 to the peroxide oxygen atom,
giving formic ester 36. On hydrolysis, 36 gives formic acid and the
D-glucose hemiacetal of D-glyceraldehyde; by alkaline hydrolysis the
latter yields D-glucose and D-glyceraldehyde. These compounds by
further reactions are converted into formic acid [225].

$$
\begin{array}{c}
\text{HC=O} \\
| \\
\text{HCOH} \\
| \\
\text{HOCH} \\
| \\
\text{HCOR} \\
| \\
\text{HCOH} \\
| \\
\text{CH}_2\text{OH} \\
\underline{27}
\end{array}
\xrightarrow{\text{H}_2\text{O}_2}
\begin{array}{c}
\text{OH} \\
| \\
\text{HCOOH} \\
| \\
\text{HCOH} \\
| \\
\text{HOCH} \\
| \\
\text{HCOR} \\
| \\
\text{HCOH} \\
| \\
\text{CH}_2\text{OR} \\
\underline{32}
\end{array}
\longrightarrow
\begin{array}{c}
\text{HCO}_2\text{H} \\
+ \\
\text{HC=O} \\
| \\
\text{HOCH} \\
| \\
\text{HCOR} \\
| \\
\text{HCOH} \\
| \\
\text{CH}_2\text{OH}
\end{array}
\xrightarrow{\text{H}_2\text{O}_2}
\begin{array}{c}
\text{OH} \\
| \\
\text{HCOOH} \\
| \\
\text{HOCH} \\
| \\
\text{HCOR} \\
| \\
\text{HCOH} \\
| \\
\text{CH}_2\text{OH} \\
\underline{33}
\end{array}
\longrightarrow
\begin{array}{c}
\text{HCO}_2\text{H} \\
+ \\
\text{HC=O} \\
| \\
\text{HCOR} \\
| \\
\text{HCOH} \\
| \\
\text{CH}_2\text{OH}
\end{array}
\xrightarrow{\text{H}_2\text{O}_2}
\begin{array}{c}
\text{OH} \\
| \\
\text{HCOOH} \\
| \\
\text{HCOR} \\
| \\
\text{HCOH} \\
| \\
\text{CH}_2\text{OH} \\
\underline{34}
\end{array}
$$

(85a)

$$
\begin{array}{c}
\text{HC}\langle^{\text{OH}}_{\text{OOH}} \\
| \\
\text{HCOR} \\
| \\
\text{HCOH} \\
| \\
\text{CH}_2\text{OH} \\
\underline{34}
\end{array}
\longrightarrow
\begin{array}{c}
\text{HC}\langle^{\text{O}}_{\text{O}} \\
| \\
\text{HCOR} \\
| \\
\text{HCOH} \\
| \\
\text{CH}_2\text{OH} \\
\underline{35}
\end{array}
\xrightarrow{\text{H}_2\text{O}}
\begin{array}{c}
\text{HCO}_2\text{H} \\
+ \\
\text{HC=O} \\
| \\
\text{HCOH} \\
| \\
\text{CH}_2\text{OH} \\
+ \\
\text{HOCH}_2(\text{CHOH})_4\text{CHO}
\end{array}
\begin{array}{l}
\xrightarrow{\text{H}_2\text{O}_2} 3\,\text{HCO}_2\text{H} \\[6pt]
\xrightarrow{\text{H}_2\text{O}_2} 6\,\text{HCO}_2\text{H}
\end{array}
$$

(85b)

5.5 Influence of Metals on the Decomposition of H_2O_2

It is well known that traces of iron salt greatly accelerate the decomposition of hydroperoxides by forming hydroxyl and peroxy radicals [232,233]. The catalytic effects of Fe^{2+} and Fe^{3+} are found to be similar. This is understandable, since in the presence of an excess of H_2O_2, an equilibrium is established between them by Eq. (65) and (66).

The catalytic effect of the iron cations does not increase regularly with their concentration. It was reported, for example, that Fe^{3+} in concentrations of $10^{-5}-10^{-6}$ mol/liter acts in the pH range 10.5-11.4 as stabilizer, while it is a catalyst at a concentration of 10^{-4} mol/liter [207]. A difference in activity of the Fe^{3+} cations depending on the quality of the water was observed (Table 2.21). The catalytic effect in tap water was pronounced, while no such effect was observed in distilled water [205].

Copper ions appear to be more effective catalysts than iron and are active at lower concentrations (Table 2.22) [205]. A promotion effect (synergism) is observed in catalytic activity when two ions are present in the solution, for example, iron and copper [218]. A mixed catalyst composed of hydroxides of Cu^{2+}, Mg^{2+}, and Fe^{2+} catalyzes H_2O_2 decomposition 100 times faster than Fe^{2+} alone [218]. $Cr^{3+} + Cu^{2+}$ are much more active than either separately, in dilute H_2O_2 solutions. They do not augment each other in concentrated solutions. In general, the catalytic activity of the heavy metal ions

Table 2.21 Influence of Fe^{3+} Ions on the Decomposition of H_2O_2 at pH 9^a

	Decomposition constant K, sec^{-1}		
Fe ions added, mg/liter	0.0	1	10
Distilled water	5×10^{-5}	6.2×10^{-5}	5.5×10^{-5}
Tap water	4.4×10^{-6}	4.1×10^{-5}	5.6×10^{-4}

[a]Conditions: temperature, 90°C; initial concentration of H_2O_2, 3.5 g/liter.
Source: Ref. 205.

increases with increase in alkalinity. Nickel is an exception and so is cobalt ion to a certain extent [207].

Copper and iron contaminations occur frequently on fabrics before the bleaching stage. Copper traces come from brass rollers and steam pipes; iron occurs as rust or swarf from machine bearings and from water and steam used in protessing; as dust of worn-out loom parts during warp and filling stop motions. Of special importance are oil and grease applied to worn out bronze bearings, either in preparatory processes or in finishing, forming copper soaps. Cu(OH), Cu(OH)$_2$, and Cu are also found in lubricants. Rusty iron fillings were shown to be the most effective catalysts for H_2O_2 decomposition (Table 2.23).

Different anions influence the decomposition of H_2O_2 in different ways (Table 2.24). For example, ferric sulfate has a greater catalytic effect than the equivalent concentration of $FeCl_3$. It is seen in Table 2.24, that in the absence of catalysts, iodide, carbonate, sulfate, and nitrate affect the decomposition of H_2O_2 more strongly than the hydroxyl ions. On the other hand, stabilizing effects are evident for phosphate, pyrophosphate, citrate, tartrate,

Table 2.22 Influence of Copper Ions on the Decomposition of H_2O_2 Solutions in Pure Distilled Water^a

Cu^{2+}, mg/liter	1.7	1.1	0.78	0.56	0.44	0.22	0.11
$K, sec^{-1} \times 10^3$	1200	810	300	161	85	23	8.3

[a]H_2O_2, 10 ml/l; NaOH, 10 g/l.
Source: Ref. 205.

Table 2.23 The Catalytic Effect of Metals on the Stability of the Peroxide Solution[a]

Types of metals[b]	Residual concentration (vol.)	Decomposition H_2O_2 (%)
Blank	1.58	21
Aluminium foil	1.56	22
Stainless steel turnings	1.48	26
Lead, pure foil	1.30	35
Nickel foil	1.23	38.5
Iron filings	1.15	42.5
Brass sheet	1.11	44
Monel metal	1.07	46.5
Sheet lead composition	1.06	47
Tin, granular	1.02	49
Copper foil	0.92	54.5
Rusty iron filings	0.09	95.5

[a]Two vol. H_2O_2 with 10 g/liter sodium silicate for 2 hr at 80°C.
[b]The metal having the least effect is aluminium showing why peroxide is transported in vessels of this metal.
Source: Ref. 234.

and borate. Iodide and nitrate probably attack H_2O_2. However, it is assumed that most of the anions interact with the cations to increase or decrease the activity of the latter. Silicate, citrate, and tartrate ions form complexes with heavy metal cations and maintain these ions in solution. Cyanide is inert in the presence of sodium cations but forms stable complexes with heavy metal cations and inhibits the H_2O_2 decomposition.

5.6 Catalytic Degradation of Cotton in the Presence of Metals

The catalytic effect of metal ions is not limited to the decomposition of H_2O_2. Simultaneously the fibers are attacked, and the DP and the tensile strength decrease considerably. For example, when a fabric is contaminated with grease which contains copper soap, its

Table 2.24 Stability of Alkaline Peroxide in the Presence of
Various Anions

Anion added as sodium salt 0.01 M	H_2O_2 Remaining after 1 hr at 50°C	Final pH
Hydroxyl	90	10.87
Acetate	87	10.86
Bicarbonate	77	10.50
Tetraborate	89	9.37
Bromide	87	10.90
Perborate	95	10.90
Pyrophosphate	100	10.83
Dihydrogen phosphate	97	8.15
Iodide	5	11.85
Chloride	92	10.60
Citrate	97	10.80
Nitrate	90	10.82
Tartrate	96	10.80
Sulfate	79	10.92
Carbonate	59	11.21
Cyanide	88	10.80
Silicate	91	10.75

[a]0.15% H_2O_2; deionized H_2O × 0.01 N C.P. caustic soda.
Source: Ref. 207.

tensile strength decreases with the increase in the concentration of
the copper in the grease (Fig. 2.52) [235]. The damage is inhibited
after removing the metal by pretreating the fabric with acid.
    The catalytic degradation of the fabrics by metallic contamina-
tion depends on the bleaching system. For continuous bleaching at
low liquor ratios, the catalytic effect will be relatively smaller if the
contaminant comes from water or steam, since in this case a smaller
amount can be carried over as compared to the higher liquor ratios
applied in the winch, in the package machine, and in the kier. If,
however, the catalytic impurity is already present in the fibers even

in trace amounts, the catalytic effect in the low liquor ratio processes will be higher due to the presence of the much higher concentrations of H_2O_2 in close contact with the fiber [236].

Localized concentrations of the catalyst accelerate the oxidation of the fabric to such an extent that "pinholes" or "razor cuts" are being formed in the fabric. The pinholes can appear as a result of local dissolution of the cellulose, due to low DP values corresponding to fluidities above 38 rhes. The "razor cuts" may be formed already at fluidities between 18 and 38 rhes [336]. It was recently found that the pinhole formation is caused by specific concentrations of metal salts. Systematic trials were undertaken in which drops (3-mm diameter) of copper and iron sulfate solutions of various concentrations were deposited on cotton fabric and dried. The fabric was subsequently bleached with alkaline H_2O_2 by a pad-steam procedure. Pinholes were found in the localized spots which contained 6 μg copper sulfate. They were not formed at the spots containing 0.6 and 60 μg [237] of the salt.

5.7 Bleaching of Cotton with H_2O_2

Mechanism of H_2O_2 Bleaching

The color in organic substances and also in unbleached fabrics is due to mobile electrons, usually in systems of conjugated double bonds, and decolorization requires the demobilization of such electrons

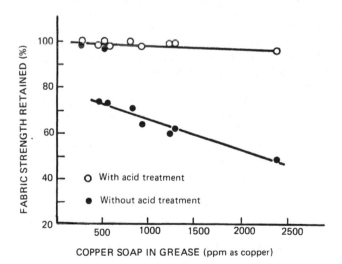

Figure 2.52 The effect of dilute sulfuric acid applied prior to bleaching with H_2O_2 on the breaking strength of fabric contaminated with grease which contains copper soap. (From Ref..235.)

[48]. The chemical structure of the faint, creamy-colored matter in the unbleached cotton is not exactly known, although some workers believed it to be associated with proteins containing aromatic groups in very small amounts [238-240]. It is probable that the structure of the chromophore involved consists of conjugated double bonds which can be attacked by free radicals. According to Cates [209, 210], the free radicals will be added to a double bond in the same manner as in free-radical addition polymerization of unsaturated monomers. For example a quinoid structure,

$$R\cdot + \quad \text{[structure]} \quad \longrightarrow \quad R\text{[structure]}\cdot \qquad (86)$$

when attacked by a free radical, will be converted to a benzenoid structure. The colored chromophores may compete for the $\cdot OH$ radicals formed in the reactions of Eqs. (65), (67), and (69) and thus inhibit the decomposition of H_2O_2. They will perform the bleaching operation according to Eqs. (87) and (88):

$$\cdot OH + \overline{\overline{S}} \rightarrow HO\overline{S}\cdot \qquad (87)$$

$$HO\overline{S}\cdot + H_2O_2 \rightarrow HO\overline{S}OH + \cdot OH \qquad (88)$$

where $\overline{\overline{S}}$ is the colored chromophore and $\overline{S}$ the oxidized chromophore [209].

This will break the chain of Eqs. (69) and (70). It is accordingly seen in Fig. 2.47 tha the decomposition of H_2O_2 in the pH range 7-11 is much smaller in the presence of an unbleached fabric. Upon increasing the concentration of the fabric in the bath (Fig. 2.53), the rate of decomposition decreases and at the same time the whiteness increases and the DP decreases, showing that the peroxide was consumed in bleaching rather by evolution of oxygen.

It is of interest to note that the maximum in H_2O_2 decomposition in the presence of fabric at pH 10.3 (Fig. 2.47) coincides with the maximum in percent reflectance (Fig. 2.54). A similar maximum, obtained by Nicoll and Smith [207] for the decomposition of H_2O_2 in distilled water, was explained by them by assuming the formation of colloidal hydroxides from the trace impurities with increase of alkalinity. These colloidal hydroxides are believed to be even more active than the metal perions which also exist in the system. With a large excess of alkali the heavy metal hydroxides dissolve and, as a result, the stability of H_2O_2 increases at high alkali concentrations [207].

It appears, therefore, that during bleaching the H_2O_2 is consumed (a) for whitening the fiber, (b) for chemically damaging the fiber, and (c) by evolving oxygen. "The proportions of H_2O_2 used for each of these routes depend on the nature, concentration, and location (i.e., on the fabric or in solution) of the impurities capable

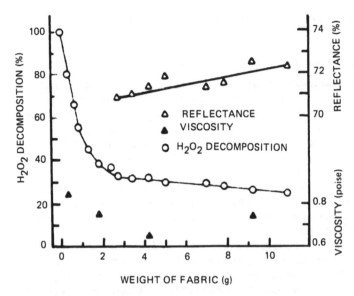

Figure 2.53 H_2O_2 decomposition, reflectance, and viscosity as a function of the weight of the fabric; pH 10.5; initial concentration of H_2O_2, 0.15 mol/l; temperature 87.7°C; time of bleaching, 10 min. (From Ref. 209.)

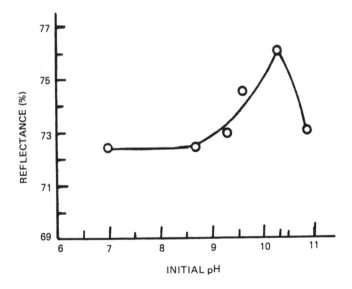

Figure 2.54 Reflectance as function of pH of the H_2O_2 bleaching solution. Initial H_2O_2 concentration, 0.15 mol/l; temperature and time as in Fig. 2.53. (From Ref. 209.)

of either possitive or negative catalysis; and the pH of the solution" [209].

Stabilization of Hydrogen Peroxide

Hydrogen peroxide solutions attain their maximum bleaching action at pH 11.0-11.5 and at temperatures up to 130°C. Both high pH and high temperature lead to decomposition of H_2O_2 and to degradation of the fibers, which, as described above, are strongly catalyzed by heavy metal ions. Considerable efforts have been made by many investigators to counteract these effects by developing stabilizers for peroxide baths.

Stabilizers act partly by providing a buffering action, which ensures that the pH is not too high. Their action is also based on their ability to complex traces of heavy metals, particularly Fe, Cu, Cr, and Mn and their salts.

The most commonly used stabilizers are sodium silicates, used in the colloidal polymerized form (waterglass), or as orthosilicate and metasilicate. Their stabilizing activity is enhanced by the presence of water hardness (Ca and Mg ions) [241]. When soft water is used, it is advantageous to add magnesium sulfate (0.6 g/liter) to the bath [234]. Colloidal magnesium silicate together with a dispersing agent is a most effective stabilizer [234,242]. There are several advantages to the use of silicates. They are cheap and are more effective than any other stabilizer. They also have a detergent action and inhibit corrosion of metal materials [216]. In addition they provide a pH buffering action, which suppresses the concentration of free sodium hydroxide [243], and this is probably a factor in reducing fiber damage. It has been suggested that colloidal silicates coat surfaces, including sharp-edged solids, which are catalytically active in decomposing peroxide [1]. However, it has been shown that nonpolymerized silicates are more effective than waterglass in inhibiting decomposition [243].

The stabilization of H_2O_2 by sodium silicate was explained by the formation of a complex [222] according to the equation:

$$H_2O_2 + X \leftrightarrow H_2O_2 \cdot X \tag{89}$$

where X is silicic acid.
The equilibrium constant of Eq. (89) is:

$$K_c = \frac{[H_2O_2 \cdot X]}{[H_2O_2] [X]} \tag{89a}$$

From Eq. (62) the dissociation constant of H_2O_2 is:

$$K = \frac{[H^+] [OOH^-]}{[H_2O_2]} \tag{62a}$$

hence,

$$[OOH^-] = \frac{[H_2O_2 \cdot X] \, K}{[H^+] \, [X] \, K_c} \qquad (90)$$

It follows from Eq. (90) that for a constant concentration of silicate the concentration of OOH^- increases with the increase in pH, which will express itself in decreased stability of H_2O_2. On the other hand the concentration of OOH^- decreases with the increase in silicic acid concentration [222].

The existence of this complex stabilizes the H_2O_2 by preventing its interaction with the metal ion catalysts [207].

According to another explanation the activity of silicate is due to its colloidal nature [245], especially in the presence of calcium and magnesium ions which enhance the stabilization effect of the silicate. The colloidal particles formed absorb the metal ions. Sodium silicate acts similarly at elevated temperatures [245].

The major disadvantage of silicate stabilizers is that their use tends to cause the formation of hard scales on processing equipment [4,244,246-249]. These interfere with the free running of the cloth through the equipment, abrade the cloth, and reduce the efficiency of heat exchangers. The problem can be especially acute in a continuous, one-state bleaching process [244], particularly during rinsing, due to a rapid fall in pH [246]. As the silicate stabilizers become dehydrated, they polymerize, losing water, and their molecular weight increases until finally a barely soluble residue of $(SiO_2)_n$ is precipitated. This can also happen when hot-impregnated fabrics are stowed for a long time. Even the time of heating with steam to the desired temperature may suffice for the precipitation [292]. In many instances the precipitates contain, in addition to silicic acid, calcium and magnesium silicates which are also difficultly soluble especially in very hard waters. There is also a tendency for deposition of silicate onto the fiber [185,242,246,248]. If this is excessive, the cloth will have a high ash content which causes poor handle. It will also cause a decrease in sewability, as evidenced by the increase in the force needed to puncture the fabric by the needles during the sewing operation [244].

The precipitates do not present problems to textiles with smooth surfaces, since subsequent treatments may remove most of them. It is different for fabrics having a structured surface such as knits, velvets, etc., which suffer a pronounced loss of hand. The precipitation usually does not occur in the cold peroxide bleaching system which can be applied for the sensitive fabrics. Problems with dyeing and printing have also been attributed to silicate deposition. This latter problem arises particularly under conditions where washing-off is difficult, for example, in package machines. In such instances additional alkali (sodium carbonate) should be added [185], or half the sodium silicate should be replaced by trisodium phosphate

[248]. The latter has little stabilizing effect when used alone. The use of triethanolamine together with silicate has been recommended [249].

Considerable attention has been given to the stabilization of hydrogen peroxide bleaching liquors with inorganic and organic sequestering agents, both in the presence and in the absence of sodium silicate. The sequestering agents are compared by their chelation value, that is, the weight in milligrams of Ca as $CaCO_3$ sequestered by 1 g of sequestering agent. A more accurate definition is based on the ability of a compound to keep in solution a metal ion complex in the presence of a precipitating agent, as shown in Eq. (91):

$$M^{n+} + L^{m-} \leftrightarrow ML^{m-n} \tag{91}$$

where M^{n+} and L^{m-} are the activities of the metal ions and of the sequestering agent, respectively. The equilibrium constant of the formation of the complex is:

$$K_e = \frac{[ML^{m-n}]}{[M^{n+}] [L^{m-}]} \tag{91a}$$

Since the activities of the ions involved are in many cases unknown, the measurements are made with solutions maintained at a constant ionic strength by addition of simple electrolytes. The activity coefficients probably remain constant in this case and may be determined using concentrations for the calculation. In order to determine the true thermodynamic equilibrium constant, K_e values at different ionic strength are measured and extrapolated to zero ionic strength. The sequestration efficiency depends to a large extent on the pH and on the concentration of the precipitating agent present in the system. The efficiency of EDTA for complexing of calcium and magnesium falls off at a pH of about 7, while copper is fully complexed at pH values above 3.

At high pH values hydroxyl ions compete with the sequestrant for the metal ions, bringing about the decomposition of the complex and the precipitation of the metal hydroxide as follows:

$$ML^{m-n} + nOH^- \leftrightarrow M(OH)n + L^{m-} \tag{92}$$

with

$$K_h = \frac{[M(OH)n] \times [L^{m-}]}{[ML^{m-n}][OH^-]^n} = \frac{[M^{n+}][L^{m-}]}{[ML^{m-n}]} \times \frac{[M(OH)n]}{[M^{n+}][OH^-]^n} \tag{93}$$

$$= \frac{1}{K_e K_s}$$

where K_s is the solubility product of the metal hydroxide.

The stability constant K_e and the solubility product K_s determine the effectiveness of the sequestering agent in alkaline solutions. For high values of K_h the equilibrium of Eq. (92) is shifted to the right and the amount of the metal hydroxide precipitated is increased [1]. The higher the stability constant the more effective the complexation of the metal.

In addition to stability the stabilizer or sequestrant "has to be compatible with the process, and has to be stable to oxidizing and reducing agents (if need be) at temperature ranges employed in the process. Also the sequestrant should possibly contribute to process efficiency and goods quality by helping in the removal of soil, its dispersion, and suspension in solution, in order to prevent soil redeposition on textile goods" [246]. In addition, it is also required that the sequestrant should minimize or prevent the formation of the insoluble precipitates. The most commonly used inorganic stabilizers are the polyphosphates [242,247]. Tetrasodium pyrophosphate and sodium hexametaphosphate are used only in soft water, since Ca and Mg ions precipitate them from solution. Moreover, they lose their effectiveness at temperatures above about 70°C and at a pH above 10.3. This is probably due to hydrolysis of trisodium phosphate, accompanied by decomposition of metal-polyphosphate complexes [191, 246,251]. Their application is thus limited to peroxide bleaching under mild conditions, as used for protein, polyamide, and cellulose acetate fibers.

Many commercial organic, nonsilicate stabilizers are available at present. Sequestering agents are active in this respect [1,5,242, 246,247,252,253,259]. Protein and protein degradation products have a stabilizing effect on peroxide bleaching baths: soyabean flour [248], gelatin, egg albumin, and degraded wool protein [1] are known to be stabilizers. Protein fatty acid condensation products are also effective. In addition gluconic and glucoheptonic acids have a sequestering effect [246], and polycarbonic acid derivatives are also used. Though soybean flour reduces the rate of peroxide decomposition, it does not prevent accelerated damage to cotton cloth in the presence of copper [1]. Commercial preparations may include more than one sequestrant, a detergent, and possibly a softening agent.

These stabilizers may be used alone [1,244,247,254,259] or in the presence of tartaric or citric acid or of a phosphate buffer such as sodium tripolyphosphate [246]. More usually they are used together with a reduced amount of sodium silicate [4,5,244,246,254]. For instance, a concentration of 10-20 g/liter sodium silicate may be replaced by 5 g/liter silicate and 5-7 g/liter of an organic stabilizer [5]. In this way problems of scaling are greatly reduced and the cloth has improved handle [244], while the bleaching effect is better than that obtained in the complete absence of silicate [5].

Several of the most important sequestering agents are listed in Fig. 2.55, and their stability constants are tabulated (Table 2.25)

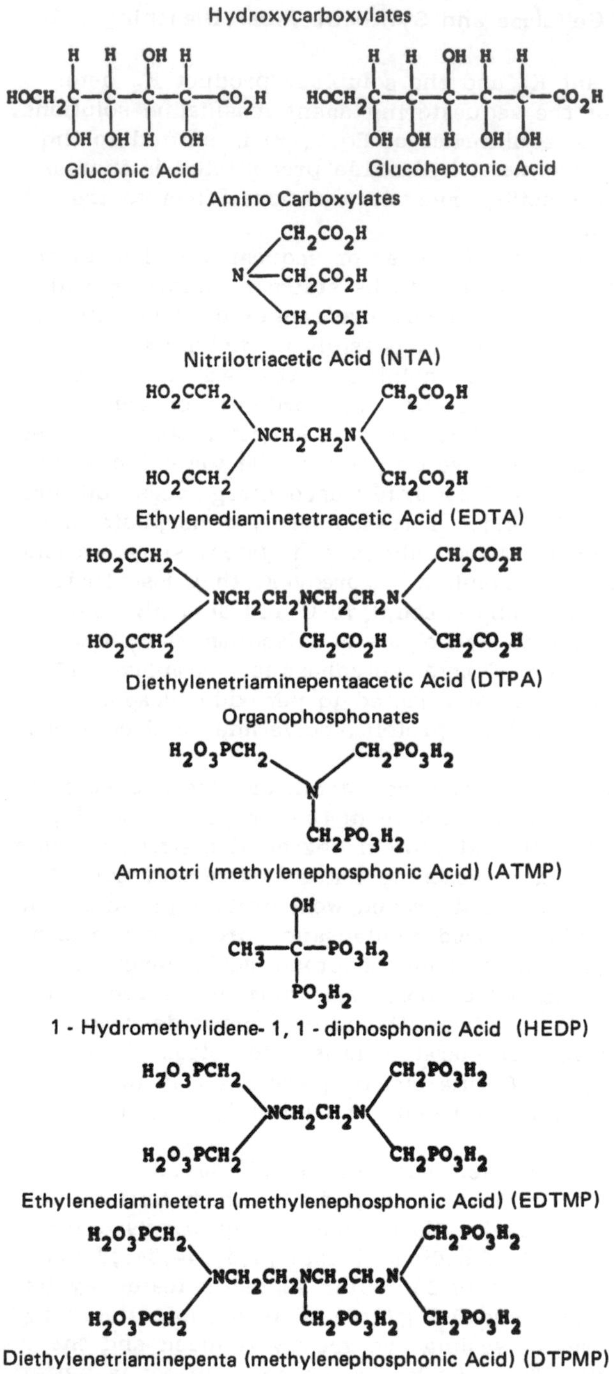

Hydroxycarboxylates

Gluconic Acid

Glucoheptonic Acid

Amino Carboxylates

Nitrilotriacetic Acid (NTA)

Ethylenediaminetetraacetic Acid (EDTA)

Diethylenetriaminepentaacetic Acid (DTPA)

Organophosphonates

Aminotri (methylenephosphonic Acid) (ATMP)

1 - Hydromethylidene- 1, 1 - diphosphonic Acid (HEDP)

Ethylenediaminetetra (methylenephosphonic Acid) (EDTMP)

Diethylenetriaminepenta (methylenephosphonic Acid) (DTPMP)

Figure 2.55 Several sequestrants used in H_2O_2 bleaching. (From Ref. 246.)

Table 2.25 Stability Constants, Log K_e, of Chelants for Ca^{2+}, Mg^{2+}, Fe^{3+}, and Cu^{2+}

Chelant	Ca^{2+}	Mg^{2+}	Fe^{3+}	Cu^{2+}
ATMP	6.9	6.5	14.6	17.4
HEDP	7.1	6.4	21.6	19.0
EDTMP	9.3	8.6	19.6	23.2
DTPMP	7.1	6.4	—	19.5
EDTA	10.7	8.7	25.1	18.7
HEDTA	8.4	5.8	19.8	17.5
DTPA	10.9	9.3	27.5	21.2
Gluconic Acid	1.2	0.7	37.2	—
NTA	6.4	5.4	15.9	12.9

Source: Ref. 246.

for several metal ions. The aminocarboxylates (EDTA, NTA, DTPA) do not act as detergents by soil dispersion or suspension. They do not prevent scale formation when applied in concentrations less than stoichiometric to the scale-producing salts. They are, however, efficient metal sequestrants. The organophosphonates (ATMP, HEDP, EDTMP, DTPMP) act similarly to the polyphosphates in reducing scale-forming precipitates as well as by sequestering and by dispersing and suspending soil. They are, however, much more stable than the polyphosphates and remain effective at elevated temperatures.

Bleaching solutions containing polyphosphates decompose much more rapidly than the solutions containing phosphonates. This opens the possibility of reducing the amount of H_2O_2 and sodium silicate in the bleaching liquor and of obtaining cotton fabrics of lower fluidity and better hand [246]. The phosphonates were also shown to be much more effective than the inorganic polyphosphates in maintaining the silicates in solution, which permits reduction of the ash content in the bleached fiber.

A comparison of the stabilizing effect of peroxide bleaching baths of a phosphonate-diethylene-triaminepentamethylene phosphonate pentasodium salt (DTPMP·5Na) and its aminocarboxylic homologue, diethylenetriaminepentasodium acetate (DTPA·5Na), is presented in Fig. 2.56. In the DTPA + $CaCl_2$ baths the decomposition of the peroxide in the absence of silicate is rapid, indicating a far-reaching oxidation of the DTPA. In the presence of silicate the effect is probably due to the calcium silicate suspension, which is capable of

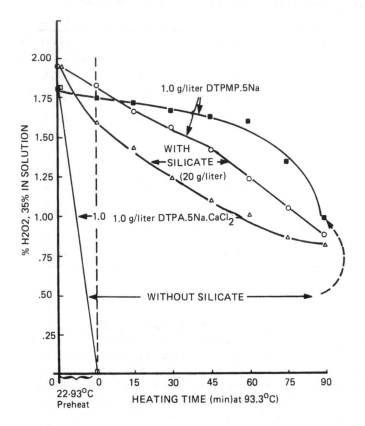

Figure 2.56 Stabilization of H_2O_2 bleaching baths. 20 g/l H_2O_2, 35%; 2.5 g/l NaOH, 50%; 1.0 ppm Fe^{2+}; 0.25 ppm Cu^{2+}. (From Ref. 246.)

tying up the heavy metals [246]. The DTPMP·5Na is seen to be highly effective even without $CaCl_2$, both in the presence and absence of silicate. The amounts of DTPMP·5Na applied depend on the concentrations of the calcium and magnesium salts present in the bleaching bath. They are used up by the phosphonate, and its amount may not be sufficient for the sequestration of the copper and iron ions. At a level of 30 ppm Ca and 10 ppm Mg, 3.56 g/liter of the phosphonate is needed for a 1:1 molar ratio usually required for efficient sequestration. However, lower molar ratios were found to yield good peroxide stabilization in the absence of silicate. Blends of the phosphonate with polyphosphates and DTPA can also be used advantageously.

Hydrogen peroxide is also used for desizing [4,257-259]. Though magnesium sulfate is added as a stabilizer, there is a greater degree of free radical peroxide decomposition here than in a stabilized bleachbath. Thus, in order to avoid excess fiber damage the concentration of hydrogen peroxide should not be higher than that required for desizing even though this might improve the final whiteness of the cloth [257].

Although magnesium salts have been introduced in special commercial stabilizer formulations used for bleaching [273] and desizing [258], and although they have been known since 1964 to inhibit the degradation of cellulose during bleaching of wood pulp with oxygen and applied for this purpose [267-272], the mechanism of the effect is not yet fully understood. Rowe [258] believes that "the stabilizing effect obtained in the presence of sodium hydroxide is due to the formation of an active magnesium compound from the magnesium salt contained in the hydrogen peroxide. For that reason additives that complex magnesium should not be included in the desizing solution." A similar view was expressed by Samuelson [54]. Robert [272] has suggested that the degradation catalysts are absorbed on or coprecipitated with a magnesium hydroxide substrate. Magnesium hydroxide is also believed to form a complex with the primary oxidized derivative of cellulose and thus to inhibit further degradation [270]. Isbell [214] suggested recently that magnesium forms a complex with the superoxide radical formed in Reactions (66a), (67), and (70) as follows:

$$Mg(OH)_2 + \cdot O_2^- \rightarrow \cdot OOMgOH + OM^- \tag{94}$$

$$\cdot OOMg(OH) + \cdot OH \rightarrow MgO + O_2 + H_2O \tag{95}$$

This complex would immobilize and even precipitate the superoxide radical, thereby breaking the chain of the production of the $\cdot OH$ radical in Eqs. (69) and (70) and effectively decreasing its concentration. It may also act as a scavenger for the $\cdot OH$ radicals (Fig. 2.57).

Other workers suggested that Mg deactivates the transition metal compounds via the formation of coordination compounds. Evidence for the formation of such a compound with 3:1 or 6:1 $Mg^{2+}:Fe^{3+}$ ratios was also brought forward for the oxidation of cotton linters with oxygen [271] and for the oxidation of glucose with hydrogen peroxide in the presence of iron salts [227]. In this case a Mg:Fe complex of 6:1, with a structure similar to ferrocyanide or ferricyanide was suggested.

It appears, therefore, that magnesium hydroxide is an active stabilizer both in the presence and in the absence of iron. The retardation of the catalytic effect of cobalt on bleaching and H_2O_2 decomposition by magnesium ions is not nearly as effective as that

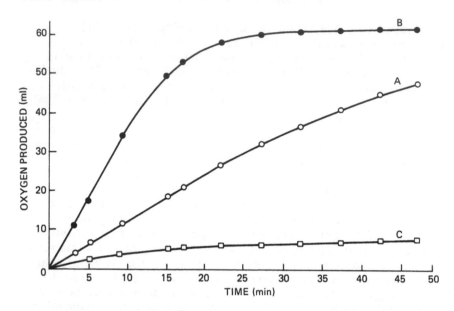

Figure 2.57 Effect of a carbonyl compound and of magnesium hydroxide on the decomposition of H_2O_2 at 40°C. Curve I: KOH 1 ml; 3 M), H_2O (1 ml), and H_2O_2 (1 ml, 30%). Curve II: same as Curve I, plus 2-deoxy-D-arabino-hexose (164 mg). Curve III: same as Curve I plus $MgSO_4$ (25 mg). (From Ref. 214.)

of iron although the degree of stabilization was found to be linear with the amount of magnesium present [271]. Additional reagents (i.e., titanium dioxide, sodium tetraborate, and silver hydroxide [271]) were reported to act as inhibitors although with lower effectiveness. Borax was also found [148] to inhibit the rate of the oxidation of glucose and cellulose by hypochlorite and hyprobromite, and it was suggested it forms complexes with the C_1 and C_2 hydroxyls of the carbohydrate, similar to the complex of Mg-gluconic acid suggested by Rapson [271].

It is of interest to note that the same relationship between viscosity and percent brightness was found when bleaching scoured as well as purified fabrics at several temperatures and in the presence and absence of a stabilizer (Fig. 2.58). Freytag [273-275] developed a quality factor, Q, which is composed of the normalized values of the degree of whiteness and of the change in the DP of the sample. The relative statistical weight of each of these parameters can be changed according to the use of the fabric. For high-white fabrics, where the DP is less important, the statistical weight of the DP change is decreased by multiplying it with a suitable factor. A different factor is applied if fabrics are intended for dyeing.

$$Q = \frac{1}{L} \tag{96}$$

and

$$L = \sqrt{(100 - w)^2 + (20S)^2} \tag{97}$$

The factor 20 can be changed in order to change the relative weight of the DP in the quality factor.

The degradation parameter is:

$$S = \frac{1}{\lg 2} \cdot \log \left(\left[\frac{2000}{DP_t} - \frac{2000}{DP_i} \right] + 1 \right) \tag{98}$$

The whiteness parameter is:

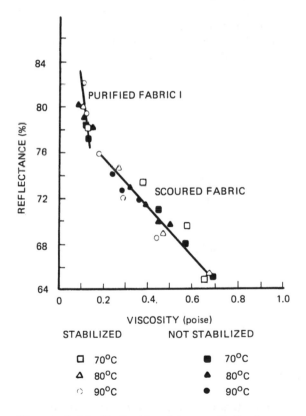

Figure 2.58 H_2O_2 bleaching: Reflectance vs. viscosity of bleached fabrics. (From Ref. 208).

$$W = 100 - \sqrt{(100 - TM)^2 + (FC)^2} \qquad (99)$$

where TM is the brightness factor,

$$TM = \frac{A + B + G}{3} \cong \frac{X + Y + Z}{3} \qquad (100)$$

and A, B, G represent reflection with the tristimulus filters measured with a photometer, and FC is the chromaticity:

$$FC = [TM - A] + [TM - B] + [TM - G] \qquad (101)$$

The application of this quality factor makes it possible to draw some interesting conclusions concerning the effects of several bleaching parameters on the quality of the bleached fabrics [273-275]. Increasing the temperature of steaming from 100°C to 134°C at pH 11, when using for the impregnation 1% H_2O_2; 4% Na_2SiO_3 36°Be, owf; and 80 mg/liter $MgCl_2$, decreases the quality factor of the bleached fabric from 68 to 50. Magnesium silicate increases the quality factor at 100°C from 62 to 68. The optimum time of the steaming was 25 min at 100°C, while at the higher temperatures 0.5-1 min were the optimal times. At 134°C magnesium silicate was not effective, and this decreased the quality factor (Fig. 2.59). The activity of the

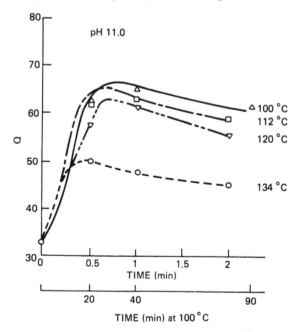

Figure 2.59 Influence of time and temperature of steaming on the quality factor, Q. Bleaching with 1% H_2O_2, 4% Na_2SiO_3 (36°Be), owf, and 80 mg/l $MgCl_2 \cdot 6H_2O$. (From Ref. 275.)

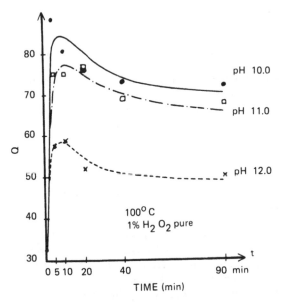

Figure 2.60 Influence of the pH on the quality factor Q at 100°C and 1% owf H_2O_2. (From Ref. 274.)

magnesium at 134°C was, however, restored when it was added in the form of a salt of EDTA. The quality factor increased with decrease in pH from 12 to 10. At pH 10 and 100°C, with 0.5% H_2O_2, a factor of 84 was reached (see Fig. 2.60).

Activation of H_2O_2

The best known activator of H_2O_2 is alkali. In pure water and in acidic solution no oxidizing activity is obtained. In order to be able to bleach under weakly alkaline, neutral, or acidic conditions (e.g., for bleaching of rayon, acetate, silk, and wool) and also for cold alkaline H_2O_2 bleaches, other activators are needed. Activation by metal ions brings about degradation of the fiber and can therefore not be applied, except for bleaching of dark-colored animal and human hair and brushes.

The prevalent activators are produced by conversion of H_2O_2 into peracids, such as peracetic acid. The latter can be obtained directly according to Eqs. (72) and (74) as an equilibrium mixture or by acetylation of H_2O_2 with acetic anhydride in the presence of a catalyst such as sodium hydroxide or EDTA at room temperature:

$$(CH_3CO)_2O + H_2O_2 \leftrightarrow CH_3COOOH + CH_3COOH \tag{102}$$

The by-product diacetylperoxide may also be formed under unsuitable conditions. This compound spontaneously decomposes:

$$(CH_3CO)_2O + H_2O_2 \leftrightarrow (CH_3CO)OO(OCCH_3) + H_2O \qquad (103)$$

Bleaching with peracetic acid is carried out under mildly acidic conditions between pH 6 and 7.5. The commercial product contains 37% paracetic acid and is fairly stable at room temperature in dilute solution. Only about 2% decomposed under these conditions [1]. Peracetic acid splits into radicals:

$$CH_3COOOH \rightarrow CH_3COO\cdot + \cdot OH \qquad (104)$$

and its decomposition is also catalyzed by transition metal ions so that the addition of stabilizers is needed.

Peracetic acid is applied to synthetic fibers, rayon, acetate, and in certain cases to cotton. It has several advantages over H_2O_2 and chlorite: it causes less swelling of the cellulose, it is less corrosive than chlorite, and it exerts a smaller effect on dyestuffs than chlorite and alkaline H_2O_2. The peracetic acid does not bleach the cottonseed husks. Impregnating scoured cotton poplin with a solution containing 2.37% peracetic acid, 0.5% sodium lauryl sulfonate wetting agent, 0.2% sodium hexametaphosphate stabilizer, and 1.25% NaOH, having a pH of 5.5, at a pickup of 100%, gave a high degree of whiteness upon 5 min of steaming at 100°C. Increasing the steaming time brought about a pronounced degradation of the fabric [283].

Another activating agent is performic acid, which is obtained in very concentrated solutions in equilibrium with H_2O_2 [277]:

$$HCOOH + H_2O_2 \rightleftarrows HCOOOH + H_2O \qquad (105)$$

It decomposes by hydrolysis with increase in pH or by evolution of CO_2:

$$HCOOOH \rightarrow CO_2 + H_2O \qquad (106)$$

Formaldehyde with H_2O_2 yields monomethylol and dimethylol peroxides:

$$H_2O_2 + HCHO \rightarrow HOCH_2OOH$$

$$HOCH_2OOH + HCHO \rightarrow HOCH_2OOCH_2OH \qquad (107)$$

The dimethylol peroxide decomposes yielding formic acid and formaldehyde:

$$HOCH_2OOCH_2OH \rightarrow 2HCOOH + H_2 \qquad (108)$$

$$HOCH_2OOCH_2OH \rightarrow HCOOH + HCHO + H_2O \qquad (109)$$

Based on similar principles a large number of additional activators could be developed, and many such acylating compounds, that is, organic compounds which form percarbonic acids with H_2O_2, have been prepared [279]. These include: carbonic acid anhydride, chloride, and ester [278]; acylation products based on the group N$-$CO$-$R$-$ [279], such as acylamides (e.g., N,N,N',N'-tetraacetyle-thylenediamine and acylated hydroxylamine); acylated heterocyclic materials based on diketopiperazine, imidazolinon, thiazolinon, and 1,3,4,6-tetraacetylglycoluril (TAGU). Compounds with a smaller molecular weight such as the esters and acylation products of in-organic acids (e.g., H_2CO_3, H_3PO_3, H_3PO_4) are described in the patent literature as activators. Similarly the peracids and perdi-acids of sulfuric and phosphoric acids are described as activating agents [259]: H_2SO_5, $H_2S_2O_8$ [260,261], H_3PO_5, and H_4PO_8 [260, 261]. The perdisulfuric acid and its sodium salt bleach very slowly and are applied only in cold pad-batch processes.

Ammonium persulfate activates H_2O_2 in the cold bleaching of wool and silk [162,191]. Tetrapotassium perdiphosphate has a small but significant effect on whiteness and seed husk removal in a com-bined continuous desize-scour-bleach process involving steaming for 30 min.

The percompounds of the inorganic acids hydrolyze in dilute solutions and afford H_2O_2 and the acid so that their behavior is similar to that of the H_2O_2. Buffers are therefore added in order to stabilize the pH of the solution. The DP of the cellulose was found to decrease with the decrease in pH, upon bleaching with activators in the acidic pH range [242]. Bleaching at pH 7.9-7.3 gave a fabric with a brightness of 81% and a DP of 1855, while when the pH was in the range of 4.6-3.8, values of 64.3% and 505 were obtained, respectively. The largest application of the activators is in the laundering industry in which a bleaching effect at relatively low temperatures, (e.g., below 60°C) is required.

Several activators of undisclosed and proprietary compositions are being used by many bleaching plants. These activators are composed of a number of materials which fulfill several functions so that a composite effect is obtained. It is claimed that such acti-vators impart to fabrics a high hydrophilicity, a soft hand, and a decrease in hardness. They can be applied at the pH range 5-8 and can also be used for regenerated cellulose fibers, for bleaching of dyed fibers, and of cotton-polyamide blends without degradation of the polyamide [281]. Such activators appear to be free from the drawbacks of peracetic acid, such as the strong unpleasant odor, the danger of explosion due to the possible formation of diacetyl-peroxide, and the sensitivity of the bleach solutions to temperature and catalysts [281].

Peroxide Bleaching by Batch Processes

In contrast to hypochlorite, when bleaching with H_2O_2 prolonged desizing and alkaline boiling stages are not always needed. The alkali included in the bleaching formulations performs to a large extent the function of the removal of the impurities simultaneously with the bleaching action of the H_2O_2. The alkali used is usually sodium hydroxide, and the ratio of $Na_2O:SiO_2$ is of considerable importance in order to minimize the precipitation of the silica oxides. In long liquor bleaching, a part of the NaOH is replaced by Na_2CO_3. If alkaline boiling has been performed on the fabric before bleaching, alkali addition in the bleaching stage can be kept to a minimum compatible with the silicate added. The bleaching process applied will depend to a large extent on the state of the fabric and the pretreatment applied to it.

There may be interactions between residual enzymes which sometimes remain on the fabric after the desizing stage [191]. Several enzymes are known to decompose H_2O_2 catalytically. It is therefore necessary to effect a thorough rinse before the bleaching stage. Acid pretreatment before the alkaline scour has been demonstrated to have an important beneficial influence on the properties of the bleached fabrics. In continuous treatments for dwell times of up to 60 min, as much as 2% of HCl owf can be used, while for higher dwell times of about 12 hr, 0.2–0.3% are used [276]. The acid treatment decreases the mineral content by removing calcium and magnesium deposits embedded in the fibers. If not removed from the fabric, they are converted in the bleaching stage into magnesium and calcium silicates which are retained in the fabric and impart to it a hard hand [185]. Furthermore, some iron ions are removed, which decreases the rate of the decomposition of H_2O_2, increases the brightness, and decreases the fluidity (Table 2.26), [276].

Kier Bleaching with H_2O_2: The kier process is applied to flock, yarn hanks, and to a smaller extent to fabrics in batches between 250 and 5000 kg. The description of its operation and the various kier designs are given in Chap. 3 of Vol. 1, Part A.

The kiers, their pumps, and heaters are usually constructed of cast iron and mild steel. In order to prevent corrosion of the kier it should be coated with a silicate cement lining (see Sec. 5.1). The pump, heaters, and external pipes can be passivated by circulating a boiling solution containing sodium silicate and magnesium sulfate. The recommended sequence of the kier bleaching process is: desizing, scouring, bleaching, soda ash scald, and washing. Desizing is carried out by enzymes or by oxidation with persulfate, hypochlorite, bromite, or hydrogen peroxide in the case of starch sizes. For soluble sizes such as polyvinyl alcohol (PVA) and carboxymethylcellulose (CMC), hot water plus detergent accompanied by oxidation or enzyme

Table 2.26 Effect of Acid Prebleaching Treatment on the Properties of the Bleached Cotton.

Sample no.	Percent reflectance		Fluidity (Rhes)		Ash (%)	
	Acid sour	No sour	Acid sour	No sour	Acid sour	No sour
1	92.5	89.9	1.5	6.0	0.06	0.14
2	93.1	91.2	2.6	5.5	0.10	0.15
3	92.2	90.8	1.5	6.7	0.03	0.15
4	90.6	89.5	1.1	6.7	0.89	0.15
5	91.2	90.9	2.0	8.5		
6	91.7	89.4	1.1	4.4		

[a]Six cotton samples of varying constructions from commercial bleaches in two plants. Half of each construction was bleached in one plant by a sequence of desize, sour, scour, and peroxide bleach. The other half was bleached in a second plant by an identical procedure, omitting the souring stage.
Source: Ref. 276.

treatment, if some starch is present, are used. (Desizing is discussed in Chap. 1 of this volume.)

The caustic scouring is needed only in exceptional cases. In most cases the scouring is done simultaneously with the peroxide bleaching with the addition of a suitable surfactant. The chemicals applied for the bleaching stage are as follows: 35% H_2O_2, 3-5%; NaOH, 0.6-1.4% or NaOH, 0.3-0.8% plus Na_2CO_3, 0.6-1%; sodium silicate, 79°Tw, 2-3%. When the cloth has been scoured in NaOH, silicate and Na_2CO_3 only are added. The loading and steam heating of the kier is done gradually over a period of 1 hr and the temperature is raised to 85°C and maintained for 2-3 hr of bleaching. The liquor ratio in the kier is 3-5:1 [191].

The changes in alkalinity and H_2O_2 concentration during the kier bleaching are shown in Fig. 2.61. It is seen that after 3 hr of bleaching about 0.3-0.6 g/liter H_2O_2 still remains in the bath. In order to utilize this and improve whiteness and absorbency a part of the liquor is replaced by a solution of 2% Na_2CO_3 and maintained for 0.5-1 hr at 80-85°C.

Kier bleaching is also applied for dyed goods. The conditions for this are milder: the maximum temperature of the kier liquor is 80°C, and the concentration of H_2O_2 is lower (up to 4%), and only Na_2CO_3 (1-1.5%) is added together with sodium silicate [191].

Bleaching in Package Machines: The liquor ratio used in the package and beam machines is 5-10:1; these machines are used mainly for treating yarns and loose cotton and sometimes also for fabrics processed on beam machines.

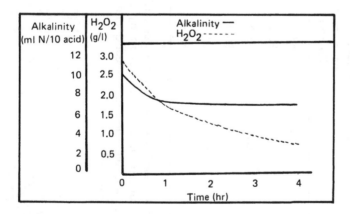

Figure 2.61 Change in concentration of alkali and H_2O_2 during a typical kier bleaching. Liquor ratio 4:1. Alkalinity determined on 10 ml of solution. (From Ref. 191.)

For cotton yarns a pretreatment is sometimes carried out, consisting of circulation through the package of a solution of sodium tripolyphosphate and some wetting agent at 95°C for 5-10 min. This treatment is believed to remove surface impurities and prevent deposition of Ca and Mg salts in the package.

A typical bleaching solution is composed of sodium silicate 79°Tw, 2-7 g/liter; NaOH, 0.5 g/liter; Na_2CO_3, 1.8 g/liter; H_2O_2, 4.5 g/liter; and wetting agent. An organic stabilizer, 0.5-2 g/liter can be applied in place of the sodium silicate in order to prevent filtration of insoluble precipitates on the surface of the yarns. The temperature is raised to 90°C in 20-30 min and maintained for 1-2 hr (Fig. 2.62) [191].

The package machines can also be used for high-temperature (HT) bleaching. In this case, 4 g/liter sodium silicate are taken with a higher concentration of NaOH, about 2-5 g/liter and Na_2CO_3, 1 g/liter. The temperature is first raised to 80°C in 20 min, the machine is closed, and the temperature is elevated to 120-130°C and maintained at this level for a further 20 min [191]. HT bleaching is usually applied for cotton-polyester (COT/PET) blends, and a temperature of 115-118°C is used.

Bleaching Cotton with H_2O_2 in the Winch Machine: Bleaching in the winch machine is done at a liquor ratio of 10-40:1; therefore little pretreatment if any is needed. The solution is composed of 1-2 g/liter of H_2O_2, 7 g/liter sodium silicate, 0.5 g/liter NaOH, and wetting agent. The sodium silicate can be replaced by 0.5-2 g/liter of an organic stabilizer. The temperature is 90-95°C and the treatment time is 1-2 hr. A similar process is applied for COT/PET.

Bleaching in the Jig Machine: Since the liquor to goods ratio in jig processing is low (3-7:1), a pretreatment consisting of a desizing and wash-off is needed. The pretreatment may consist of an enzyme desizing, a dilute mineral acid treatment, or an impregnation of the fabric with 2-7 g/liter available chlorine as NaClO together with 5 g/liter Na_2CO_3. This serves also as a partial bleaching.

The bleaching solution is prepared according to degree of whiteness required, to the type of fabric, and to the plant conditions. For a full white, usually a higher amount of H_2O_2 is taken along with higher amounts of silicate and alkali than for a fabric for subsequent dyeing. The bleaching solution is composed of 1.5-3.5 g/liter H_2O_2, 4-10 g/liter sodium silicate or 1-4 g/liter of an organic stabilizer, plus 1-1.5 g/liter of NaOH, and wetting agent. The temperature of the treatment is 80–100°C, and its duration is 1-3 hr [191].

Continuous and Semicontinuous Bleaching Processes with H_2O_2

The most important developments in bleaching processes have occurred in the last 20 years in continuous bleaching. The need to

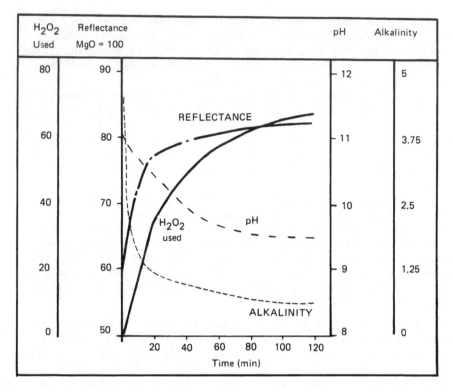

Figure 2.62 Bleaching cotton with hydrogen peroxide in package machines. Liquor ratio 10:1. Relationship of bleaching, time, alkalinity, pH, and peroxide utilization. Conditions of bleaching: temperature, 95°C; initial H_2O_2, 125 g/l; NaOH, 0.8 g/l; organic stabilizer, 2.0 g/l. (From Ref. 191.)

decrease the general cost of the bleaching operation and to save water and energy and the need to comply with the new regulations for effluent disposal and waste-water treatment have brought about the emergence of a number of semicontinuous and fully continuous processes; first, consisting of several stages, such as desizing, souring, scouring, and bleaching; then, decreasing the number of stages; and lately consisting of one stage only. These developments involved a gradual decrease in the time of bleaching and an increase in treatment temperatures and in alkali concentrations. At the same time, more sophisticated and highly efficient machinery ranges became available.

All forms of continuous bleaching have essentially similar basis, and the bleaching ranges consist of a sequential placement of stages which perform all of the bleaching functions. Each of the stages listed above consists of 3-unit operations: saturation, reaction, and removal [191]. *Saturation* refers to the application of the necessary chemical to the fabric, which can be done by a variety of methods, such as solution padding on dry or wet fabric and foam padding. *Reaction* denotes the creation of conditions under which the chemicals in the saturated fabric are activated. Such conditions are either short dwell times at high steam temperatures or longer dwell times at ambient temperatures. *Removal* consists of washing out the impurities of the fabric. These three-unit operations characterizing each stage appear several times in each bleaching range with the appropriate chemical compositions, times, and temperatures for each stage [191].

The bleaching processes may be carried out in rope or in open-width form (Table 2.27). The choice between these processes depends upon the nature of the fabric and its construction and upon production requirements. The open-width processes are preferred for heavier, higher-twist fabrics, such as poplin, duck, twills, and drills, in order to minimize wrinkles and crushing. The rope treatments are used for fabrics which do not have creasing problems such as knitgoods, sheeting, broadcloth, shirting, toweling, and lining. Very light-weight, woven fabrics and knits are also processed in jet machines, prior to the dyeing operation. Usually these fabrics require very little scour and bleach since most of them are constructed with man-made fibers.

Rope Bleaching: The rope bleaching ranges are based mainly on the J box, which is considered the workhorse for this industry. The main variations between the rope bleaching ranges are the type of J box, the number of stages, and the type of washer used. The number of stages employed varies from one to five depending on the bleaching method adopted and on the size of the plant, on the throughput required, and cost considerations. The stages are desizing, alkaline scour, acid sour, hypochlorite prebleach, and peroxide bleach. There are two types of J boxes, differing in the method of heating. In one case, the J box is insulated and heated (Becco system). In the other case, it is equipped with a preheating tube (U tube) followed by a heat-insulated J box (DuPont System).

The rope systems require fewer chemicals and permit a higher production rate than the open-width machines. The large J boxes can store up to 2000 kg fabric. Dwell times of 30-60 min are possible, and speeds of production of about 180 m/min can be reached.

Table 2.27 Rope versus Open-Width Bleaching

	Rope	Open-width
Fabric weight (yd/lb)	>2.5	≤2.5
Twist	Low twist, soft yarns	High-twist yarns, hard surface
Construction	Loose	Tight weave
Surface	Soft	Pile
Subsequent treatment	Finish white or print base	Dyeing
Applications	Sheeting	Poplin
	Printcloth	Corduroy
	Broadcloth	Duck
	Shirtings	Twills
	Toweling	Drills
	Linings	
	Knit goods	

Source: Ref. 4.

The rope systems are suitable for a large variation of fabric width, and the equipment involved is simple and efficient.

An interesting jet system for continuous rope bleaching of knits has recently been applied. The jet machine (CRT, Argathen) [282a] makes it possible to carry out the prescouring, the underliquor hypochlorite prebleaching and H_2O_2 bleaching, and afterwashing continuously in one machine, while the fabric is moved through several compartments. Each of the compartments can be serviced separately and may hold up to 70 kg of fabric. The liquor ratio is 1:12-1:20, the temperature is up to 100°C, and the speeds are 10-100 m/min.

Open-Width Bleaching: Semicontinuous Systems: In the *pad-batch* system the fabric is impregnated in open width. It is then preheated by passing through a steam or infrared chamber and made into a circular batch consisting of about 2000 m on a rotating roller. It is then held in a steam-heated reaction chamber for a given time prior to washing off. The retention time is usually 1-3 hr at a temperature of 80-100°C. The amounts of chemicals used in typical runs are 0.5-0.8% NaOH, 2-3% sodium silicate, 0.8-1.2% H_2O_2, all owf.

In the pad-batch process as in all other processes involving impregnation, the liquor ratio is very low and, therefore, the concentration of chemicals is high. A strong stabilization of the H_2O_2 is needed; thus, only a part of the silicate can be replaced by an organic stabilizer. The pad-batch system is also carried out *cold*. It requires a multidip padder, batching on A frames that rotate slowly for 8-16 hr, allowing flexibility in production scheduling. The fabric is passed through a small atmospheric steamer (80-100 m) before going through the final washer. The danger of catalytic degradation is lower in the process, but the brightness is also lower than when working at higher temperatures. The short steaming stage improves the brightness. The amounts of H_2O_2 and NaOH applied are understandably higher in order to compensate for the low temperature. The impregnating bath is composed of 1.4-1.8% H_2O_2 (100%); 1.4-2.0% sodium silicate, 38°Be; 1-1.5% alkali; and 0.8-1.0% organic stabilizer at 100% pickup. (For COT/PET the pickup is usually 70%, and 1.0-1.3% H_2O_2, 1.0-1.5% sodium silicate, and 0.8-1.0% sodium hydroxide are taken.) This system is being used to a large extent in European mills.

In another modification of this system which is also widely used in Europe, the pad-roll process is transformed from a semicontinuous to a fully continuous process: *the rebatching process* [175]. In this process two rolls are simultaneously employed in the steam chamber. On each roll two superimposed layers of fabric are wound up. The first inner layer stems from the second roll after partial aging. The outer layer came directly from the saturator. Upon completion of the batch on the first roll, the rotation of the roll is reversed, and the outer layer that came from the saturator is wound on the second roll, while the inner layer is led to the washing stage. The production speed is about 120 m/min.

Fully Continuous Open-Width H_2O_2 Bleaching: The open-width bleaching processes have been developing in recent years at a rapid pace. These processes are of particular importance for the COT/PET blends which are very sensitive to creasing during finishing treatments. There are at present several processes (Fig. 2.63) which differ mainly in the time and conditions of steaming and in the steaming machines used.

Chemical formulations and processing conditions for two open-width processes, the *rapid bleach* and the *minute bleach*, are summarized in Table 2.28 and compared to the standard rope system. It is evident that with the decrease in steaming time, the amounts of hydrogen peroxide, sodium hydroxide, and sodium silicate are increased. The severity of the scour is also increased [4-7]. For the rapid bleach, smaller, 200-kg J boxes or continuous pile storage steamers, in which the fabric is plaited down onto conveyor belts, are used in order to minimize fabric weight on sensitive fabrics. The effect of the shortening of the reaction time on H_2O_2 and NaOH concentration as well the time-temperature relationship in the process has been systematically investigated [276,284,286]. In the minute-bleach system the fabric is tightly guided on guide rollers through a continuous roller steamer containing 200 m of fabric, so as to avoid any crease marks (Figs. 2.64 and 2.65)

Another rapid system which is being used by a number of plants around the world is *pressure bleaching*. It is to be expected that at a temperature of 130-140°C in a saturated steam atmosphere, the reaction will be considerably speeded up, and an extremely high

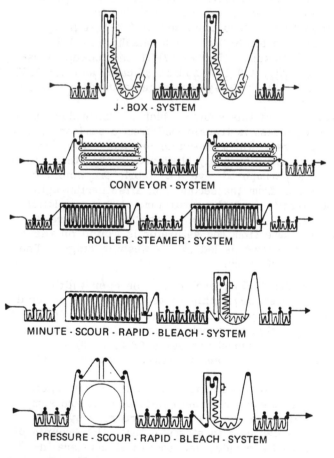

Figure 2.63 Standard and combined open-width bleaching ranges. Units between the major equipment consist of saturators, wash and rinse boxes, and similar auxiliary equipment. (From Ref. 280.)

Table 2.28 Typical Conditions for Continuous Bleaching

	Standard rope	Rapid open-width	Minute open-width
Scour			
Steaming time (min)	60	8-15	2
Temperature (°F)	200-210	200-210	212
Caustic soda (%)	3-4	4-6	6-9
Trisodium phosphate (%)	0.25-0.50	0.25-0.50	0.25-0.50
Surfactant (%)	0.1-0.2	0.1-0.2	0.1-0.2
Sequestering agent (%)	0.1-0.2	0.1-0.2	0.1-0.2
Bleach			
Steaming time (min)	60	8-15	2
Temperature (°F)	200-210	200-210	212
Sodium silicate (%)	0.75-1.5	1.5-2.0	1.5-2.0
Caustic soda (%)	0.1-0.2	0.2-0.6	1.0-2.0
Hydrogen peroxide (%)	0.35-0.60	0.60-0.90	1.0-1.4
Sequestering agent (%)	0.10-0.20	0.10-0.20	0.10-0.20
Tetrasodium pyrophosphate 9%)	0.10-0.20	0.10-0.20	0.10-0.20
Surfactant (%)	0.10-0.20	0.10-0.20	0.10-0.20
Epsom salt (%)	—	—	0.05

Source: Ref. 4.

level of absorbency will be reached in about 1 min, especially if the
fabric is reimpregnated inside the pressure chamber [7] (Fig. 2.66
and 2.67). The process is carried out in two stages [285]. The
first stage is an alkaline scour with 80 g/liter NaOH and an anionic
detergent. The temperature of steaming is 142°C and the time is
40 sec. The peroxide bleaching stage is carried out under similar
conditions for 45 sec. The composition of the impregnating solution is: 10 g/liter H_2O_2; 9 g/liter Na_2SiO_3, 35°Be; 1 g/liter NaOH;

Figure 2.64 An open-width bleaching range. (Courtesy of Messrs. Benninger.)

and 9 g/liter organic stabilizer. The fabric is passed after impregnation in the pressure chamber through special seals via inflatable Teflon lips that maintain the pressure inside the chamber. The fabric can be reimpregnated in a special saturator located in the chamber. It is guided through the chamber on a bed of slowly rotating rollers so that 30 m of fabric are contained in the chamber at any time. It was stated after a trial production period of 18 months [285] that the pressure system gave good results: a brightness of about 84% and a DP of 1500-2000. No yellowing was obtained on aging. These properties were constant throughout any given batch of fabric.

It was, however, also stated that the "plants which use this system have considerably more maintenance problems than would be tolerated on an open-width preparation range. The main problem areas are bearings inside the machine as well as inlet and exit seals which have to be replaced frequently" [7].

As distinct from the above range, which is produced by Kleine-wefers, in the Mather and Platt "Vaporloc" range (Fig. 2.68) the steaming time is about 2 min at 140-145°C, and the amount of fabric in the chamber is over 150 m. This is achieved by the fabric being plaited down [7,281].

In another modified pressure system the bleaching is carried out at 110°C and reduced pressure [7]. There are also ranges in which the fabric is submerged in the solution during bleaching. It is claimed that the fabric floating through a liquid medium will be more open and less prone to crease marks. Such a process is called "wet heel."

A number of combined bleaching ranges such as the *pressure-scour and rapid bleach*, and the *pressure scour and minute bleach* are also being used now. Their operating conditions are summarized in Table 2.29.

Figure 2.65 A continuous steamer. (Courtesy of Messrs. Mortensen.)

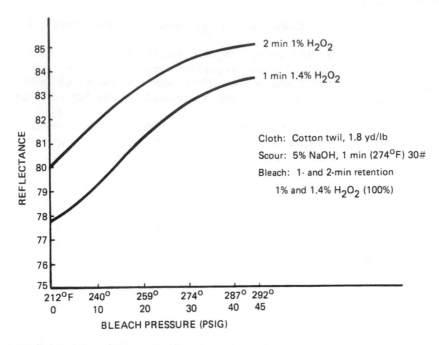

Figure 2.66 Effect of time, temperature, and peroxide concentration on the reflectance attained in the laboratory continuous, open-width pressure unit. (From Ref. 280.)

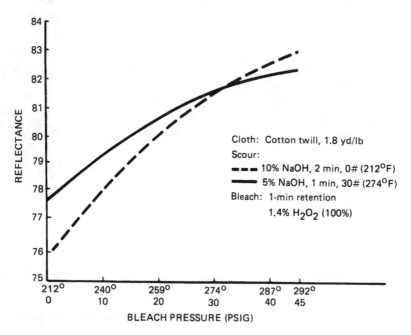

Figure 2.67 Effect of bleach temperature and caustic soda scour on the reflectance attained in the laboratory open-width, continuous pressure unit. (From Ref. 280.)

Figure 2.68 Vaporloc high-temperature bleaching machine showing roller bed. (Courtesy of Messrs. Mather and Platt.)

Single-Stage Peroxide Bleaching: Considerable efforts were made in recent years to decrease the number of bleaching stages. Stage combination saves a large amount of energy and water. Important progress has been made toward a one-stage process. Rope ranges are preferred for a single-stage bleaching, since they allow long steaming times in their large J boxes. Single-stage bleaching is being

Table 2.29 Summary of Operating Conditions of Continuous Processes[a]

	Caustic scour stage			Peroxide bleach stage		
	Time (min)	Temperature (°F)	Yd cloth in process	Time (min)	Temperature (°F)	Yd cloth in process
Large J-box system	60	200-205	6,000-10,000	60	200-205	6,000-10,000
Rapid bleach	8-15	200-205	800-1,200	1-25	200-205	800-1,200
Minute bleach	2	210-212	200	2	210-212	200
Pressure bleach	2	280-292	150	2	280-292	150
Pressure scour + rapid bleach	2	280-292	150	8-15	200-205	800-1,200
Pressure scour + minute bleach	2	280-292	150	2	210-212	200

[a]Data obtained in laboratory and confirmed in mill runs except for "Pressure bleach" which is only laboratory data.
Source: Ref. 280.

Table 2.30 Direct Comparison of the Optimized KPP and C/P Processes with the Conventional Plant Process[a] (West Point Pepperell Fabrics)

Fabric	Process	Whiteness degree	Absorbency (sec)	Noncotton content (%)					Breaking strength (lb)
				Water extracts	Enzyme extracts	Solvent extracts	Total extracts		
5/5S	Conventional	77.3	10.2	3.4	0.2	1.2	4.8		148
	KPP	77.0	7.0	0.1	0.1	0.7	0.9		154
	C/P	82.4	5.4	0.2	0.2	0.4	0.8		—
5/5T	Conventional	80.3	<1	2.5	0.1	0.5	3.1		416
	KPP	80.0	2.5	0.2	0.1	0.2	0.5		406
	C/P	75.1	1.5	0.4	0.2	0.5	1.1		—
100T	Conventional	69.2	<1	0.1	0.7	0.6	1.4		261
	KPP	75.3	<1	2.4	0.1	1.0	3.5		267
	C/P	65.7	<1	0.6	0.1	0.6	1.3		—

[a]Conventional method utilized timed 30-45 min dwell in each J-box steamer.
Source: Ref. 6.

conducted now in several mills on COT/PET and 100% cotton fabrics. It requires higher alkalinity for seedhusk penetration and wax removal, and 25-30% more H_2O_2 than in similar multistage formulations. Absorbency is poorer, and the seedhusks are sometimes not fully removed. Prewashing with water is alleviating the problem [4-7]. Similar problems are encountered in mills carrying out open-width, single-stage bleaching using roller steamers which permit short, 2-min steaming.

A comparison between two new experimental single-stage and the conventional three-stage processes, carried out recently in a study by the Southeastern Section of the AATCC, is shown in Table 2.30. In one process, designated KPP, the following chemicals were employed: H_2O_2 (35%), 5.0%; Triton X-100, 0.1%; Varsol 18, 0.8%; Na_2SiO_3 42°Be, 2.0%; NaOH, 0.8%; tetrapotassium peroxydiphosphate (KPP), 1.0%. All percentages are owf. The pad dwell time was 18 sec; steam time, 45 min; wet pickup, 100%; wash bath temperature, 80°C.

The chemicals used in the second process, designated C/P, were applied from foam: H_2O_2 (50%), 3.0%; NaOH, 1.0%; Na_2SiO_3, 1.0%; Grifferge NMT-66 (surfactant), 1.0%; Griffwet PA835 (wetting agent) plus chelate ACK, 1.0%. The wet pickup was 100-105% and the steam time, 60 min.

It can be seen (Table 2.30) that for all three treated fabrics the single-stage processes performed in a more satisfactory way than the conventional process, and a saving of about 20%, chiefly in energy, was achieved.

6. BLEACHING OF REGENERATED CELLULOSE AND BAST FIBERS

6.1 Rayon, High-Wet-Modulus, and Polynosic Fibers

Regenerated cellulosic fibers, normal viscose rayon, or high-wet-modulus (HWM) fibers are more vulnerable than cotton to chemical damage, this being particularly so with normal viscose rayon. They must therefore be bleached under conditions milder than those used for cotton. Moreover, the raw fibers are whiter than those of un-bleached cotton, and their colored impurities are more easily removed. Sodium hypochlorite, hydrogen peroxide, sodium chlorite, and peracetic acid can all be used for viscose rayons [1,162,185,286-290]. However, cupraammonium rayon can be bleached under normal conditions only with sodium chlorite [191].

Filament viscose rayon can be bleached at the production stage. The cakes are washed to remove occluded coagulation-bath liquors, desulfurized (if this is required), and then treated with a dilute solution of sodium hypochlorite. This is followed by washing with dilute hydrochloric acid to destroy residual bleach liquor [291].

Hydrogen peroxide bleaching at this stage, together with a sequestering agent, has also been described [292]. However, the majority of rayons are produced in staple form, and these are less likely to have been bleached at the production state [1].

Recommended conditions for sodium hypochlorite bleaching are 1 hr in a cold liquor containing 1.5 g/liter of available chlorine at pH 10-11 (sodium carbonate) [185]. For peracetic acid, bleaching is done for 1 hr at 65°C with a 0.3% solution of peracetic acid (36-40%) containing sodium hexametaphosphate at near neutral pH [185,287]. A cold-pad batch process with peracetic acid has been developed in which desizing is claimed to accompany bleaching [293]. With sodium chlorite, conventional conditions are used: 30-45 min at 80°C and pH 3.8-4.0 (formic acid and sodium dihydrogen phosphate) [1,162, 185,286]. The properties of rayon staple fiber for nonwoven packaging material are improved by bleaching the wet slurry with sodium chlorite in the hollander beater trough, before wet laying [294].

The most common bleaching technique is probably with hydrogen peroxide [1,185,286-288]. For the bleaching of yarn or fabric in a long liquor, treatment is with 0.5-1.0 vol. hydrogen peroxide for 1 hr at 50-70°C. Sodium silicate, caustic soda, soda ash, and a detergent are added. As with all cases of peroxide bleaching, it is often advantageous to add a sequestering agent. Alkaline peroxide bleaching of normal and HWM rayon cloth for 1 hr at 88°C in the presence of sodium silicate drastically reduces the average molecular weight of both types of rayon. The effect is uniform throughout the filament [295]. Similarly the rapid bleach and the minute bleach processes can be applied.

Peroxide bleaching of rayon and HWM cloth is also adaptable to cold-pad-dwell processing [287]. The fabric is padded with warm alkaline (soda ash and sodium silicate) peroxide solution, so as to achieve a pickup of 0.4% hydrogen peroxide (35%) owf, and is then batched. For this process, preliminary desizing is usually required. Peroxide bleaching can be combined with dyeing in a continuous pad-steam process. Chloramine can replace peroxide here as the bleaching agent [296].

Bleaching processes for both normal and HWM staple fibers have been compared [297]. Generally, HWM fibers bleach to a higher whiteness than normal rayon, and they are degraded less. For both types of fiber, chlorine dioxide was marginally superior to hydrogen peroxide with respect to both efficiency of bleaching and to the amount of degradation. The latter was measured by the fall in degree of polymerization. Both the above bleaching agents were superior to sodium hypochlorite in all respects. In all cases bleaching increased the crystallinity, and this was particularly the case with sodium hypochlorite. The water retention capacity of the fibers also rose on bleaching, this being more marked with HWM rayon. The recommended optimum conditions were peroxide bleaching at pH 10 and 85°C for 1 hr in the presence of 0.6-1.0 g/liter hydrogen peroxide, using sodium silicate as stabilizer.

Rayon fibers may be bleached at 70°C with sodium dichloroisocyanurate at a concentration of 0.5-2.0 g/liter and at pH 5.6 for 30-45 min [298]. The use of urea peroxyhydrate [299,300] and sodium perpyrophosphate [300] has also been described.

6.2 Cellulose Acetate and Triacetate

Strongly basic conditions will hydrolyze the acetate groups in cellulose acetate fibers. Nevertheless, they can be bleached satisfactorily with sodium hypochlorite and hydrogen peroxide at a pH lower than that used for cotton bleaching. Bleaching under mildly acidic conditions with peracetic acid is also possible.

Sodium Hypochlorite

Bleaching of cellulose acetate is at room temperature for 45-90 min with a sodium hypochlorite solution containing 0.5 g/liter available chlorine to which dilute acetic acid has been added to lower the pH to not greater than 10. This is followed by an antichlor treatment with sodium bisulfite (30 min at 70°C) [185].

Hydrogen Peroxide

A 0.5- to 1-vol. hydrogen peroxide solution is applied for about 1 hr at 65-75°C. The alkalinity is adjusted to pH 9-9.5 either with sodium silicate, tetrasodium pyrophosphate (TSPP), ammonia, or one of the commercial peroxide stabilizers recommended for wool [185,301]. Contamination with heavy metals should be avoided [1]. Cellulose acetate may be bleached at 75°C during 1 hr in the presence of 7 g/liter sodium silicate without suffering fiber damage. However, with triacetate, a concentration of 10 g/liter can be used safely [301]. When bleached in the presence of soda ash (0.5 g/liter) and sodium silicate (1.2 g/liter) at 80°C or below, damage is confined to the fiber surface [302].

Peracetic Acid

Both acetate and triacetate fibers may be bleached with peracetic acid under conditions similar to those used for polyamide, but at a lower temperature. Bleaching is for 1 hr at 65°C, pH 5-6, in the presence of sodium hexametaphosphate (0.5 g/liter) [185].

Other Bleaching Agents

Sodium chlorite is suitable for bleaching both acetate and triacetate [1,191]. An acidity of pH 3.6 (formic acid) has been used [301].

Solutions of di- and trichloroisocyanuric acids can also be used, at room temperatures and pH 3-9, with an active chlorine content of 2-3 g/liter [303]. An acetate textile has been bleached at 50°C for 1 hr using sodium percarbonate (5 g/liter) in the presence of acetic anhydride (3.3 g/liter) [304].

6.3 Flax and Linen

Retted flax contains only about 70% α-cellulose (dry basis), and the high proportion of noncellulosic matter (relative to cotton) complicates the bleaching process [1,185,305-307]. About two-thirds of these impurities are hemicelluloses with a low degree of polymerization. These are soluble in strong alkali so that bleaching under conditions used for cotton can cause an unacceptably large loss in weight, up to 25%. Milder conditions, with a longer treatment time, are used than with cotton, and the presence of residual hemicellulose is an advantage [307]. Unbleached linen has a much deeper color than has raw cotton, and a two-stage (combination) bleaching process is usually applied, often with sodium chlorite followed by hydrogen peroxide [306-308]. The more severe the conditions used, the higher the degree of bleaching, but the greater the loss in weight. The degree of bleaching and scouring applied to the fiber is thus adjusted to suit the purpose for which the cloth is intended [1]. If a soft hand is required, that is, if the fabric will be used for apparel, it is necessary to effect a far-reaching removal of the noncellulosic materials. In this case an alkaline extraction is imperative. If, however, the fabric is intended for tablecloth and similar uses and a hard hand of the linen is desired, the extraction is performed with sodium carbonate or omitted altogether in order to minimize the amount of the impurities removed.

A major difference between flax and cotton is the presence of a small amount (about 2%) of dark-colored lignins is retted flax. These have to be removed completely if a high level of whiteness is to be achieved. Also, efficient scouring to remove all traces of "sprit" (i.e., remants of the woody core of the flax stem) is essential. Kier boiling with lime has been recommended for this latter purpose, but caustic soda-soda ash is now used [307].

While a cotton fiber is unicellular, the cellulosic component of flax is composed of bundles of relatively long fibers (20-25 mm) cemented together both laterally and longitudinally by the middle lamella, which is composed of pentosans, pectin, and polyuronides [305]. Bleaching, like other preparatory processes, such as retting, scutching, and particularly scouring, breaks down the bundles into ultimate (unicellular) fibers. This process is termed "cottonization." A fully bleached linen is composed almost wholly of ultimate fibers [313].

Hypochlorite

The traditional method of bleaching linen involved treatment at room temperature with acidified sodium hypochlorite solution, followed by a mild peroxide or alkaline hypochlorite [307]. Hypochlorite is applied at pH 2.5-4.5 [286,309-311]; chlorine water or acidified bleaching powder has also been used. Alkaline hypochlorite bleaching is done at about pH 10.5 [306,311,312,314]. Acid hypochlorite chlorinates the lignin fraction, which then becomes soluble on subsequent alkaline scouring. However, no bleaching of the lignin fraction occurs [306,315]. Hypochlorite bleaching causes marked fiber degradation, as shown by a rise in fluidity [312]. Bleaching at pH 4 gives higher whiteness, but also causes a higher rise in fluidity than at pH 10.5 [311].

Hydrogen Peroxide

Typical conditions for the subsequent peroxide bleach are 4-5 hr at about 75°C in the presence of sodium silicate, soda ash, and magnesium sulfate [185,308]. A higher peroxide concentration (2% on weight of goods) is recommended when bleaching cloth than when bleaching yarn. Yarn may be bleached in hank form, using 0.5% peroxide, on weight of goods, and with constant agitation [1,286]. Soft water should be used for peroxide bleaching [306,316], and the presence of trace heavy metals should be avoided [1].

Sodium Chlorite

The use of sodium chlorite instead of hypochlorite gives good results with less fiber degradation [185,263,286,311,317]. Sodium chlorite not only solubilizes two-thirds of the flax lignins, but also bleaches them, unlike hydrogen peroxide or peracetic acid [306,315]. It is therefore not necessary, with sodium chlorite bleaching, to remove all the lignin. In contrast to chlorine, the action of chlorite on lignin is one of oxidation, not of chlorination. A mild alkaline scour before bleaching, for example, with sodium carbonate or with a 3:1 mixture of this with caustic soda [135,307,318,319], is essential in order to obtain good whiteness without excessive fiber damage. This is at 80-90°C together with a detergent. A similar scour, with sodium carbonate or caustic soda [313], should follow the chlorite bleaching stage. Yarn which has been chlorite-bleached and then scoured is composed almost wholly of ultimate fibers, but bleaching after scouring is less effective in reducing strands to ultimate fibers. Thus the bleaching agent apparently attacks the fiber cement and renders it soluble in alkali. Recommended conditions for chlorite bleaching are [185,286]: sodium chlorite (80%), 1.5-3% on weight of goods; sodium dihydrogen phosphate (anhydrous), 0.25-0.5% on weight of goods; wetting agent; and formic acid, to pH 3.8-4.2.

Soft water should be used [316]. The temperature is brought slowly to around 85°C and kept there for 2-3 hr. This is followed by peroxide bleaching using the conditions described above [1,185, 318,319]. A chlorite bleach sandwiched between two peroxide bleaches is particularly effective [316]. Chlorite bleaching following treatment with hypochlorite has also been described [320]. Sodium chlorite bleaching at pH 2 (i.e., bleaching by chlorine dioxide) gives the best whites, but this causes more fiber damage than at pH 4 [263]. A one-step pad-steam process for the bleaching of linen, hemp, and jute utilizes a sodium chlorite solution buffered to pH 6.8 and stabilized by the addition of hydrogen peroxide. Steaming is for 3 hr at 90°C [306,321]. Linen can be bleached from the gray state if a high concentration (20-30 g/liter) of chlorite is used [307]. Polyester linen blends (50:50) may be bleached with sodium chlorite at 95°C for 1 hr [322].

Other Bleaching Agents and Processes

Peracetic acid is suitable for bleaching flax [191]. The use of sodium dichloroisocyanurate [323] and cyanuryl chloride [306,324] have been described for linen bleaching. Solid peroxide compounds were found to be more effective than hydrogen peroxide [325]. The effect of bleaching on the permanent press finishing of linen cloth has been investigated [319].

6.4 Bleaching of Jute

Jute, together with flax, ramie and hemp, are bast fibers, i.e., fiber bundles which are structural components of plant stems and are organized in a concentric bast layer around the woody stem of the jute stalk. Jute differs from linen in its high lignin content, about 11-12% dry weight, in the small dimensions of the ultimate fibers, approximately 2 millimeters, as well as in the composition of the hemicellulose fraction [305]. The lignin, together with the hemicellulose, provide inter-cellular binding material known as middle lamella, holding the fibers together [305]. Attempts to remove it, by bleaching or by treatment with alkali, cause a marked loss in strength and change in properties [305,306,327]. In contrast to linen it is neither desirable nor necessary to delignify jute. For example, though white jute is effectively bleached by hydrogen peroxide most of the lignin remains [326].

A consequence of the lignified nature of bleached jute is that its whiteness is not fast to light, and there is a gradual reversion to the original color (photo-yellowing) [306]. Though the best varieties of jute are sufficiently white to allow dyeing without bleaching [291], it is difficult to achieve light-fast dyeings in pale and bright shades [306, 328]. This has lead to the development of light-fast bleaching treatments.

As with linen, the greater the bleaching effect the greater also the loss in fiber weight. A combination (two-stage) bleach is usual: sodium hypochlorite as the first stage followed by treatment with hydrogen peroxide, potassium permanganate, or sodium chlorite [327, 329,330]. Bleaching with sodium hypochlorite (using 1-3 g/liter active chlorine) should be in a neutral or slightly alkaline medium; in an acid bath there is a strong tendency for chlorine retention by the lignin [327]. Optimum conditions for bleaching with sodium chlorite (7 g/liter) are for 80 min at 65-70°C and at pH 4. This causes the minimum reduction in degree of polymerization. At higher temperatures bleaching is more rapid [306,330]. Jute has been bleached by hydrogen peroxide at 75-85°C for 2 hr in the presence of sodium silicate and soda ash. Under these conditions no degradation of the α-cellulose is observed at concentrations of 1 vol. and below [326]. With peroxide concentrations of up to 2 vol., dry strength of yarn is slightly increased, but wet strength falls by about 25% [326].

Combination hypochlorite/peroxide and hypochlorite/chlorite bleaching have been compared [331]. For the particular conditions used in these experiments, both processes gave practically the same whiteness, but bleaching with hydrogen peroxide caused a markedly lower loss in weight (10.0% as against 17.9%) and a slightly lower loss in strength. Sodium chlorite removes more lignin than does hydrogen peroxide, and the former reagent has been used for the complete delignification of jute [326,332].

A number of patents describe lightfast bleaching treatments [306]. Treatment with chlorine and water, or with sodium hypochlorite at pH 6, causes surface delignification only. Inner portions of the fiber are not affected appreciably, and loss of strength is satisfactory. The fibers are then extracted with alkali or sulfite, washed and bleached by conventional procedures [333]. This process is claimed to give a smooth fiber and makes possible the preparation of dyed and printed fabrics [306]. Another lightfast bleaching process involves treatment with an acidified solution of potassium permanganate followed by discharge of excess permanganate using a mild reducing agent [334].

7. BLEACHING OF SYNTHETIC FIBERS AND BLENDS

7.1 Polyester Fibers

Polyester fibers may be bleached using an acidified solution of sodium chlorite [1,185,338-343]. This does not cause damage to the fiber and, indeed, there is a slight increase in the degree of crystallization [343]. Hydrogen peroxide or sodium hypochlorite have little or no bleaching effect, though they do not damage the fiber. These fibers are naturally of good color and bleaching is often not necessary; fluorescent whitening alone may suffice for white goods, and it is not usually necessary to bleach before dyeing. However, for a "super-white" effect, sodium chlorite bleaching may be combined with fluorescent whitening [338-340]. (See Chap. 3, this volume.)

A standard recipe for making up 100 liters of bleaching solution is [338]: 2-4 kg sodium chlorite; chlorite stabilizer, according to manufacturer's recommendations; formic acid, to pH 3.5-3.8; 1.5-2 kg sodium nitrate; wetting and dispersing agents.

Acetic acid may replace formic acid. Bleaching in the presence of nitric acid at pH 2-3 has also been described [185]. Bleaching temperatures of 90-120°C are used for 30-60 min. At 90-95°C fiber penetration is slow, and better whites are obtained at higher temperatures, though this entails the use of closed equipment with provision for venting chlorine dioxide. As an alternative to the above long-liquor treatment, foam processing may be used [344]. A pad-roll method also gives good results [341,345]. With fabric woven from rotor-spun yarn, a lower liquor concentration is used in the padding bath than with ring-spun yarn [341]. Discoloration during bleaching may result from residual Mn or Sb polymerization catalysts. This can be prevented by pretreatment with a reductive bleaching agent [346] or by including hydrogen peroxide in the chlorite bleaching bath [347]. An advantage is claimed for adding an organic hydroperoxide and ethyl butyrate to the bath [348].

Polyester fibers have also been bleached with solutions of di- and trichloroisocyanuric acid at pH 3-9 during 30-45 min at 60-65°C [303].

7.2 Polyamide Fibers

Bleaching of polyamide is often not necessary, fluorescent whitening being sufficient [185]. Bleaching should not only remove natural color but also yellowing caused by heat setting. Prescouring is essential. Polyamide, like wool, is sensitive to alkaline conditions. Sodium hypochlorite is not used, having very little bleaching action. In neutral and alkaline solutions it causes damage to the fiber and at slightly acid pH it chlorinates amino groups, necessitating a subsequent antichlor treatment [1,185]. Yellowing caused by heat or photodegradation can be bleached using sodium perborate, percarbonate, or peracetic acid [350].

Sodium Chlorite

Most polyamide bleaching is done with sodium chlorite [185], which probably gives the best whites. It does not degrade the fiber; indeed some fiber strengthening [351] and partial recrystallization [343] occur. A standard recipe for 100 liters of bleaching solution is [185]: sodium chlorite, 0.05-0.2 kg; acetic or formic acid, to pH 3.8-4.2; sodium nitrate, 0.1-0.2 kg; wetting agent. A buffer, such as tetrasodium pyrophosphate, may be added. A liquor ratio of 20:1 to 40:1 is used. Processing requires 30-60 min at 80-85°C, and is followed by rinsing. The goods are finally treated with sodium bisulfite (antichlor) at 20-25°C for 15 min. It is possible to bleach together with a suitable fluorescent whitening agent [353], or together with scouring, using an acid stable surfactant [185]. Addition of sulfur compounds, such as tetramethylthiuram disulfide, zinc dimethyldithiocarbamate, benzothiazoles, and imidazolinethiols, is stated to improve the bleaching effect [352].

Peracetic Acid

This is a major use for peracetic acid as a bleaching agent [1]. Recommended quantities (for 100 liters) are: peracetic acid (36-40%), 0.3 kg; sodium pyrophosphate (anhydrous), 0.025 kg; wetting agent, 0.1 kg; and dilute NaOH, to pH 6-7. The goods are entered cold, the temperature raised to 80-85°C during 30 min and then kept for 30 min at that temperature. There is some fiber damage, which is increased by working at lower pH or higher concentrations [185,351]. Trace metals (Cu,Fe) should be avoided since they catalytically decompose peracetic acid. Peracetic acid bleaching causes a decrease in subsequent dye uptake [185], but inclusion of sodium chloride and other halide ions in the bath counteracts this [354].

Hydrogen Peroxide

Hydrogen peroxide under cotton bleaching conditions causes significant damage [351,355], and even under wool bleaching conditions (pH 8-9.5 and at 50°C or below) a fall of 50% in tensile strength can occur [185]. Acid peroxide bleaching has been practiced in the past [356], including in the presence of oxalic acid and tetrasodium pyrophosphate buffer, or with a stabilizer.

Agents for protecting nylon from fiber damage during alkaline peroxide bleaching are marketed. The best known is Proventine-7 (Degussa) [353,357-359] which is stated to contain a biguanidine derivative [358]. In the presence of this agent, bleaching may be carried out at pH 10-11 and 90°C during 2-4 hr without excessive damage to the fiber. It is used in the presence of sodium silicate and magnesium sulfate, and can be used under HT conditions (120°C for nylon 66, 110°C for nylon 6) in the presence of a peroxide stabilizer [353]. When using Fiber Protecting Agent PA2 (Peroxid-Chemie), bleaching is for 90 min at 95°C [162]. A number of Japanese patents have described nylon protecting agents [360]. An alternative is to incorporate an additive into the polyamide polymerization stage [361].

Reductive Bleaching

Stabilized sodium hydrosulfite is a less effective bleaching agent for polyamide than is sodium chlorite. However, Blankit D (BASF), which contains zinc hydroxymethylsulfinate together with activators, is claimed to be superior to sodium chlorite [362]. It is used in a long liquor at pH 2.6 (formic acid) for 20-30 min at 80°C or, together with hydrosetting, for 15 min at 120°C. Other methods recommended for use with this product are a pad-steam continuous process (100-102°C/ 10-20 min) or a cold or warm pad-batch process [363]. Another reductive bleach which has been recommended for polyamide is Clarite PS (Ciba-Geigy) [353], which contains sodium hydrosulfite together with anhydrous sodium pyrophosphate [185].

7.3 Acrylic Fibers

Though some acrylic fibers are markedly yellow, they are generally of good color. Bleaching is therefore usually necessary only for goods of high whiteness and sometimes for those in pale shades. When a high degree of whiteness is required, bleaching is almost always accompanied by fluorescent whitening [185,364-367]. White acrylics are generally bleached in yarn form, including high-bulk yarns [364,367]. Acrylic fibers yellow when boiled in a solution at pH > 7 [185], so that neither sodium hypochlorite nor hydrogen peroxide is applicable. Bleaching is almost invariably with sodium chlorite, typically for 1 hr at 90°C, but if an FWA (fluorescent whitening agent) is present, it is preferable to bleach at near the boil, for 30-45 min [364,367]. However, there are many different types of acrylic fibers, and while some, for example, Dralon [162], can be bleached under standard conditions (in the presence of formic acid at pH 3.6), others, such as Courtelle [370], are stated to be incompatible with sodium chlorite. In the latter case a mild reductive bleach with sodium hydrogen sulfite, together with fluorescent whitening, is recommended [370]. The conventional procedure, using formic or acetic acid at pH 3.5-4, is not always effective, the whiteness obtained being unstable to heat and light. Instead oxalic acid is used [185,364-366]. After the treatment, the bath should be cooled slowly with continual movement to prevent creasing, to 50-60°C over 20-25 min [264,267]. This may be accompanied by a prolonged rinse, or excess chlorite may be removed subsequently with sodium bisulfite at 50°C [267].

An alternative is to work with sodium chlorite or hypochlorite at pH 3 or below, which is achieved by adding phosphoric or nitric acid [191,368]. Sodium ultraphosphate ($Na_2O:P_2O_5 < 1:1$) can also be used to achieve pH 3 [369].

Acrilan 1656, which contains basic groups for acid dyeing and has poor color, benefits from a preliminary acid steep at pH 2 and 50°C [185]. For chlorite bleaching of similar fibers, ammonium acetate can be added to the bath [370].

Bleaching agents which have been proposed as alternatives to sodium chlorite are chloroisocyanuric acids [371] and permonosul-

furic or peroxyphosphoric acid at pH 2 [372]. Treatment with urea
peroxide in the presence of sodium silicate as a preliminary step to
dyeing, has also been advocated [373], as has peroxide bleaching at
pH 6.5 in the presence of glycolic or citric acid [374].

7.4 Fiber Blends

Two general rules apply to the bleaching of fiber blends:

1. The severity of the bleaching conditions should be no more
    than the most sensitive component can stand [191]. Thus
    the most sensitive component largely determines the condi-
    tions of bleaching. However, if this component represents
    only a small percent of the mixture it will withstand rather
    harsher conditions than would the pure fiber [375].
2. When the components have similar sensitivities, bleaching is
    according to the component present in the largest amount
    [191]. Nevertheless, for mixtures of natural with man-
    made fibers, it is often only necessary to bleach the former.

The bleaching of wool blends is described elsewhere (This volume
Chap. 4).

Polyester-Cotton

It is usually sufficient, when preparing cloth for dyeing or for medium
whites, to bleach the cotton component only. For this, conventional
cotton bleaching methods are used, largely involving hydrogen peroxide [349,376-399], and this process may be carried out in the cold
[392]. Sodium hypochlorite bleaching is also possible [378,391]. If,
for high whites, the polyester component also requires bleaching,
sodium chlorite is used, since it is effective with both fibers. For
the very highest white effects, as required for shirtings, hypochlorite or chlorite bleaching may be followed by a second, cotton, bleaching stage with peroxide [378,391,392]. Heat setting of the polyester
can cause some yellowing of the cotton component, so that bleaching
should always be after heat-setting [391,393,394]. Fluorescent whitening of the polyester component can accompany chlorite bleaching,
and the cotton can be fluorescently whitened together with peroxide
[390-395]. Continuous processes are preferred where sufficiently
large quantities are treated [376].

Thus, the dominant bleaching agent for polyester-cotton blends
is hydrogen peroxide [349]. With the introduction of improved pad-
bake fluorescent whitening processes for polyester (cf. Chap. 3,
this volume), less chlorite bleaching has been performed. This is
in spite of the fact that chlorite bleaching gives the cloth a blueish
tinge which enhances the whiteness effect [349].

Peroxide Bleaching: Essentially, conventional methods of cotton peroxide bleaching are used (see [349,379,381,383-387]).   Polyester
cannot withstand prolonged treatment with alkali, as used in kier
boiling, and time and temperature limits have been determined for
treatments with NaOH of various concentrations [276].   Bleaching
in rope form can cause creasing of the polyester.   However, "rapid-
bleach" techniques can be used, with dwell times of about 8-10 min
[376] or less [386].   Also, tightly woven fabrics, such as for rain-
wear, can be treated in rope form [376].

The conventional cotton bleaching methods recommended for
polyester-cotton include: a pad-roll system with batching at 80-90°C
and rotating for 2 hr, continuous high-temperature treatment, or
treatment at 100°C according to the "Autobleach" semicontinuous
system [376].   Continuous open-width pad-steam processes are particularly suitable [28,383], as in an open-width J box [385,386].   It
may be advantageous to combine bleaching with scouring, or the
two together with desizing [387,388].   Another rationalization is to
combine dyeing with mild bleaching at pH 9-10 an at 98-100°C in the
presence of a carrier and a scouring detergent [396].   The importance of good scouring before bleaching has been stressed [385].
For practical purposes this can be followed by bleaching on a range
working at 100°C, rather than under pressure [385].   However, the
concentration of peroxide and sodium hydroxide must be adjusted according to the working temperature [381,385].

Peroxide bleaching of COT/PET blends can also be performed
advantageously in the presence of the less alkaline sodium orthosilicate (Na_4SiO_4) using magnesium phosphate as stabilizer [380].   The
effect of various pretreatment sequences on color strength of dyeing
has been investigated.   Heat setting before scouring apparently
reduces color strength [399].   Peroxide bleaching in the presence
of organic solvents has been claimed [388,400].

Other Bleaching Processes: Apart from sodium chlorite and hypochlorite, bleaching of COT/PET with a chlorine monoxide-air mixture
in the presence of sodium carbonate has been described [401].   Chlorite bleaching of COT/PET has been performed in the presence of
isopropyl formal as activator [402], and in the presence of hydroxylamine salts [403].

Cotton-Rayon Blends: Cotton is blended with viscose and with high-
wet-modulus (HWM) or polynosic rayons [405].   Viscose rayon is the
most sensitive to alkali, while polynosics approach the stability of
cotton [405].   These blends are bleached with any of the methods
used for cotton, but with the stringency of the conditions modified
to suit the nature of the rayon component and its proportion in the
blend [185].   Bleaching of cotton blended with viscose rayon or with
Vincel 64 (a HWM fiber) is done preferably with 1-2 g/liter sodium
chlorite for 1-1.5 hr at 85-90°C, together with formic acid at pH 3.6

[386,405]. This is adaptable to continuous J-box operation (at pH 3.9).

Blends of polynosic and cotton are usually bleached with NaClO$_2$, 2% owf, for 2 hr at 90°C in the presence of NH$_4$HF$_2$, 2%; HCOOH (85%), 3 ml/liter; NaNO$_3$, 6 g/liter; and a wetting agent. However, satisfactory results can also be achieved with hydrogen peroxide, in the presence of sodium silicate, tetrasodium pyrophosphate, a protein-based stabilizer, and a sequestrant, at 90°C [386]. Peroxide bleaching in the presence of sodium hydroxide has also been described [305, 406]. Nevertheless, bleaching of cotton-polynosics at 90°C is with a reduced peroxide concentration (3.65-4 g/liter) and is generally not recommended unless the blend contains 50% or more cotton [405,407]. For viscose rayon-cotton the recommended conditions are 4 g/liter hydrogen peroxide at 65°C, plus 0.3 g/liter NaOH; 1 g/liter Na$_2$CO$_3$; and Na$_2$SiO$_3$ 79°Tw, 3.5 g/liter. This treatment reduces the sulfur content of the viscose component remaining from the viscose manufacturing process [287]. A cold pad-batch peroxide bleach in the presence of sodium silicate and soda ash is a simple, economical process [409]. Sodium dichloroisocyanurate is a satisfactory bleaching agent for the viscose rayon component [298,405]. It is carried out with 2 g/liter active chlorine at 70°C without much degradation.

Cotton-Polyamide: Since polyamide is effectively bleached with sodium chlorite, and since the latter is superior to peracetic acid for cotton, sodium chlorite gives the best effects on cotton-polyamide mixtures. This also applies to mixtures of polyamide with other cellulosics. A conventional method is used: 1.5-2 hr at 90°C and at pH 4. This is followed by treatment with a warm solution of sodium carbonate. Pad-roll and continuous processes are used [185]. Hydrogen peroxide bleaching is possible in the presence of borax-boric acid (90 min at 95°C) [404], lithium hydroxide (98°C, 1 hr) [404], or acetic anhydride at pH 6 [408]. Sodium dichloroisocyanurate may also be used at 90°C and pH 7-9.5 [409].

Acrylic Blends with Cotton and Cellulosics: Alkaline conditions are not compatible with acrylics; thus sodium chlorite is the best choice, since it bleaches both components. Chlorite bleaching in the presence of formic acid and sodium pyrophosphate (1 hr at 80-90°C) has been recommended [185]. For mixtures with not more than 20% acrylics content, the cotton alone may be peroxide-bleached in the presence of sodium silicate (1 hr at 90°C) [185]. Acrylic-rayon blends may be bleached with sodium dichloroisocyanurate, this being followed by treatment with sodium bisulfite [410].

Polyester-Rayon Blends: Polyester blended with Vincel 64 is most suitably bleached with sodium chlorite at pH 4 for 1 hr at 95°C. This can be combined with fluorescent whitening of both components, with a suitable carrier for the polyester [386]. The rayon component can be bleached with hydrogen peroxide, under the conditions used for Vincel-cotton blends [386]. A high-temperature continuous pad-steam process for peroxide bleaching of rayon-polyester has been claimed [411].

Other Mixtures: Mixtures of elastomers with cotton or cellulosic fibers can be peroxide bleached, and this may be followed by reductive bleaching [185]. Peroxide bleaching of cotton-cellulose acetate has been described [406].

Other Polyester Blends: For polyester blends with *polyamide, polyacrylonitrile, cellulose-acetate,* and *triacetate* fibers, sodium chlorite is the agent of choice, since it is the only one which bleaches all of these fibers and does not degrade them [412]. Hydrogen peroxide can be used for polyamide, but this is far less effective and is only possible in the presence of a polyamide-protecting agent. Reductive bleaching, with hydrosulfite or sulfoxylate derivatives is possible, but caution is required since these cause yellowing with some polyester fibers. The acrylic fiber Courtelle is stated to be incompatible with sodium chlorite [412]. Suitable FWAs and carriers can be added to the chlorite bleach-bath.

The following conditions are recommended for chlorite bleaching: for polyamide blends, 30-45 min at 90-115°C; for polyacrylonitrile blends, 30 min at 96-98°C; for acetate blends, 30-45 min at 75-80°C; for triacetate blends, 30-45 min at 90-125°C.

In all cases, formic acid is added to pH 3.5-3.8 and a chlorite stabilizer is used.

Various Fiber Blends: A series of Japanese patents deals with the bleaching of polyacrylonitrile-protein graft copolymers [413]. A polyurethane elastomer-polyamide fiber mixture may be bleached with sodium hydrosulfite [414]. As a general rule, mixtures containing elastomeric yarns cannot be bleached with sodium hypochlorite or chlorite, though hydrogen peroxide can be used [185]. Reductive bleaching (sodium hydrosulfite or bisulfite) is possible, though not always very effective. However, reductive bleaching is the only possibility where blends of elastomers with polyamide or acrylic fibers are concerned [185].

8. TREATMENT OF BLEACHING EFFLUENTS

Scouring and bleaching are carried out either together in one stage or in two consecutive stages, and the former process contributes the major proportion of organic matter to the effluent from a bleaching plant [415-417]. Effluent from textile wet processing is very variable in content, and it is usually balanced by mixing in a lagoon or tank, and before this the effluent flow is neutralized to near neutral pH. It is then usually treated biologically before discharge into a watercourse, though occasionally it may be sent to a local sewage plant to be treated together with domestic wastes. In either case, toxic materials must not be allowed to arrive at the treatment plant, since they will interfere with the microorganisms which are used to break down the organic matter [415-417].

Spent bleaching liquors and wash waters from a bleaching operation contain unused bleaching agent together with stabilizers, alkali or acid, and surfactants, and they also contain fibers, particularly if natural fibers are being treated [421]. Hydrogen peroxide is not detrimental, but residual sodium hypochlorite can cause trouble [415]. It may be destroyed during the balancing stage by its reaction with excess oxidizable material. Otherwise it must be pretreated with sodium bisulfite [417], or slowly bled into the main stream [416]. Sodium chlorite is also toxic, and effluent containing this may have to be pretreated with a reducing agent [415] or highly diluted before disposal [418]. In connection with chlorite bleaching, it should be noted that the BOD of formic acid is five times that of acetic [418]. There is thus, in conclusion, a premium attached to peroxide bleaching, in that it entails easier effluent treatments [415].

A description has been given of the treatment method for a cotton bleaching plant producing 4000 m^3 of wastewater a day [419]. The effluent has a BOD of 500-600 mg/liter, and suspended solids seldom exceed 100 mg/liter. Treatment is by sedimentation, aided by the addition of a polymer, and then with activated sludge. This reduces the BOD to 25-30 mg/liter, suitable for discharge. Thickened sludge is dewatered by centrifugation before disposal.

The volume of effluent requiring treatment can be reduced by water reuse, though this does not lessen the organic load. Thus, peroxide cotton-bleaching wash liquors can be collected and reused, for example, for rinsing the cloth after scouring [417] or for making up process liquors [416]. Such recycled liquors must be monitored for their content of salts, FWAs, and cationic surractants. The introduction of FWAs has reduced the volume of bleaching effluents since it allows milder bleaching conditions [418]. However, fluorescent whitening only supplements chemical bleaching and cannot replace it.

Wastewater from the bleaching of cellulosic fabric can be treated by passage through columns of γ-Al$_2$O$_3$. This only reduces the BOD to 300-400 mg/liter and further treatment is necessary [420,421], but color is removed and foam formation is avoided.

ACKNOWLEDGMENT

The author wishes to thank Dr. R. Levene of the Israel Fiber Institute for assistance in the preparation of sections 6, 7, and 8 of the manuscript.

REFERENCES

1. R. H. Peters, *Textile Chemistry*, Vol. II, Elsevier Publ., 1967.
2. B. K. Easton, *Ciba Geigy Rev.* 3, 3 (1971).
3. J. S. Sconce, Ed. *Chlorine, Its Manufacture, Properties and use*. Reinhold Publishing Co., New York (1962).
4. B. A. Evans, *Text. Chem. Color. 13*, 254 (1981).
5. B. K. Easton, *Text. Chem. Color. 13*, 252 (1981).
6. G. R. Turner, *Text. Chem. Color. 13*, 246 (1981).
7. S. W. Poser, *Am. Dyest. Rep. 71*, 13 (1982).
8. N. E. Hauser, J. C. Martin, and M. White, Jr., *Am. Dyest. Rep. 70*, 19 (1981).
9. W. Kothe and N. Leppert, *Textilveredlung 13*, 333 (1978).
10. K. Dickinson, *J. Soc. Dyers Colour. 95*, 119 (1979).
11. G. Roesch, *Melliand Textilber. 62*(11), 790 (1982).
12. U. Kirner, *Melliand Textilber. 51*, 1089 (1970).
13. K. H. Rucker and H. Fuchs, *Textil Prax. Intern.*, 274 (1973).
14. Tennant, British Patent 2319 (1799).
15. B. M. Baum, J. H. Finley, J. H. Blumberger, E. J. Elliott, F. J. Scholer, and H. L. Wooten, *Othmers Encyclopedia of Chemical Technology*, 3rd Ed. Vol. 3, 938 (1978).
16. C. W. Dence, *The Bleaching of Pulp*, Tappi Monograph Series No. 27 (1963).
17. M. Lewin, *Tappi 48*, 333 (1965).
18. M. Lewin and M. Avrahami, *J. Am. Chem. Soc. 77*, 4491 (1955).
19. R. L. Whistler and R. Schweiger, *J. Am. Chem. Soc. 80*, 5701 (1958).
20. K. W. Young and A. J. Allmand, *Can. J. Res. 27*, 318 (1949).
21. N. V. Sidgwick, *Chemical Elements and Their Compounds*, Oxford University Press, London (1950).
22. B. P. Ridge and A. H. Little, *J. Textile Inst. 33*, T33, T59 (1942).
23. J. A. Epstein and M. Lewin, *Bull. Res. Counc. of Israel 9A*, 3-4 (1960); *9B*, 3 (1961).
24. J. A. Epstein and M. Lewin, *J. Polymer Sci. 58*, 991 (1962).
25. M. Lewin and J. A. Epstein, *J. Polymer Sci. 58*, 1023 (1962).
26. M. Lewin and M. Albeck, Mild Oxidation of Cotton, Final Report FG-IS-101-58, submitted to ARS, USDA, Jerusalem, 1963.
27. W. A. Roth, *Z. Phys. Chem. Abt. A, 145*, 289 (1929).
28. J. J. Bernard and H. J. Bolker, *Chem. Rev. 76*, 487 (1976).
29. A. Meller, *Pure Appl. Chem. (Australia) 6*, 40 (1956).
30. A. Meller, *Holzforschung 14*, 78, 129 (1960).
31. C. Birtwell, D. A. Clibbens, and B. P. Ridge, *J. Textile Inst. 16*, T13 (1925).
32. D. A. Clibbens and B. P. Ridge, *J. Textile Inst. 18*, T135 (1927); *18*, T148 (1927); *19*, T389 (1928).

33. H. A. Rutherford, F. W. Minor, A. R. Martin, and M. Harris, *J. Res. Nat. Bur. of Standards 29*, 131 (1942).
34. G. F. Davidson, *J. Textile Inst. 43*, T291 (1952); *29*, T195 (1938); *39*, T65 (1948).
35. C. Birtwell, D. A. Clibbens, A. Geake, and B. P. Ridge, *J. Textile Inst. 23*, T185 (1932).
36. A. R. Urquhart, W. Bostock, and N. Eckersall, *J. Textile Inst. 23*, T135 (1932).
37. A. R. Martin, L. Smith, R. L. Whistler, and M. Harris, *J. Res. Nat. Bur. of Standards 27*, 449 (1941).
38. O. Samuelson and C. Ramsel, *Svensk Papperstidn 53*, 155 (1959).
39. O. Samuelson and L. A. Wikstrom, *Svensk Papperstidn 63*, 543 (1960).
40. H. A. Turner, G. M. Nabar, and F. Scholefield, *J. Soc. Dyers Col. 53*, 5 (1937).
41. L. F. McBurney *Cellulose and Cellulose Derivatives*, Part 1. (Ott, Spurlin, and Grafflin, Eds.). 2nd Ed. Interscience, New York (1954).
42. D. A. Clibbens and A. H. Little, *J. Textile Inst. 27*, 285 (1936).
43. C. Birtwell, D. A. Clibbens, A. Geake, and B. P. Ridge, *J. Textile Inst. 21*, 85 (1930).
44. C. G. Schwalbe, *Z Angew. Chem. 21*, 1921 (1908); *22*, 197 (1909); *Ber. 40*, 1347 (1907).
45. C. Birtwell, D. A. Clibbens, and B. P. Ridge, *J. Textile Inst. 16*, 13 (1928).
46. P. Van Fossen and E. Pacsu, *Textile Res. J. 16*, 163 (1946).
47. M. Lewin, Thesis, Hebrew University, Jerusalem, 1947.
48. G. Holst, *Chem. Rev. 54*, 169 (1954).
49. O. Theander, *Svensk Papperstides 61*, 581 (1958); *48*, 105 (1965).
49a. B. Lindberg and O. Theander, *Svensk Papperstidn. 57*, 83 (1954).
50. J. Schmorak, D. Mejzler, and M. Lewin, *J. Polymer Sci. 49*, 203 (1961).
51. H. Kaufman, *Z. Angew. Chem. 37*, 364 (1924); *43*, 840 (1930).
52. G. Nabar and H. A. Turner, *J. Soc. Dyers Color. 61*, 258 (1945).
53. P. N. Agarwal, *J. Sci. Ind. Research 21D*, 65 (1962).
54. O. Samuelson, *Das Papier 24*, 671 (1970).
55. I. Norstedt and O. Samuelson, *Svensk Papperstidn. 68*, 565 (1965).
56. B. Alfredsson and O. Samuelson, *Svensk Papperstidn. 77*, 449 (1974).
57. S. I. Andorsson and O. Samuelson, *Cell. Chem. Techn. 10*, 209 (1976).
58. J. A. Epstein and M. Lewin, *Textile Res. J. 30*, 652 (1960).
59. J. Schmorak and M. Lewin, *Anal. Chem. 33*, 1403 (1961).
60. M. Lewin, *Methods in Carbohydrate Chemistry*, Vol. 6. (R. L. Whistler and J. N. Bemiller, Eds.). Interscience Publ. New York, p. 76 (1973).

61. P. O. Bethge and P. T. Nevell, *Svensk Papperstidn.* 62, 236 (1959).
62. W. K. Wilson and A. A. Padget, *Tappi* 38, 274 (1955).
63. G. F. Davidson and T. P. Nevell, *J. Textile Inst.* 48, T356 (1957); 39, T102 (1948); 46, T407 (1955).
64. E. C. Ellington and C. B. Purves, *Can. J. Chem.* 31, 801 (1953).
65. R. L. Mitchel, *Ind. Eng. Chem.* 38, 843 (1946).
66. R. J. E. Cumberbirch and W. G. Harland, *Shirley Inst. Mem.* 31, 199 (1958).
67. R. I. Colbran and T. P. Nevell, *J. Textile Inst.* 49, T333 (1958).
68. M. Lewin and S. Weinstein, *Textile Res. J.* 37, 751 (1967).
69. M. Lewin and A. Ettinger, *Cell. Chem. Technol.* 3, 9 (1969).
70. M. Lewin and H. Ben-Bassat, Int. Techn. Textile Cotoniere, 1er, Paris, (CIRTEC), 1969, p. 535.
71. AATCC Test Method 82-1968: Fluidity of Dispersions of Cellulose from Bleached Cotton Cloth.
72. O. A. Battista, *Ind. Eng. Chem. Anal. Ed.* 16, 351 (1944).
72a. R. H. Marchesault and B. G. Ranby, *Svensk Papperstidn.* 62, 230 (1959).
73. G. F. Davidson and H. A. Standing, *J. Textile Inst.* 42, T141 (1951).
74. G. F. Davidson and T. P. Nevell, *J. Textile Inst.* 47, T439 (1958).
75. D. Mejzler, J. Schmorak, and M. Lewin, *J. Polymer Sci.* 46, 289 (1960).
76. T. P. Nevell, *J. Textile Inst.* 42, T91 (1951).
77. E. D. Kaverzneva, *Doklady Akad. Nauk S.S.S.R.* 68, 865 (1949).
78. J. Pinte and M. P. Rochas, *Bull. Inst. Textile France* 3, 139 (1947).
79. H. S. Isbell, *J. Res. NBS* 32, 54 (1944); *Ann. Rev. Biochem.* 12, 205 (1943).
80. M. Lewin, Mild Oxidat on of Cotton, Project FG-Is-169, Final Report submitted to the Agricultural Research Service, U.S.D.A. Jerusalem 1968.
81. R. L. Colbran and G. F. Davidson, *J. Textile Inst.* 52, 773 (1961).
82. H. Richtzenhain and B. Abrahamsson, *Svensk Papperstidn.* 57, 538 (1954).
83. O. Samuelson and A. Wennerblom, *Svensk Papperstidn.* 57, 827 (1954).
84. D. W. Haas, B. F. Hrutfiord, and K. V. Sarkanen, *J. Appl. Polymer Sci.* 11, 587 (1967).
85. M. Albeck, A. Ben-Bassat, J. A. Epstein, and M. Lewin, *Textile Res. J.* 35, 836 (1965); *Israel J. Chem.* 1(3a), 304 (1963).

86. M. Albeck, A. Ben-Bassat, and M. Lewin, *Textile Res. J. 35*, 935 (1965).
87. M. Lewin, *Textile Res. J. 35*, 979 (1965).
88. M. Lewin, Oxidation of Cellulose with Hydrogen Peroxide, Proc. Inter. Symposium on Delignification with Oxygen, Ozone, and Peroxide, Raleigh, NC. Tappi, Abstracts of Papers.
89. I. Ziderman, A. Basch, A. Ettinger, and M. Lewin, *Israel J. Chem. 7*, 90 (1969).
90. J. Bel Ayche, I. Ziderman, and M. Lewin, 41st Ann. Meeting of the Israel Chemical Society, Abstract of Papers, p. 178 (1971).
91. I. Ziderman, J. Bel Ayche, A. Basch, and M. Lewin, *Carbohydrate Res. 43*, 255 (1975).
92. M. Lewin, I. Ziderman, N. Weiss, A. Basch, and A. Ettinger, *Carbohydrate Res. 62*, 393 (1978).
93. I. Ziderman, *Carbohydrate Res. 81*, 196 (1980).
94. M. Lewin and I. Ziderman, *Cell. Chem. Technol. 14*, 743 (1980).
95. I. Ziderman, *Cell. Chem. Technol. 14*, 703 (1980).
96. M. Lewin and L. G. Roldan, *Textile Res. J. 45*, 308 (1975).
97. A. Meller, *Appita Proc. 7*, 263 (1953).
98. H. W. Giertz, *Svensk Papperstidn. 48*, 317 (1945).
99. I. H. Spinner, *Tappi 45*, 495 (1962).
100. H. E. Virkola, Y. Hentola, and H. Sihtola, *Papperi ja Puu 40*, 635 (1958).
101. I. D. Dean, C. M. Fleming, and R. T. O'Connor, *Textile Res. J. 22*, 609 (1952).
102. J. F. Haskins and M. J. Hogsed, *J. Org. Chem. 15*, 1264 (1950).
103. W. M. Corbett *Recent Advances in the Chemistry of Cellulose and Starch*. (J. Honeyman, Ed.). Heywood and Co., Ltd., London, pp. 106-134 (1959).
104. E. W. Montroll and R. Simha, *J. Chem. Phys. 8*, 721 (1940).
105. H. Wakeham, *Cellulose and Cellulose Derivatives, Part III*. (E. Ott, H. M. Spurlin, and M. W. Grafflin, Eds.). Interscience Publishers, New York, p. 1331 (1955).
106. E. B. Marum, *Tappi 39*, 390 (1956).
107. N. E. Virkola and O. Lehtikoski, *Paper and Timber 42*, 559 (1960).
108. N. E. Virkola and H. Sihtola, *Norsk Skogindustrie 12*, 87 (1958).
109. H. Sihtola, B. Antoni, Y. Hentola, I. Palenius, and N. E. Virkola, *Paper and Timber 40*, 11; 579 (1958).
110. I. Jullander and K. Brune, *Acta Chem. Scand. 11*, 570 (1957).
111. W. H. Rapson, C. B. Anderson, and G. F. King, *Tappi 41*, 442 (1958).
112. W. H. Rapson and K. H. Hakim, *Pulp and Paper Mag. Canada 58*, 7, 151 (1957).
113. W. H. Rapson and J. H. Spinner, *The Bleaching of Pulp*, 3rd Ed. (R. P. Singh, Ed.). Tappi Press, p. 357 (1979).

114. F. L. Browne, Theories of the Combustion of Wood and its Control, Forest Products Laboratory, Rept. No. 2136, Madison Wisconsin, 1958.

115. A. Meller, *Svensk Papperstidn.* 65, 629 (1962).

116. W. R. Rasch, *J. Res. NBS* 7, 465 (1931).

117. I. P. White, M. C. Taylor, and G. T. Vincent, *Ind. Eng. Chem.* 34, 782; 978 (1942).

118. G. A. Richter, *Ind. Eng. Chem.* 23, 371 (1931).

119. J. C. Tongren, *Paper Trade J.* 107, 76 (1938).

120. J. W. McIntire, *Paper Trade J.* 109, 317 (1939).

121. H. Sihtola, *Paper and Timber* 45, 71 (1963).

122. V. I. Ivanov and E. D. Stakheeva-Kaverzneva, *Usp. Khim.* 13, 281 (1944); cited in R. L. Whistler and I. N. BeMiller, *Adv. Carbohydr. Chem.* 13, 289 (1958).

123. O. Theander, *Acta Chem. Scand.* 12, 1887 (1958).

124. J. Defaye, H. Driguez, and A. Gadelle, *Carbohydrate Res.* 38, C4-C6 (1974).

125. Yu. A. Zhdanov. V. I. Minkin, R. M. Minjaev, I. I. Zacharov, and Yu. E. Alexeev, *Carbohydr. Res.* 29, 403 (1973).

126. R. L. Whistler and J. N. BeMiller, *Adv. Carbohydr. Chem.* 13, 289 (1958).

127. E. R. Garret and J. F. Young, *J. Org. Chem.* 35, 3502 (1970).

128. G. Machel and G. N. Richards, *J. Chem. Soc.* 1932 (1960).

129. P. S. Fredricks, B. O. Lindgren, and O. Theander, *Svensk Papperstidn.* 74, 597 (1971); *Acta Chem. Scan.* 21, 2895 (1967); *Tappi* 54, 1, 87 (1971).

130. G. M. Nabar, F. Scholefield, and H. A. Turner, *J. Soc. Dyers and Colour.* 53, 3 (1957).

131. S. H. Mhaire, G. M. Nabar, and G. M. Vyas, *Proc. Indian Acad. Sci.* 31, 234 (1950).

132. W. B. Achwal and S. A. Bhatt, *Indian J. Technol.* 5, 124; 129 (1967).

133. V. A. Shenai and J. K. Bhatt, *Indian J. Technol.* 7, 82 (1969).

134. V. A. Shenai and R. B. Prasad, *Textile Res. J.* 42, 603 (1972); 44, 591 (1974).

135. G. M. Nabar, V. A. Shenai, and V. G. Kulkarni, *Indian J. Technol.* 7, 360 (1969).

136. V. A. Shenai and O. P. Singh, *J. Soc. Dyers Colour.* 87, 228 (1971).

137. V. A. Shenai and K. D. Shah, *Textile Res. J.* 44, 214 (1974).

138. V. A. Shenai and K. K. Sharma, *J. Appl. Polymer Sci.* 26, 377 (1976).

139. V. A. Shenai and A. G. Date, *J. Appl. Polymer Sci.* 20, 385 (1976).

140. V. A. Shenai and A. S. Patil, *J. Appl. Polymer Sci.* 23, 123 (1979).

141. B. A. S. F. Ratgeber, *Cellulosefasern*, 1977.

142. R. L. Derry, *J. Soc. Dyers Colour.* 71, 884 (1955).

143. H. Borsten, *Text. Recorder 82*, No. 974, 71 (1964).
144. I. Rushnak, I. Kovacs, B. Losenczi, and J. Morgos, *Textil-veredlung 14*, 442 (1979).
145. P. B. Merkel and D. R. Kearns, *J. Am. Chem. Soc. 94*, 7244 (1972).
146. J. F. Rabek and B. Ranby, *Polymer Eng. Sci. 15*, 40 (1975).
147. R. Bloch, M. Lewin, and L. Farkas, *J. Am. Chem. Soc. 71*, 1988 (1949).
148. T. P. Nevel, *J. Textile Inst. 42*, T185 (1961).
149. M. Lewin, *Bull. Res. Council of Israel 2*, 101 (1952).
150. R. O. Griffin, A. McKeown, and A. G. Winn, *Trans, Faraday Soc. 28*, 101 (1932).
151. H. A. Liebhaftsky, *J. Am. Chem. Soc. 56*, 1500 (1934).
152. G. Jones and S. Backstrom, *J. Am. Chem. Soc. 60*, 490 (1938).
153. E. Shilov, *J. Am. Chem. Soc. 60*, 490 (1938).
154. L. Farkas and M. Lewin, *J. Am. Chem. Soc. 72*, 5766 (1950).
155. *C.R.C. Handbook of Chemistry and Physics.* (R. C. Weast and M. J. Astle, Eds.). 61st Ed. (1981-1982).
156. W. P. Lawrence, *Paper Trade J. 124*, 38 (1947).
157. M. C. Taylor, J. F. White, and G. P. Vincent, *Techn. Assoc. Papers 23*, 251 (1940).
158. H. W. Giertz, *Tappi 34*, 209 (1951).
159. M. Lewin and J. A. Epstein, 3rd Ann. Report on Project FG-Js-101-58, Mild Oxidation of Cotton, submitted to the Agricultural Research Service, U.S. Dept. of Agric., Jerusalem, Israel (1961).
160. A. Treinin, M. Sc. Thesis, Hebrew University, Jerusalem (1955).
161. R. Bloch, K. Goldschmidt, P. Goldschmidt, J. Schnerb, British Patent 596,192 (1948); 596,193 (1948); 615,604 (1949) (to Palestine Potash, Ltd.).
162. W. Sebb, *Textil Praxtis Intern. 163*, 266 (1981).
163. C. Kujirai and I. Fujita, *J.S.D.C. 78*, 80 (1959).
164. H. Baier, *Melliand Textilber. 32*, 141 (1951).
165. G. Holst, *Svensk Papperstidn. 31*, 23 (1945).
166. M. Hefti, *Textile Res. J. 30*, 861 (1960).
167. R. S. Higginbottom and R. A. Leigh, *J. Textile Inst. 53*, 312 (1962).
168. E. Doerfel, *Melliand Textilber. 38*, 285, 413 (1957).
169. R. Soila, O. Lehfikosti, and N. E. Virkola, Finnish Pulp and Paper Res. Inst. No. 287, 1963; *Svensk Papperstidn. 65*, 17, 632 (1962).
170. H. Hefti, *Textil Rundschau 11*, 82 (1956).
171. R. A. Leigh and L. Chesner, British Patent 1,071,494 (1967), (to Laporte Chemicals, Ltd.)
172. K. Fischer and S. Rosinger, German Patent 2,140,645 (1973) (to Hoechst A-G).

173. J. Balland, French Patent 4,141,685 (1979); *Teintex 44* (6-7), 13 (1979).
174. M. Tanemoto and H. Fukushima, Japanese Patent 76,82,080 (1976).
175. P. Buschman, *Melliand Textilber. 35*, 304 (1954).
176. M. G. Shiker, L. A. Rybkina, and N. P. Bolova, *Chem. Abs. 69*: 97641.
177. K. Ascik and K. Jablonska, *Przgl. Wlok. 30*, 33 (1976); Polish Patent 97,834 (1978).
178. J. Tourdot and J. Breiss, French Patent 1,466,436 (1967).
179. J. Haeusermann and E. Hablutzel, French Patent 1,541,184.
180. M. Uehara, Japanese Patent 74 96,982 (1974).
181. F. R. W. Sloan, H. C. Boyle, and A. Greenwood, British Patent 1,539,017 (1979).
182. Y. Sando and E. Nakano, Japanese Patent 76 67,472 (1976).
183. T. Shitamura and S. Hagihara, Japanese Patent 7003,311 (1970).
184. P. Mosse, *Teintex 19*, 811 (1954).
185. E. R. Trottman, *Textile Scouring and Bleaching*, Griffin Co., London (1968).
186. S. N. Soldushenkov and A. E. Shilov, *Zhur, Fizicheskoi Khimii 19*, 105 (1945).
187. J. Hirade, *Bull. Chem. Soc. Japan 10*, 92 (1935).
188. H. Taube and H. Dodgen, *J. Am. Chem. Soc. 71*, 3350 (1949).
189. F. Lenzi and W. H. Rapson, *Pulp and Paper Mag. Can. Tech. Soc. 63*, T442 (1962).
190. T. P. Forbath, *Chem. Eng. 180* (1961).
191. *Interox Technical Manual* (1979).
192. A. Agster, *Melliand Textilber. 59*, 11 (1978).
193. O. Deschler, *Melliand Textilber. 49*, 1301 (1968).
194. N. Warbel, U.S. Patent 2,711,363 (1955).
195. Belgian Patent 449,839 (to Degussa A. G.).
196. F. Kocher, *Bull Inst. Text. France 6, No. 31*, 19 (1952); *No. 32*, 19, (1952).
197. W. H. Rapson, *Paper Mill News 78*, 88 (1955); *Tappi 39*, 284 (1956).
198. W. H. Rapson, *The Bleaching of Pulp*. Tappi Monograph Series No. 27 (1963).
199. M. Sihtola and W. E. Virkola, *Paper and Timber 41*, 35 (1959).
200. O. Samuelson, *Das Papier 24*, 671 (1970).
201. W. H. Rapson and C. B. Anderson, *Transactions Techn. Sec. C.P.P.A. 3*(2), 52 (1977).
202. W. H. Rapson and G. B. Strumila, *The Bleaching of Pulp*. (R. P. Singh, Ed.). 3rd Ed. Tappi Press (1979).
203. W. B. Achwal and G. Shenkar, *J. Appl. Polymer Sci. 16*, 1791 (1972).
204. I. Ziderman, Abstr., 46th Ann. Meeting Israel Chem. Soc., Jerusalem, 1979, p. 84.

205. W. Ruttiger and U. Kirner, *Melliand Textilber.* *51*, 1075 (1970).
206. W. Triselt, *Melliand Textilber.* *51*, 1094 (1970).
207. W. D. Nicolls and A. F. Smith, *Ind. Eng. Chem.* *47*, 2548 (1955).
208. D. M. Cates and W. M. Cranor, *Textile Res. J.* *28*, 708 (1958); *30*, 848 (1960).
209. A. M. M. Taher and D. M. Cates, *Text. Chem. Color.* 7, 220 (1975).
210. W. G. Steinmiller and D. M. Cates, *Text. Chem. Color.* 8, 14 (1976).
211. F. Haber and J. Weiss, *Naturwissensch.* *20*, 948 (1932); *Proc. Roy. Soc. (Lond.), Ser. A 147*, 932 (1934).
212. W. G. Barb, J. H. Baxendale, P. George, K. R. Hargrave, *Trans. Faraday Soc.* *47*, 462 (1951).
213. H. J. H. Fenton, *J. Chem. Soc.* *65*, 899 (1894).
214. H. S. Isbell, M. L. Frush, R. Naves, and P. Suntracharden, *Carbohyd. Res.* *90*, 111-122 (1981).
215. L. Erdey, *Acta Chem. Sci. Hung.* *3*, 95 (1953).
216. W. C. Schumb, C. N. Satterfield, and R. L. Wentworth, *Hydrogen Peroxide*, Reinhold Publ. Co. (1955).
217. H. S. Isbell and H. Frush, *Carbohyd. Res.* *28*, 295 (1973).
218. A. Krause, *Ber.* *72*, 161 (1938).
219. W. C. Schumb, *Ind. Eng. Chem.* *41*, 992 (1949).
220. G. Bergman, *Textile-Praxis 23*, 261 (1968).
221. S. A. Simon and A. Derlich, *Textile Res. J.* *16*, 609 (1946).
222. N. Hartler, E. Lindahl, C. G. Moberg, and L. Stockman, *Tappi 43*, 806 (1960).
223. V. I. Ivanov, E. Kaverzneva, and Z. I. Kuznetsova, *Bull. Acad. Sci. U.S.S.R., Div. Chem. Sci.* *2*, 374 (1953); *Textil Prom.* *14*, 31 (1954).
224. H. S. Isbell, M. L. Frush, and E. T. Martin, *Carbohyd. Res.* *26*, 287 (1973).
225. H. S. Isbell and R. G. Naves, *Carbohyd. Res.* *36*, Cl (1974).
226. H. S. Isbell, H. L. Frush, and Z. Orhanovic, *Carbohyd. Res.* *36*, 283 (1974).
227. H. S. Isbell, E. W. Parks, and R. G. Naves, *Carbohyd. Res.* *45*, 197 (1975).
228. H. S. Isbell, H. L. Frush, and E. W. Parks, *Carbohyd. Res.* *51* C5 (1976).
229. H. S. Isbell and H. L. Frush, *Carbohyd. Res.* *59*, C25 (1977).
230. H. S. Isbell and H. L. Frush, *Carbohyd. Res.* *72* 301 (1979).
231. W. S. Wood and K. W. Richmond, *J. Soc. Dyers Color.* *68*, 357 (1952).
232. J. G. Wallace, *Hydrogen Peroxide in Organic Chemistry.* E. I. duPont de Nemours & Co., Wilmington, Del. pp. 60-72 (1962).
233. W. A. Prior, *Free Radicals.* McGraw-Hill, New York, pp. 136-138 (1966).

234. A. S. Ramadan, *J. Textile Inst.* *51*, T215 (1960); *J. Soc. Dyers Color.* *75*, 384 (1959).

235. E. P. Bayha, L. H. Hubbart, and W. H. Martin, Report 32, Inst. of Textile Technology, Charlotteville, Va., 1964; *Intern. Dyer* *131*, 529 (1964).

236. D. V. Parikh, V. G. Agnihotri, and S. S. Kalra, *Text. Ind.* *132*, 251 (1969).

237. Y. Kobliakov, I. Ziderman, J. Bel-Ayche, and M. Lewin, Israel Fiber Institute, unpublished results.

238. M. Chilikin, *Bull. Feder. Int. Assoc. Chem. Text. Color.* *5*, 367 (1935); *Melliand Textilber.* *18*, 365 (1937).

239. E. Kornreich, *Melliand Textilber.* *19*, 61 (1938).

240. H. R. Mauersberger, *Matthews Textile Fibers*, 5th Ed. Wiley and Sons, New York, p. 266 (1952).

241. V. S. Strel'tsov, M. V. Korchargin, and G. A. Bogdanov, *Teknol. Tekst. Prom.* *(6)* 81 (1975); *Chem. Abs.* *84*; 106948q (1977).

241a. D. R. Solani, *Man-Made Text. India 21*, 590; 608 (1978); *22*, 39 (1979).

242. W. Ney, *Textil Praxis Int.* *29*, 1392, 1552 (1974).

243. N. A. Soldatkina, Z. N. Dymova and G. A. Bogdanov, *Tekst. Prom.*, (Moscow) *1*, 56 (1975); Chem. Abs. 126471r (1975).

244. E. Gottlieb, *Am. Dyestuff Rep.* *69*, 20 (1980).

245. A. Sigg, *Textil Rundschan 11*, 391 (1956).

246. X. Kowalski, *Am. Dyestuff Rep.* *68*, 49 (1979); X. Kowalski *Text. Chem. Color.* *10*, 215 (1978).

247. J. F. Leuck, *Am. Dyestuff Rep.* *68*, 49 (1979).

248. L. Chesner, *J. Soc. Dyers Color.* *79*, 140 (1963).

249. L. Chesner and G. C. Woodward, *J. Soc. Dyers Color.* *74*, 531 (1958).

250. J. V. Butcher, *Rev. Prog. Color.* *4*, 90 (1973).

251. V. V. Safonov, G. A. Bogdanov, and M. V. Korchagin, *Teknol. Tekst. Prom.* *(2)*, 84 (1973); *Chem Abs.* *79*; 93308a.

252. O. Oldenroth, *Seifen, Ole, Fette, Wachse 93*, 371 (1967).

253. M. Rupin, Book of Papers, AATCC National Techn. Conf., 197 (1979).

254. G. Teppich, *Textilverdl. 11*, 156 (1976).

255. W. Bachman, *Melliand Textilber.* *49*, 449 (1968).

256. E. P. Frieser, *Spinner Weber Textilveredl.* *82*, 32 (1964).

257. K. Dickinson and T. J. Thompson, *Am. Dyestuff Rep.* *67*, 19 (1980).

258. M. Rowe, *Text. Chem. Color.* *10*, 215 (1978).

259. AATCC Southeastern Section, *Text. Chem. Color.* *14*, 10 (1982).

260. K. Dickson, R. R. Kindron, and P. Curzons, Book of Papers, AATCC Natl. Techn. Conf., 100 (1980); U.S. Patent 3,026,266 (1962).

261. L. A. Sitver, B. K. Easton and R. E. Yelin, *Text. Chem. Color. 5*, 79 (1973); U.S. Patent 3,765,834 (1973).

262. British Patent 1,184,940 (1966).
263. J. K. Skelly, *J. Soc. Dyers Colour.* 76, 469 (1960).
264. L. Chesner and R. A. Leigh, *Textil. Rundsch.* 20, 217 (1965).
265. C. Garrett, *Text. Manuf.* 91, 51 (1965).
266. Japanese Patent 81-73,166 (1979), *Chem. Abs.* 95: 134325e.
267. A. Robert, P. Perolle, A. Viallett, and O. Martin-Bovat, *ATIP Bull.* 18 151-176 (1964).
268. A. Robert, A. Viallet, P. Perolle, and J. P. Andreoletti, *Revue ATIP* 20 207 (1966); *Paper Trade J.* 152, 49 (1968).
269. M. Manouchehri and O. Samuelson, *Svensk Paperstidn.* 76, 486 (1973).
270. J. Defaye, H. Driguez, and A. Gadelle, *Appl. Polymer Symp.* 28, 555 (1976).
271. A. F. Gilbert, E. Pavlova, and W. H. Rapson, *Tappi* 56, 95 (1973).
272. A. Robert, P. Traynard, and O. Martin-Borret, U.S. Patent 3,384,533 (1968); A. Robert and A. Viallet, *ATIP* 25, 238 (1971).
273. R. Freitag, *Colourage* 35 (1975).
274. R. Freitag, *Melliand Textilber.* 51, 72 (1970).
275. R. Freitag and J. Diemunsch, *Deutsche Textiltech.* 21, 162 (1971).
276. I. L. Moore, *Am. Dyestuff Rep.* 57, 27 (1968).
277. O. Schmidt, *Z. Ges. Textilind.* 66, 849 (1964).
278. U.S. Patent 2,955,905; British Patent 836,988 and 839,715
279. German Patent 1,162,967.
280. I. E. Lynn, *Am. Dyestuff Rep.* 58, 20 (1969).
281. W. Kueppers, *Melliand Textilber.* 51, 1069 (1971).
282. D. Better, *Intern. Dyer Printer* 292 (1978).
283. A. M. Sookne, C. A. Rader, U.S. Patent 3,416,879 (to Union Carbide Corp.) (1968).
284. W. Kueppers, *Melliand Textilber.* 51, 1069 (1970).
285. J. Despierre and J. P. Hauger, *Melliand Textilber.* 49, 1187 (1968).
286. *Bleaching Manual*, Laporte Industries, Ltd. (1968).
287. V. G. Agnihotri, *Colourage* 23, 49 (1976).
288. N. F. Crowder and W. A. S. White, *J. Soc. Dyers Colour.* 71, 764 (1955).
289. A. G. Fisher, *Text. Chem. Color.* I. 1079 (1969).
290. C. E. Harrison and H. I. Welch, Book of Papers, AATCC Natl. Tech. Conf. 307 (1978).
291. E. R. Trotman, *Dyeing and Chemical Technology of Textile Fibres*. Griffin Publ., London (1970).
292. Japanese Patent 75-31,204.
293. German Patent 1,292,128.
294. E. Gruber, *Melliand Textilber.* 60, 435 (1979).
295. J. Dyer and L. H. Phifer, *J. Polym. Sci. (Part C)* 36, 103 (1971).

296. L. A. Gotovtseva, N. L. Zabotina, L. I. Konovalova, and V. Y. Demidova, *Chem. Abs.* *84*, 137068g (1974).

297. F. Kocevar and H. Talovic, *Melliand Textilber.* *54*, 1064 (1973).

298. A. A. Burinskaya, V. M. Bel'tsov, and E Akim, *Tekhnol. Tekst. Prom.* *4*, 78 (1974); *6*, 82 (1974).

299. A. P. Lazareva, L. A. Gotovtseva, M. A. Ogurova, N. A. Zabavina, I. V. Trifonova, and V. G. Ermilov, *Tekst. Prom. (Moscow)* *12*, 67 (1976).

300. G. A. Bogdanov, Z. N. Dymova, S. V. Petrov, B. V. Emel'yanov, and A. F. Shiskma, *Chem. Abs.* *83*, 11985q (1973).

301. B. Kudziene, *Chem. Abs.* *68*, 88108d (1966).

302. Z. K. Naryshkina, F. M. Fisher, and E. I. Raiskina, *Tekst. Prom. (Moscow)* *32 (1)*, 77 (1972); *Chem Abs.* *76*, 142254d.

303. V. M. Bel'tsov and I. V. Kalaus, *Chem. Abs.* *78*, 44898p.

304. Japanese Patent 72-33, 034, *Chem. Abs.* *78*, 98988m (1972).

305. M. Lewin, *Tappi 41*, 8 (1958).

306. R. R. Mukherjee and T. Radhakrishan, *Text. Progr.* *4*, 54-63 (1972).

307. F. R. W. Sloan, *Rev. Progr. Color.* *5*, 12 (1974).

308. A. Y. Bedina, O. G. Vishnyakova, and G. F. Chankova, *Tekst. Prom. (Moscow) (9)*, 64 (1974).

309. E. Weidemann and L. T. Van der Walt, SAWTRI Techn. Rep. No. 434 (1978).

310. E. Weidemann and H. Grabherr, SAWTRI Techn. Rep. No. 365 (1977).

311. M. Van Lancker, *Ann. Sc.: Text. Belg.* *21*, 278 (1973).

312. E. Bonte, *Bull. Inst. Text. France 24*, 443 (1970).

313. K. Poklewska, *Prace Inst. Przem. Wlok.* *14*, 223 (1968).

314. Japanese Patent 73-30,987 (1973); U.S. Patent 3,281,202 (1966).

315. V. I. Lebedeva, *Tekhnol. Tekstil. Prom. (1)*, 113 (1969); Vi. I. Lebedeva and P. V. Moryganov, *Tekhnol. Tekstil Prom. (3)*, 90 (1968).

316. E. Bonte and R. Croain, *Rev. Inst. Techn. Ronbaisien*, *58*, 31 (1967); *59*, 31 (1968); *Chem. Abs.* *72*, 134001w.

317. J. Zielinska, *Przegl. Wlok.* *28*, 489 (1974).

318. K. H. Rucker, *Melliand Textilber.* *51*, 1083 (1970).

319. I. Lambrinoû, *Melliand Textilber.* *52*, 1184 (1971); *51*, 815 (1970).

320. British Patent 1,539,017 (1979).

321. British Patent 1,266,896 (1968).

322. J. P. Bruggeman, *Bull. Inst. Text. France 25*, 875 (1971).

323. V. M. Bel'tsov, A. M. Zadevalkina, and M. A. Sekushina, *Tekhnol. Tekstil Prom. (4)*, 98 (1973); *Tekstil Prom. (Moscow) (5)*, 59 (1976); V. M. Bel'tsov, A. Zadevalkina, and M. Kalaus, *Teknol. Tekstil Prom. (5)*, 84 (1974).

324. British Patent 1,181,850 (1970).

325. B. V. Emel'yanov, A. I. Kuznetova, L. M. Klotatova, and M. V. Osipova, *Tekhnol. Tekstil Prom. (1)*, 70 (1978); *Chem. Abs. 89*: 112051m.

326. D. B. Das, M. K. Mitra, and J. F. Wareham, *J. Text. Inst.* *43*, T443; 449 (1952).
327. F. Tucci, *Ciba Review 108*, 3911 (1955).
328. P. S. Patro, *Text. Dyer Printer 4*, 57 (1971).
329. G. Rukhana, *Text. Dyer Printer 1*, 41 (1968).
330. M. H. Rahaman and M. M. Huque, *Pakistan J. Sci. Indust. Res. 13*, 303 (1970).
331. M. A. Salam, A. B. M. Abdullah, and N. B. Khan, *Bangladesh J. Jute Fibre Res. 13 (1-4)*, 181 (1978); *11 (1)*, 57 (1977); *Chem. Abs. 90*:138961d; *92*:60237s.
332. M. Kabir and N. G. Saha, *Bangladesh J. Sci. Ind. Res. 12 (1-2)*, 91 (1977); *Chem. Abs. 87*:169092.
333. U.S. Patent 3,521,991 (1967).
334. U.S. Patent 3,472,609 (1968); 3,384,444 (1964).
335. H. Uzdowsk; *Przegl. Wlok. 21*, 567 (1967); *Chem. Abs. 68*: 79714w.
336. Japanese Patent 77-121,579 (1977).
337. A Venkateswaran, *Tappi 48*, 191 (1965).
338. Anon., *Bayer Farben Revue S16*, 11 (1977).
339. A. Liddiard, *Textilveredl. 5*, 93 (1970).
340. G. Schlichtmann, *Textilveredl. 13*, 227 (1978).
341. H. Beckmann, Chemifasern, *Textilind. 28*, 180 (1978); E 43 (1978).
342. R. K. Narkar and A. K. Narkar, *Text. Dyer Printer 7*, 45 (1974).
343. A. Wlochowitcz and Z. Malinowska, *Przegl. Wlok. 32*, 584 (1978).
344. J. Skoufik, *Am. Dyestuff Rep. 68*, 20 (1979).
345. F. Dziwisz and L. Skrobecki, *Przegl. Wlok. 29*, 448 (1978).
346. Japanese Patents 73-01, 988; 73-01, 989 (1973).
347. Japanese Patent 71-26,428 (1971).
348. Japanese Patent 73-02,912 (1973).
349. I. V. Butcher, *Rev. Prog. Color. 4*, 90 (1973).
350. V. G. Agnihotri, *Colourage 22*, 25 (1975).
351. S. Shaw and W S. Wilson, *J. Soc. Dyers Colour 71*, 857 (1955).
352. Japanese Patent 73-02,913 (1973).
353. W. Schürings, *Textilveredl. 5*, 81 (1970).
354. British Patent, 1,148,374 (1969).
355. H. U. Schmidlin, *Preparation and Dyeing of Synthetic Fibres*. Chapman and Hall, London (1963).
356. H. Hantge and K. G. Rucke, *Textilveredl. 1*, 307 (1966).
357. K. Kithmar, *Melliand Textilber. 48*, 341 (1967).
358. M. Tourdot, *Teientex 37 (2)*, 67 (1972).
359. French Patent, 1,577,134 (1969) (Degussa).
360. Japanese Patent, 70-38,083; 71-37,383 to 71-37,385 (1971); 72-26,990 to 72-26,996 (1972); 74-30,229 (1974); 74-87,868 (1974); 75-14,876 (1975).

361. Japanese Patent 72-19,612; 72-19,613 (1972).
362. O. Schmidt, *Textil-Ind.* 73, 588 (1971); *Colourage 24 (18A)*, 55 (1977).
363. BASF Techn. Leaflet, M5431e (1975).
364. H. Hefti, *Fluorescent Whitening Agents: Environmental Quality and Safety*, Supp. Vol. IV. (F. Coulston and F. Korte, Eds., R. Anlike and G Muller, Guest Eds.). Thieme Publishers, Stuttgart; Academic Press, New York, pp. 53-55 (1975).
365. H. J. Weber, *Textilveredl.* 5, 114 (1970).
366. W. Blume, *Textilveredl.* 4, 88 (1969).
367. W. Guth, *Melliand Textilber.* 8, 715 (1981).
368. A. Decorte, *Text. Chim.* 27 (4), 4 (1971); *Chem Abs. 75:* 37735t.
369. Jap. 72-42,874 (1973); U.S. 3,790,343 (1974).
370. *Bayer Farben Revue*, Special Edn. 16, 38 (1977).
371. U.S. 3,579,287 (1971); (Monsanto); V. M. Bel'tsov, A. A. Burinskaya, and L. G. Shibleva, *Chem. Abs. 90*, 1054912 (1977).
372. German Patent 1,277,797 (1968) (Degussa).
373. Japanese Patent 73-20,975 (1973).
374. Japanese Patent 73-01,472; 73-01,473 (1973).
375. M. Drewniak, *Am. Dyestuff Rep. 68*, 45 (1979).
376. J. W. Davis, *J. Soc. Dyers Colour. 89*, 77 (1973).
377. G. Dierkes, *Textilbetrieb 99 (12)*, 56 (1981).
378. H. G. Smolens, *Text. Ind. 130 (9)*, 164 (1966).
379. J. Diemunsch and R. Freytag, *Bull. Sci. Inst. Text Fr. 3 (9)*, 37 (1974).
380. W. A. Millsaps, Book of Papers, Nat'l. Tech. Conf. AATCC, 72 (1977).
381. U. Kirner, *Tinctora 68*, 397 (1971).
382. B. K. Easton, *Text. Chem. Color 1*, 592 (1969).
383. T. S. Sambasivan, *Colourage 20 (7)*, 30a (1973).
384. B. D. Baehr, *Textilveredl. 11*, 145 (1976).
385. M. H. Rowe, *Text. Chem. Color 3*, 170 (1971).
386. B. K. Easton, *Am. Dyestuff Rep. 51* 499 (1962).
387. H. Beckman and R. Braun, *Textilveredl. 11*, 336 (1976).
388. Japanese Patent 79-27,098; German Patent 2,022,929 (1970); 2,148,702 (1972); 2,157,061 (1973); U.S. Patent 3,599,603 (1971).
389. G. Weiss, *Textiltechnik 23*, 438 (1972).
390. P. Gruenig, *Dtsch. Farber-Kal. 80*, 48 (1976).
391. J. Mazenauer, *Ciba-Geigy Rev.* 1973 (3), p. 36.
392. H. Beckman, *Chemiefasern Textilind. 28*, 180 (1978); E 43 (1978).
393. Anon., *Ciba Rev.*, 1966/3, p. 42.
394. C. Schork and R. Jenny, *SVF Fachorgan 30*, 33 (1965).
395. W. Schürings, *Textil-Prax. Int. 29*, 1570, 1581 (1976).
396. N. E. Houser, J. C. Martin, and M. White, *Am. Dyestuff Rep. 76*, 19 (1981).

397. A. A. Vaidya and R. R. Shah, *Text. Dyer, Printer* 7, (1) 49 (1973).
398. B. L. Neal, *Am. Dyestuff Rep.* 66, 36 (1977).
399. A. Hebeish, A. Kantouch, A. Bendak, A. Z. Moursi, and A. El-Torgoman, *Cell. Chem. Technol.* 11, 675 (1977).
400. French Patent 2,168,980 (1973).
401. German Patent 2,327,771 (1973).
402. Japanese Patent 74-75,870 (1974).
403. German Patent 2,129,772 (1972).
404. Japanese Patent 73-01,990; 73-01,991 (1973).
405. J. Varghese, D. M. Pasad, and P. B. Patel, *Colourage 24*, 23 (1977).
406. G. W. Madaras and C. G. Robinson, *J. Soc. Dyers Colour.* 89 317 (1973).
407. L. Martin and G. Oehler, *Textile Asia 4 (6)*, 38 (1973).
408. S. Bartczak, *Prezegl. Wlok.* 24, 68 (1970).
409. V. M. Bel'tzov, L. G. Shibleva, A. A. Burinskaya, and V. N. Miyasnikova, *Chem. Abs.* 89:130911w (1978).
410. V. M. Bel'tzov, A. A. Burinskaya, L. G. Shibleva, and G. P. Ivanova, *Przegl. Wlok.* 33, 685 (1979); *Tekst. Prom.* (Moscow), (11), 59 (1979); *Chem Abs.* 92 60239u.
411. German Patent 2,242,380 (1974).
412. Anon., *Bayer Farben Revue, Special Ed.* 16, 20 (1977).
413. Japanese Patent 76-43,481; 72-27,828, 72-27,827; 73-03,358, 73-03,335.
414. A. Nikolova and D. Todorova, *Tekst. Prom.* (Sofia) 20, 146 (1971); *Chem. Abs.* 75:152828n.
415. A. H. Little, *Water Supplies and the Treatment and Disposal of Effluents.* Textile Inst. Monograph Ser. No. 2, The Textile Institute, Manchester (1975).
416. A. H. Little, *J. Soc. Dyers Colour.* 83, 268 (1967).
417. S. S. Gopujkar, *Int. Dyer, Printer 144*, 47 (1970).
418. E. A. Kleinheidt, H. Theidel, and R. Aenishaenslin, *Fluorescent Whitening Agents, Environmental Quality and Safety*, Supplement Vol. IV. (F. Coulsten and F. Korte, Eds., R. Anlike and G. Miller, Guest Eds.). Thieme, Stuttgart; Academic Press, New York, pp. 47-50 (1975).
419. C. Cole, S. Corr, and J. Albert, *Am. Dyestuff Rep.* 66, 30 (1977).
420. H. Foerster, *Umweltforschung.* (H. Matthöfer, Ed.). Umschau, Frankfurt, pp. 108-118 (1976). (*Chem. Abs.* 90:76090h).
421. A. Zahn, *Melliand Textilber.* 56, 477 (1975).

3

THE FLUORESCENT WHITENING
OF TEXTILES

RAPHAEL LEVENE / Israel Fiber Institute, Jerusalem, Israel

MENACHEM LEWIN / Israel Fiber Institute and Hebrew University ersity, Jerusalem, Israel

1. INTRODUCTION

Fluorescent whitening agents (FWAs) [1,2] act as fluorescent dyes, though they are colorless. They absorb light in the near ultra-violet (UV) region of the spectrum, below about 400 nm, and re-emit the light, as fluorescence, in the violet-blue visible region (Fig. 3.1, curve 1) [3]. Near-white polymeric substrates, e.g., bleached cloth (curve 2), possess a yellowness caused by absorption in the blue region. When a FWA is applied, the blue fluorescence complements the yellowness and adds a bluish hue to the substrate, which the eye appreciates as a brilliant white. The whiteness thus obtained is much superior to that achieved by chemical bleaching, even if the latter is carried out to the limit where the cloth is weak-ened by chemical degradation, or if chemical bleaching is supplement-ed by blue tinting (curve 3). However fluorescent whitening com-plements chemical bleaching and cannot replace it. A well-bleached substrate is essential for fluorescent whitening to be effective.

The chief uses for FWAs continue to be in washing powders and on paper, which together account for about 75% of the total [4]. FWAs are universally incorporated into washing powders to compen-sate for the small loss that occurs during washing. However, their use in the finishing of white textile goods, particularly cotton, is

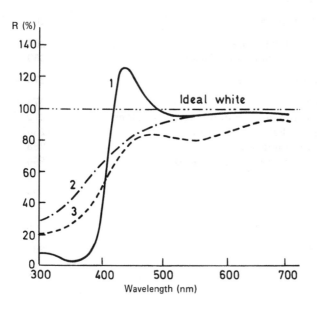

Figure 3.1 Reflectance and fluorescence of chemically bleached cloth treated with FWA (curve 1). Reflectance of chemically bleached (curve 2) and chemically bleached and blue-tinted cloth (curve 3). (From Ref. 3.)

very important. Though this usage represents only about 20% by weight of the total, the percentage calculated on a monetary basis is much higher. About one third of all textile goods are whites. Over 60% of this third were estimated in 1971 to be cotton [5]. Goods to be dyed or printed, some two thirds of the whole, are not usually treated with FWA. A small additional use of FWAs is in the mass whitening of synthetic fibers during manufacture or during melt or solvent spinning.

FWAs resemble dyes in that they must be diffused into and fixed to the fiber being treated. Otherwise the effect will not be fast to household washing. More important, however, unless the FWA is substantive to the fiber, little or no enhancement of the whiteness is achieved. This explains the multiplicity of FWAs for textile use. As with dyes, only distinct chemical types of FWAs can be applied to a particular fiber. For cotton and regenerated cellulose fibers, the equivalent of direct, anionic dyes is universally used. Acrylic fibers are whitened with cationic FWAs, while for polyester and cellulose acetate nonionic disperse-type whiteners are applied. The amount applied is much less than for dyes. Maximum whiteness can usually be achieved with less than 5 mg FWA per gram of fabric, and more than this will often spoil the whitening effect.

All the different techniques of dyeing are also applied to fluorescent whitening. These techniques are adapted to the treatment of yarn or of woven or knitted cloth, and to batchwise or continuous processes. Each process requires different properties of the FWA to be used, leading to a further variety of types. In contrast to dyeing, where the process is adjusted to give optimal results for a particular dye-fiber combination, fluorescent whitening is almost always carried out together with another process. An FWA may be applied together with chemical bleaching, or it may be incorporated in a resin finishing bath. Thus a further requirement of FWAs for textile use is that they be compatible with chemicals used in these other processes.

The designation *fluorescent whitening agent* (FWA) will be used here [6,7], though names such as optical brighteners or bleaches (OBs), fluorescent brightening or bleaching agents (FBAs), and optical whiteners are in common use.

FWAs have been the subject of recent monographs [1,2,8,9]. The first of these references includes a bibliography covering all aspects up to 1973.

2. PHYSICAL BASIS OF ACTION OF FWAs

2.1 Fluorescence

When a body is viewed under white light, it may in theory reflect all the incident light, in which case it will appear to be white

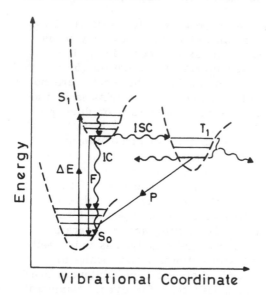

Figure 3.2 Fluorescence and vibrational relaxation. F = fluorescence, ISC = intersystem crossing, IC = internal conversion, P = phosphorescence. Radiative transition →. Nonradiative transition ⤳→.

(Fig. 3.1). However, an ideal white body cannot be achieved: materials absorb part of the incident light to a greater or lesser extent, and by reflecting the rest, appear colored.

When a dye molecule absorbs a photon of light, an electron is excited from the singlet ground state, S_0, to a higher electronic energy level, particularly the first level, S_1. This is shown in Fig. 3.2, which is schematic only and does not accurately represent energy levels. If the difference between the molecular electronic states S_1 and S_2 is ΔE, the wavelength (λ) of the absorbed light is given by hc/λ where h and c have their usual meanings.

In the simplest case of a diatomic molecule A−B, excitation of one of the pair of bonding electrons causes a decrease in electron density between the atoms accompanied by a lengthening of the bond. According to the Franck-Condon principle, the electron transition is so fast that there is no time for an adjustment in bond length. The molecule is thus in a vibrationally excited (compressed) state (Fig. 3.2). However, vibrational relaxation takes place very rapidly (in about 10^{-12} sec), and the molecule settles into the vibrational ground state of S_1, losing thermal energy to its surroundings. Fluorescence occurs when the electron returns to the ground state S_0 by a reversal of the process of energy absorption. The energy emitted will be less than that absorbed, due to the vibrational relaxation which occurred in the excited state, and also since the molecule may not return at once to the vibrational ground state of S_1. Thus the fluorescent emission is of higher wavelength than that of the absorption (Stoke's Law).

Quenching of Fluorescence

Fluorescence is not the main route for de-excitation. More usually the energy absorbed by a dye molecule is dissipated as heat among the surrounding molecules, the overall process being the absorption of light. The molecule in its excited state S_1 is deactivated by non-radiative processes leading to fluorescent quenching. In one such process, *internal conversion* (IC), the molecule relaxes from S_1 by a transition into a very high energy vibrational level of $S_0(S_1 \rightarrow S_0)$. (Fig. 3.2). Another process, termed *intersystem crossing* (ISC) into the triplet state (T_1), occurs when the spin of the excited bonding electron becomes reversed $(S_1 \rightarrow T_1)$. According to Hund's rule, the triplet state T_1, in which the spins of the electrons are parallel, is of lower energy than the corresponding singlet state S_1, in which the spins are in opposite directions. Radiative relaxation $(T_1 \rightarrow S_0)$ from the triplet state is a slow process (phosphorescence), so that a nonradiative relaxation, involving the dissipation of thermal energy, is usually preferred. These nonradiative transitions involved in fluorescent quenching are not only associated with transfer of energy to vibrational and rotational energy levels within the excited molecule itself; they may also involve, particularly where the long-lived triplet state is concerned, a long-range energy transfer process involving collision with neighboring molecules. The rate of this process is diffusion controlled, and it is active over distances up to about 5 nm [10,11]. The relation between the energy absorption by a molecule to that emitted as fluorescence is called the fluorescent quantum yield or fluorescent efficiency.

Molecular Structure

In order that the light energy absorbed in the transition $S_0 \rightarrow S_1$ should be in the near UV region of the spectrum, the molecule must contain an extended system of conjugated double bonds [8,13]. Absorption of light by the FWA molecule causes excitation of a π electron from the ground state to a low-lying excited energy state. For high fluorescent efficiency, this ππ* transition should have lower energy than that of an nπ* transition followed by intersystem crossing to the triplet state [12]. An nπ* transition involves a nonbonding electron, from a nitrogen atom or a carbonyl group in the molecule.

A requirement for fluorescence is that the extended conjugated system (the chromophore) should lie in one plane. Restrictions to free rotation should not therefore render it difficult for the chromophore to assume a planar configuration. Molecular factors affecting fluorescent efficiency have been investigated in relation to fluorescent dyes used in dye lasers [14,15]. A desirable feature for high fluorsecent efficiency is complete rigidity of the planar structure. This is particularly well illustrated by comparing the dianion of phenolphthalein (nonfluorescent) with the dianion of fluorescein (highly fluorescent). The oxygen bridge in the latter ensures molecular rigidity [16]. Further, it has been shown that factors which reduce the flexibility, when in solution, of an essentially planar molecule, enhance the fluorescent effect [14,15]. Reduced mobility of the molecule reduces the possibility of fluorescent quenching associated with vibrational relaxation. Both longitudinal and rotational modes of vibration are involved, the two being coupled.

Penetration into the Fiber

A rigid molecular structure is nevertheless, not found to be an important factor for an FWA molecule itself. However, fixation of the FWA molecule to the polymer substrate is essential in order to achieve optimum fluorescence and whiteness. In direct dyes which are similar in their constitution and action to FWAs used for cotton, fixation occurs by attachment of the dye, of linear structure and in a planar conformation, parallel to the flat linear cellulose chains [17]. Planarity is thus not only important in relation to fluorescent efficiency but also ensures high affinity to the substrate.

When an FWA for cellulose is padded onto cotton cloth by impregnation and the cloth then dried immediately, very little fluorescent whitening occurs. Optimum whiteness develops only on holding the cloth in the wet state for a period of time, the time required being shorter at higher temperatures [18-20]. During this period the FWA molecules penetrate into the fiber and become attached to the cellulose chains. Whatever the mechanism of the fixation process [17], it seems likely that once the FWA molecule is fixed to the substrate it is held there in a fairly rigid planar conformation, thus retarding fluorescent quenching associated with longitudinal and rotational modes of vibrational relaxation processes [18]. Similarly, fluorescence and whiteness enhancement associated with the deposition of an FWA for cellulose on cellophane film [21] and in the presence of polyvinyl alcohol [19,21] or a surface-active agent [22] may be attributed to increased rigidity in a planar conformation, of the FWA molecule.

In the whitening of polyester, penetration of the FWA into the fiber is also important for complete development of whiteness [23-25]. The mechanism of fluorescent enhancement is possibly similar to that with cellulose. When polyester fibers are mass-whitened at low concentrations with the FWA 4,4'-bis(benzoxazol-2-yl)stilbene, the fluorescent emission is polarized [24,26]. This is explained by assuming that the fluorescent molecules align themselves preferentially along the polymer chains in the amorphous regions of the polymer [26]. The importance of good penetration of FWA into the fiber has also been stressed in the case of polyacrylonitrile [29]. The behavior of FWAs in this respect is in contrast to that of dyes. Though good penetration of the fiber by dyes ensures good washfastness and fastness to rubbing, the effect on color yield is relatively marginal, particularly in the case of natural fibers, which are opaque [28].

External Quenching Effects

Quenching of fluorescence can be caused by long-range energy transfer from an excited state of the fluorescent (donor) molecule to an acceptor (absorbing) molecule [10,11] if the transmission spectrum of the donor chromophore overlaps with the absorption spectrum of the acceptor chromophore. Good chemical bleaching of the substrate is essential for the whitening effect of the FWA to be efficient. Residual yellowness causes quenching by providing molecules which absorb energy at the wavelength of the fluorescent emission. In addition, yellow discoloration on the fabric has an absorption tail extending into the near ultraviolet region of the spectrum (Fig. 3.1), and this absorption will compete with the FWA for the irradiating energy. Any material which is yellow, or which absorbs in the near UV region of the spectrum [22] must be avoided in a process of fluorescent whitening.

2.2 Optical Considerations

Pastel Shades

Strongly dyed fabrics are unaffected by FWA since dye absorption in the UV and blue regions of the spectrum are sufficiently strong to cause quenching of the fluorescence. It is not usual to include FWA pretreatment on cloth which is to be dyed. However pastel shades may have only a small effect on the fluorescent process. Since all washing powders contain FWA, home washing of a pale-colored garment can cause an undesirable change of shade due to the blue fluorescent emission. This may be avoided by preliminary fluorescent whitening of the cloth [29], or by the use of mass-whitened synthetic fibers [30,31], when dying with pastel shades is contemplated. A possible change in hue with respect to the same pastel shade when dying on a nonfluorescent fabric should be taken into account [37]. FWAs improve the brightness of pastel shades [30].

Greening Effect

An unpleasant green discoloration is obtained if an FWA is badly applied under conditions where it is precipitated on the fiber without

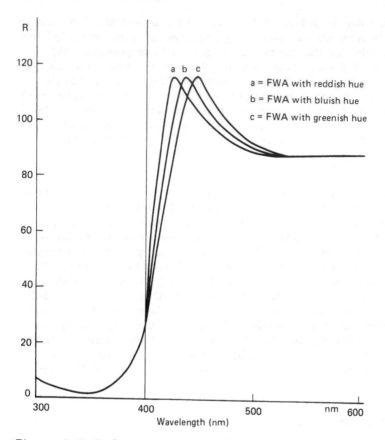

Figure 3.3 Reflectance curves of three FWAs of different shades. (From Ref. 33.)

penetration [23,32]. This occurs in the case of cellulose when the FWA is applied at too low a pH, and alkaline scouring will then cause the full whiteness to develop [32]. The green effect is attributed to aggregation of FWA molecules.

Hue

The wavelength of maximum fluorescence of the FWA affects the hue of whitened cloth. The hue varies between reddish for the shorter wavelength emissions, through a neutral blue (the most preferred), to a greenish tinge at longer wavelengths of emission (Fig. 3.3) [33]. Manufacturers sometimes include the letters R or B in their trade names to indicate the reddish or blueish shade which the FWA imparts. FWAs of a greenish hue are less in demand, though when used in conjunction with a reddish-hued whitener, a neutral blue shade is obtained. Blue-violet shading dyes are often used in conjunction with FWAs, thus producing an even higher level of whiteness [34,35].

Fluorescence-Concentration Relationship

When an FWA is applied under standard conditions to a textile substrate at technically useful rates of application, the fluorescent intensity is found to be directly proportional to the log of concentration (c) of the FWA on the fabric. Fluorescent intensity is usually measured as the total radiance factor (TRF), which is the sum of the reflectance of the cloth and the fluorescent emission. Thus

$$TRF = a \log c + b \qquad (1)$$

where a,b are arbitrary constants [36]. Under strictly standard conditions of whitening and illumination of a particular bleached fabric, using the same FWA, whiteness is directly proportional to fluorescent intensity.

Equation (1) does not hold at very low rates of application, below about 0.2 mg/g for cotton whitened with FWA (Fig. 3.4). At high levels (above about 5 mg/g cotton), a state of saturation is achieved where further increase in concentration causes no further increase of fluorescent intensity. At this stage all the energy of the ultraviolet irradiation, at the wavelengths of absorption, is absorbed

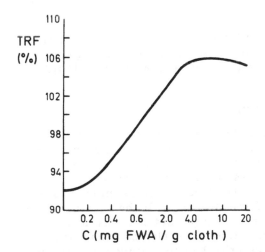

Figure 3.4 Dependence of total radiance factor (TRF), measured using filter 11 of Zeiss Elrepho photometer, on concentration (c) of FWA on cotton cloth.

and utilized by the FWA on the fabric [37] so that only an increase in intensity of illumination can increase fluorescence. At even higher application levels, whiteness begins to fall (Fig. 3.4) partly because the slight yellow self-color of the FWA causes quenching of fluorescence.

Theoretical treatments [36,38-40] have led to equations whose predictions agree well with experimental observations. However, these equations are difficult to apply, and Eq. (1) remains of practical importance. The overall process of excitation and emission of fluorescence is complicated [38,39]. Except at saturation concentrations, part of the incident, exciting UV radiation is reflected. Of that which is absorbed, some is absorbed by the substrate itself. The remaining light energy is absorbed by the FWA molecule, and a fraction, determined by the absolute fluorescent efficiency of the FWA, is transformed to fluorescent energy. Only a fraction of this emitted light finds its way out of the substrate without being reabsorbed. This fraction is a function of both the yellowness of the cloth and of the FWA itself [38,39].

Reflectance-Concentration Relationship

The dependence of reflectance, at the exciting wavelength, on the concentration (c) of FWA on the cloth is given by the relationship, based on that of Kubelka-Munk:

$$Ac = \frac{(1 - R)^2}{2R} - \frac{(1 - R_0)^2}{2R_0} \tag{2}$$

where R is the reflectance of the whitened cloth and R_0 that of the unwhitened cloth, both measured at the wavelength of absorption, where there is no fluorescent emission, and A is a constant [33].

3. THE MEASUREMENT OF WHITENESS

A manufacturer of white cloth must be able to estimate how white his product is. This refers equally to fluorescently whitened cloth or to that which has only been chemically bleached. There is a particular need to measure whiteness when a new bleaching or fluorescent whitening process is being developed; it is essential to relate changes in the parameters of the process with the whiteness obtained. Whiteness measurement is thus important not only in connection with fluorescent white substrates, but also concerns non-fluorescent ones. The methods described in this section are equally applicable in both cases. However, the ubiquitous use of FWAs has made whiteness measurement both more necessary and more difficult. The subject is also important in connection with the use and testing of washing powders, all of which contain FWAs. A bibliography on whiteness has been compiled [41].

Two methods are used to determine whiteness: visual estimation and instrumental measurement.

3.1 Visual Estimation of Whiteness

Comparing a sample with a test standard to check if it is white enough presents few problems. However, it is often necessary to arrange a series of samples in order of increasing whiteness, often done by comparing pairs of samples in the series (pair comparison). It then becomes difficult to relate the results to other series of samples, or to results obtained by other observers.

Standard Whiteness Scales

To overcome this difficulty standard whiteness scales have been devised [13], consisting of a series of samples of increasing whiteness. Each member (step) of the scale is given a number, and the whiteness of a sample is estimated quantitatively by pair comparison with the members. If the whiteness lies between two steps, then it is estimated by interpolation. This allows the whiteness to be expressed quantitatively as a number which has meaning to different observers. The assumption is here made that whiteness is a scalar quantity expressable by a single number. This is so in most cases met with in commercial practice, but is not strictly correct when comparing near-white samples of markedly different hue [42].

The Ciba-Geigy Plastic White (CGPW) Scale: The most successful and commonly used white scale was developed in the laboratories of Ciba-Geigy [13,43,44]. It is based on a series of 12 melamine plastic plates increasing in whiteness, in steps of about 20 units each, from -20 CGPW units for plate 1 to 210 CGPW units for plate 12. The plates are washable, an advantage over standard sets made from cloth or paper which become irretrievably soiled during handling. The matt side of each plate is used for estimating the whiteness of textile samples. Plates 1-4 of the scale contain decreasing amounts of a yellow dye; plates 6-12 contain increasing amounts of an FWA with a neutral (blue) hue.

Estimation of Whiteness by Pair Comparison: With training, the eye is capable of detecting very small differences in whiteness (down to about 5 CGPW units) when comparing two samples held side by side [45,46]. To estimate the whiteness of a textile sample, it is held alongside one of the plates and tilted away from the observer to reduce gloss. The samples are switched from hand to hand to aid in deciding which is the whiter. The comparison is best made against a gray background. The textile sample should be ironed flat (through a sheet of nonfluorescent paper) into sufficient layers to render the sample opaque, and so as to be the same size as one of the plates [13].

Hue Preference: Results of whiteness estimation are most reliable when performed by a panel of viewers and the results averaged. All observers rate a yellowish article as being of low whiteness. A "normal" observer will consider it whiter if it has a blue hue. Such an observer is considered to have a neutral hue preference. In addition, some observers have a green or a red hue preference: they give particular weight in their whiteness ratings to an article with a greenish-blue or a reddish-blue hue (tint) [45]. This subject is discussed further in Sec. 3.2 (see "Hue Preference Angle").

When near-white samples of similar hue are being compared, there is good agreement between the members of a panel. However, when samples of markedly different hue are involved, ratings can vary [47,48].

The Illuminant

The most important factor which must be controlled when comparing near-white fluorescent samples is the illuminant, though this is much less critical with nonfluorescent samples. The critical factor is the proportion of energy in the illuminant, in both the UV and the violet region of the spectrum, responsible for stimulating fluorescence [45]. If this proportion is increased, then the apparent whiteness of both members of a pair of fluorescent samples (one of which may be a plate of the CGPW scale) will rise. However, they may not increase in whiteness by the same extent because a difference in the intrinsic yellowness of the two substrates will cause quenching to a different extent. Thus samples will appear of different whiteness, when compared among themselves or with plates of the CGPW scale, according to the illuminant used [13,45,49]. This is an example of color metamerism and has been used [50,51] in order to compare the proportions of the near UV component in illuminant sources.

Standard Illuminants: Because of the crucial importance of the illuminant, standard illuminants have been defined by the Commission International de l'Eclairage (CIE). These illuminants are defined in tables showing the relative spectral energy (irradiance) as a function of wavelength [52]. For viewing or measuring nonfluorescent (including colored) samples, illuminant C is used, equivalent to overcast-sky daylight. Illuminant D65, should be used for fluorescent samples. It is based on the energy distribution of diffuse sunlight, without glare, and is well reproduced by a xenon lamp fitted with special filters.

Though good results in visual estimation can be obtained through a north-facing window on a cloudless day, best reproducibility, independent of latitude, time of day, and the nature of the window glass, is obtained using a viewing cabinet. It should be noted, how-ever, that while the diffuse strip lighting used in commercially avail-able cabinets adequately copies the D65 specification in the visible part of the spectrum, this lighting is rather deficient in energy in the crucial UV region [46,50,53,54]. A xenon lamp has been fitted to a viewing cabinet [48].

3.2 Instrumental Measurement of Whiteness

Colorimetry

Visual estimation of whiteness suffers from the lack of objectivity of the observer, and depends on his hue preference and the trade in which he works [45,55]. Visual estimations made by a single observer are not necessarily reproducible by another, while a panel must be organized. Thus instrumental methods of measurement have been de-vised. Clearly the arbiter for deciding the success of an instru-mental method must be the eye, since whiteness is a visual impres-sion. Ideally, a number given to whiteness on the basis of instru-mental reflection measurements should agree with that obtained by a panel of observers of neutral hue preference, and should at least give the same order of ranking within a series of samples.

Color Matching Functions: Whiteness is a form of color and is meas-ured using the same principles [13,52,56]. Unlike whiteness, which is a scalar quantity, a color is specified by three parameters. In the CIE color system all colors are defined in terms of a mixture of three primaries suited to the sensitivities of the three receptors - blue, green, and red - in the eye of a defined "standard observer," with average normal vision. The wavelength-sensitivity relationships of these receptors are shown in Fig. 3.5 as the wavelength-dependent functions $\bar{x}$, $\bar{y}$, and $\bar{z}$ (the spectral tristimulus or color matching functions).

The green receptor $\bar{y}$ is chosen to be sensitive to the bright-ness or luminous reflectance of a substrate. The $\bar{x}$ (red) receptor has part of its sensitivity in the blue region of the spectrum. The functions were first defined in 1931 for an observer with a 2° field of viewing. They were supplemented in 1964 for an observer with a larger (10°) field. It has been reported that the use of the 1964 10° observer functions gives better correlation of instrumental meas-urements with visual estimation [57,58].

Tristimulus Values: The sensation of color perceived by the brain is a synthesis of signals from the three color receptors in the eye. The signals depend, first, on the spectral distribution of the light leaving the illuminant, which is then modified by the reflection spec-trum of the substrate, and, finally, in accordance with the light

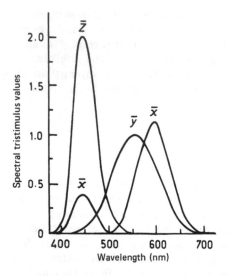

Figure 3.5   CIE spectral tristimulus functions (1964, 10° observer).

sensitivities of the eye receptors [56]. Thus the brain signals can be defined by multiples of three functions, all dependent on wavelength (λ):

1. The spectral energy distribution $E(\lambda)$ of the *illuminant*
2. The spectral reflectance factor $R(\lambda)$ of the *substrate* when illuminated by the same illuminant
3. The CIE spectral tristimulus functions $\bar{x}(\lambda)$, $\bar{y}(\lambda)$, $\bar{z}(\lambda)$ of the standard observer

The color of a substrate in the CIE system is thus defined by three CIE tristimulus values X, Y, Z, which are the values of the above multiples summed over the visible spectral region:

$$X = k \int E(\lambda)\bar{x}(\lambda)R(\lambda)\, d\lambda$$

$$Y = k \int E(\lambda)\bar{y}(\lambda)R(\lambda)\, d\lambda \qquad\qquad (3)$$

$$Z = k \int E(\lambda)\bar{z}(\lambda)R(\lambda)\, d\lambda$$

These are definite integrals usually between the limits 380 nm and 780 nm, and k is a normalization constant. The reflectance is measured as a factor, with respect to the 100% reflectance of an ideal white body.

For a fluorescent substrate, the illuminant used is usually D65 [52]. The energy reaching the eye is here dependent not only on the spectral reflectance factor $R(\lambda)$ of the substrate, but also on a fluorescent component. These two factors together are called the *total spectral radiance factor*, which replaces $R(\lambda)$ in Eq. (3).

To measure the tristimulus value X of a substrate, it is necessary to measure the reflectance or total radiance factor at wavelength intervals in the visible region, to multiply each reading by the factors $E(\lambda)\bar{x}(\lambda)$ at each wavelength, and to sum the results over the whole wavelength range. The same applied to Y and Z. A modern color spectrophotometer performs this automatically. Simpler tristimulus reflectance photometers are equipped with three tristimulus filters whose optical characteristics are suited as far as is possible to the color matching functions. The three measured photometer readings A, G. B are then transformed to the tristimulus values X, Y, Z. For the D65 illuminant and CIE 1964 (10°) colorimetric observer, the transformations are:

$$X = 0.768A + 0.180B$$

$$Y = G \tag{4}$$

$$Z = 1.073B$$

CIE Chromaticity Diagram: For practical purposes the values X, Y, Z are transformed into three *chromacity coordinates* x, y, z:

$$x = \frac{X}{X + Y + Z} \qquad y = \frac{Y}{X + Y + Z} \qquad z = 1 - (x + y) \tag{5}$$

A color in CIE color space is defined by the numerical values of three coordinates, which are the chromaticity coordinates x, y together with the luminosity, Y. By convention, the coordinates x, y define a horizontal plane with the coordinate Y normal to that plane. A horizontal section, in the x, y plane, through CIE color space is shown in Fig. 3.6 [3]. This is called the CIE chromaticity diagram. Real colors are confined within the spectrum locus shown as a thick envelope in the diagram. Along this locus lie the color points of spectrally pure monochromatic colors, said to have 100% color saturation of purity. The wavelengths of these colors, some of which are marked on the diagram, are called *dominant wavelengths*. The point of zero color saturation at the center is called the achromatic point. This is the color point (chromaticity coordinates) of the perfect diffuser (A = G = B = 100) and is marked on the diagram for illuminant D65.

Moving towards the achromatic point from the spectrum locus along a dominant wavelength line defines progressively paler colors of the same hue. The color points of near-white colors lie in a small area approximately marked at the center of the diagram (Fig. 3.6) around the achromatic point, and shown enlarged in Fig. 3.7.

A point in color space can thus be defined by:

Y = luminance factor or brightness
λd = dominant wavelength, the wavelength of the spectrum color
    obtained by extending the line (the dominant wavelength

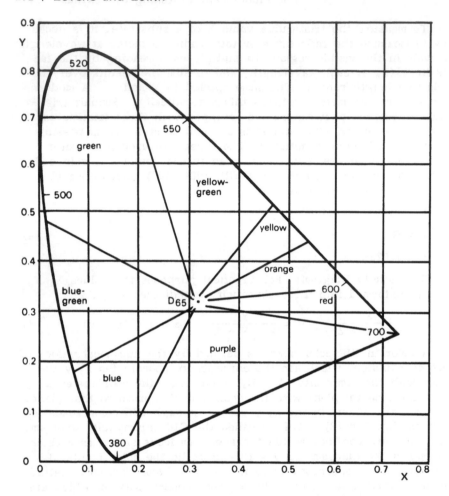

Figure 3.6 CIE Chromaticity diagram with achromatic point D65, 2° observer. (From Ref. 3.)

line) from the achromatic point through the color point until it cuts the spectrum locus

p = color purity or saturation which is the distance of the color point from the achromatic point as a ratio of the total length of the dominant wavelength line (usually %)

Near-white bodies have high values of Y and low values of p.

The disadvantage of the CIE chromaticity diagram is that equal color differences as perceived by the eye do not represent equal distances on the diagram. More complex coordinate systems replacing

Eq. (5) have been devised to overcome this problem which is most important for instrumental color matching. However, in the region of near-whites (Fig. 2.7), the x, y chromaticity diagram is linear for practical purposes [59], and no advantage is gained by using these more complex systems, such as that of Hunter [13].

Color Measuring Instruments: Reflectance photometers for measuring fluorescent substrates are equipped with a D65 source in the form of a filtered xenon lamp. In order to ensure the measurement of diffuse reflectance, instruments have an integrating sphere painted white on its inner wall. The illuminant source, the sample, and the detector are placed at ports in the wall of the sphere. The instrument must be calibrated against a near-white standard which itself may be calibrated by a standards laboratory appointed by the ISO [60].

CGPW Scale Plates: Figure 3.7 shows the central region of the CIE chromaticity diagram where the color points of near-white substrates are found. The diagram shows the color points, marked 1-12, of the plates of the CGPW scale. Plates 5-12 all have the same neutral blue hue and lie along a line pointing in to the blue-violet region of the diagram with dominant wavelength 470 nm [44]. The achromatic point D65 lies close to plate 6. The luminance Y of the plates rises from plate 1 to 12. However, the color points shown in Fig. 3.7 are

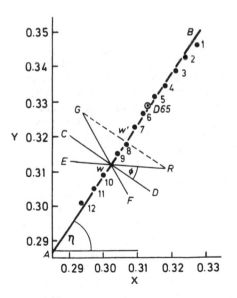

Figure 3.7 CIE Chromaticity diagram showing color points of CGPW scale for illuminant D65 (1931, 2° observer). (From Ref. 44.)

Table 3.1 AGB Whiteness Formulae

Name	Formula	Implied hue preference
Taube [65,66]	W = 4B - 3G	Neutral-red
Berger [67]	W = G + 3(B - A)	Green
Stephansen [13,45]	W - 2B - A	Neutral-green
Ganz [74]	W = 3B - 1.5G - 0.5 A	Neutral
Croes [13,45]	W = G + B - A	Green
Blue reflectance [6b]	W = B	Neutral
Hunter [65]	W = L - 3b	Neutral-red
Stensby [62]	W = L + 3a - 3b	Red

Source: Ref. 73.

a projection on the x, y plane, and do not indicate changes in Y. The chromaticity specifications (x,y,Y) of the plates vary according to the measuring instrument [43,44,45,48,49,61].

Whiteness Formulae

Many whiteness formula have been devised which attempt to express whiteness as a single number calculated from reflectance readings [13,45,46,59,62-64]. Those which are not based on tristimulus measurements have little hope of emulating the visual impression of whiteness. The most commonly used formulae are based on tristimulus photometer readings A, G, B (Table 3.1). Of these probably the most popular are those attributed to Taube [65,66], Berger [67], and a modified form of one originally proposed by Stephansen in 1933. All these formulae give W = 100 when A = G = B = 100. Similar formulae (Table 3.1) are based on Hunter coordinates [13,62,65]. Whiteness formulae embodied in standard methods for measuring whiteness include that of Taube [66] and also the formula W = B [6b].

A number of more recent formulae [13,45] based on colorimetric specifications of the substrate have been checked by correlation with visual estimations [46,48,56,61,64,68-72]. These formulae suffer either from the complexity of the calculations involved in their use, or from the necessity to use them in conjunction with a nomogram or chart. Correlation between instrumental whiteness prediction and visual estimation may be improved by using regression analysis [45, 70].

Linear Whiteness Formulae: Ganz [33,34,45,59,74-76] has developed
a series of generalized formulae of which two types will be considered:

$$W = \alpha A + \beta B + \gamma G + C_1 \tag{6}$$

$$W = DY + Px + Qy + C_2 \tag{7}$$

where α, β, γ and D, P, Q are adjustable coefficients (parameters).
Equation (6) is a generalized form of the formulae in Table 3.1.

The coefficients of Eqs. (6) and (7) can be fixed by multiple
regression analysis using the visually estimated whiteness values of
a set of near-white (fluorescent or nonfluorescent) samples and their
instrumentally measured chromaticity specifications [45,49,73]. This
is quite laborious. Moreover, a random sample set in which hue,
purity, and luminance all vary independently may not be available.
A simplified method, described later, uses the plates of the CGPW
scale. All formulae are adjusted so that W = 100 for the perfect
diffuser.

Hue Preference Angle: It has been shown [48] that for a given
value of Y there is a line of optimum preferred whiteness, defined
by substrates of neutral blueish tint, shown as AB in Fig. 3.7.
This line has a dominant wavelength near 470 nm in the blue region
and one near 573 nm in the yellow. As the hue or tint of a substrate deviates to either side of this line, the perceived whiteness
falls off. Samples with greenish tint have their color points (e.g.,
G) above the line, and those with reddish tint appear below it (e.g.,
R, Fig. 3.7).

For samples of equal luminance factor Y, equiwhiteness lines,
cutting the line AB, may thus be defined. Those defined by formulae based on chromaticity coordinates [Eq. (7)] are straight, while
for photometer reading formulae [Eq. (6)] the lines are flat curves
[74] which may be approximated by straight lines. These lines
(e.g., GF, ER) made an angle ϕ (Fig. 3.7) with the normal CD to
the preferred whiteness line. Angle ϕ is called the *hue preference
angle* of the formula. A formula with a green-hue preference angle,
$-\phi$, predicts a whiteness value W (Fig. 3.7) for a substrate of color
point G with a greenish tint, whereas a neutral hue preference formula ($\phi = 0$) would predict a lower value, W'. The same applies to
a formula with a positive hue preference angle ϕ with respect to a
substrate with a reddish tint (color point R).

The hue preference angle of a formula based on chromaticity
coordinates is fixed by the relationship [45]

$$\frac{Q}{P} = \tan(\phi + \eta) \tag{8}$$

where η is the angle made by AB with the x-axis and P, Q appear in
Eq. (7). The implied hue preference of the AGB formulae appears in
Table 3.1.

It will be clear from Fig. 3.7 that the hue preference angle becomes more important the more distant is the color point from line AB, that is, the greater the hue. Samples of neutral hue whose color points lie on AB are insensitive to the hue preference of the prediction formula.

Determination of the Coefficients of a Prediction Formula: The coefficients of Eq. (7) can be fixed using the plates of the CGPW scale. The fluorescent plates 5-12 are used with a prediction formula for fluorescent samples [45], while for nonfluorescent samples plates 1-5 may be used [73]. The hue of plates 5-12 is nonvariable (Fig. 3.7), and also brightness Y is not independent of saturation, so only simple regression can be used. It is necessary to fix reasonable values for two of the independent variables in Eq. (7).

Experience has shown that normal observers have a hue preference angle ϕ of 5-15° [48,49,75], and a value of 10° seems reasonable. The line AB may be defined by the plates of the CGPW scale [47,49]. This corresponds to $\eta = 50°$ for illuminant D65 with the 1964 observer [48,75] and to 55° for the 1931 observer [74,75]. However, it is preferable to calculate η for a particular measuring instrument from the measured values x, y, Y for the plates of the scale. The prediction formula is not sensitive to small changes in ϕ, η. For the coefficient D, values of 1 or 2 have been taken [59, 75].

Separation of variables in Eq. (7) gives the linear regression equation [47,73]:

$$W - DY = P(x + \frac{Q}{P}y) + C_2 \tag{9}$$

The coefficients P, Q, and C_2 may then be calculated using Eqs. (8) and (9) and simple linear regression between the assigned whiteness values W of plates 5-12 (or 4-12) of the CGPW scale and their measured values of x, y, Y. The prediction equation thus obtained has been shown to be reliable in predicting whiteness of textile samples met within routine commercial work whose hue does not deviate widely from neutral due to the presence of discoloration or tinting dyes [45,49]. This is so even when a nonstandard illuminant is used [49]. The values x, y, Y are calculated from readings taken with the same photometer as that used to measure these values on the CGPW plates.

It is possible similarly to determine values for the coefficients in Eq. (6), but a prediction based on Eq. (6) is less successful than one based on Eq. (7) [49,59].

The adjustable coefficients in Eq. (7) may be calculated with the aid of a small programmable calculator, so that a whiteness formula with coefficients adjusted to a particular reflectance photometer may be employed routinely in the commercial determination of whiteness. Simple graphical methods have also been devised [45,49,75].

Table 3.2 Constants for Fixed Coefficient Whiteness Formulae,
Eq. (12)

Hue preference	Values of coefficients		Values of C_W (illuminant D65)	
	P	Q	2° observer	10° observer
Neutral	-800	-1700	-809.5	-813.6
Green	-1700	-900	-827.7	-831.3
Red	800	-3000	-736.8	-741.7

By carrying out the procedure on a relfectance photometer at intervals, errors due to the aging of the xenon lamp or yellowing of the
walls of the integrating sphere [77] can be avoided [45].

For nonfluorescent samples, with coefficients determined against
CGPW plates 1-5, a prediction formula, Eq. (7), with a green hue
preference ($\phi + \eta = 30°$) was found to be necessary [73]. This is
probably a result of the slight reddish tint shown by plates 1-4 of
the CGPW scale.

Individual Perception: Individual hue preference angles have been
determined [45,49] and were found in a significant proportion of observers to vary markedly from the neutral [42,45,59,78].

The determination of equiwhiteness lines for individual observers
is a formidable task, involving the preparation of series of samples
with differing hues but constant luminance factor. The indications
are [42,48] that the lines curve away from the achromatic point. The
approximation of these to straight lines in the narrow region of white
on either side of the preferred whiteness lines is justified by experience.

Fixed Coefficient Formulae: Standard linear whiteness formulae with
fixed coefficients have been proposed by Ganz [59]:

$$W = Y + Px + Qy - C_w \qquad (10)$$

where $C_w = Px_0 + Qy_0$, and x_0, y_0 are the chromaticity coordinates
of the achromatic point for the illuminant used. A single prediction
equation cannot match the hue preference shown by all observers, so
that in addition to a standard formula of neutral hue preference, two
further equations were proposed of green and red hue preference.
The values of P and Q in Eq. (10) are shown in Table 3.7, together
with the values of C_w when using illuminant D65 (2° and 10° observer)
[59]. A proposal has been made to the CIE to adopt these formulae
for standard industrial use.

Results obtained using these equations are inferior to those of Eq. (7) with coefficients adjusted to the particular measuring instrument used [73]. However, the results are better than those predicted by the simpler formulae of Table 3.1.

Tint Evaluation: A whiteness tint prediction formula has been devised with coefficients determined by linear regression of visual assessments against instrumental readings [45]. Two standard forms with fixed coefficients have been proposed [59].

4. THE TEXTILE APPLICATION OF FWAs

4.1 Structural Types

The chemistry of FWAs, from the point of view of the synthesis of the various structural types, has been adequately and repeatedly reviewed [2,4,8,79-86]. These reviews survey the patent literature through 1973, and a recent review [4] has brought the coverage up to 1976. A bibliography covers the nonpatent literature through 1973 [1]. In the present text the main structural types will be presented, with the aim of demonstrating the connection between their chemical structures and their dyeing properties. Apart from FWA 1 (Fig. 3.8) which is CI Fluorescent Brightener 28 (constitution no. 40622), the structural formulae of very few brand name products in common use for textiles are unequivically known.

Nomenclature of Structural Types

Since FWAs are widely used, attempts have been made to codify them by a letter designation, according to structural types, followed by a number [7,88,89]. The ASTM list [7] is limited to FWAs used in washing powders in the United States. A later list [88], though much more comprehensive, unfortunately differs in its numbering, and partly in its lettering, from that of the ASTM. Table 3.3 shows the basic structures of the major types [13] together with the letter designation proposed in the above-mentioned two systems.

General Considerations

FWAs for cotton and regenerated cellulose (DASC, TS, DSBP) are anionic. They are all stilbene derivatives, contain sulfonic acid groups, and are equivalent in their use and action to direct dyes. In contrast, nonpolar polyester is whitened with neutral FWAs (DSBP, BO, NTS, naphthalimides) which are applied as aqueous dispersions, exactly as for disperse dyes. The same applies to FWAs (of type BO, C, P) for acetate and triacetate. The equivalent of basic dyes are used to whiten polyacrylonitrile, but nonionic (disperse) FWAs

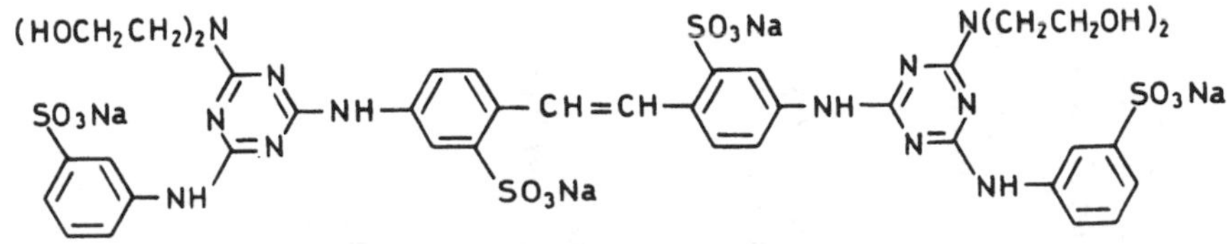

1

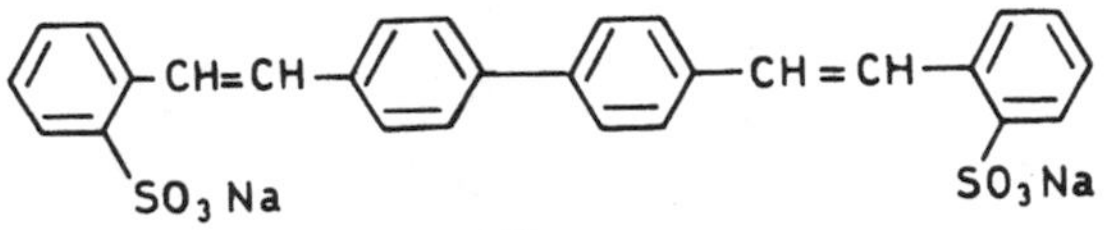

2

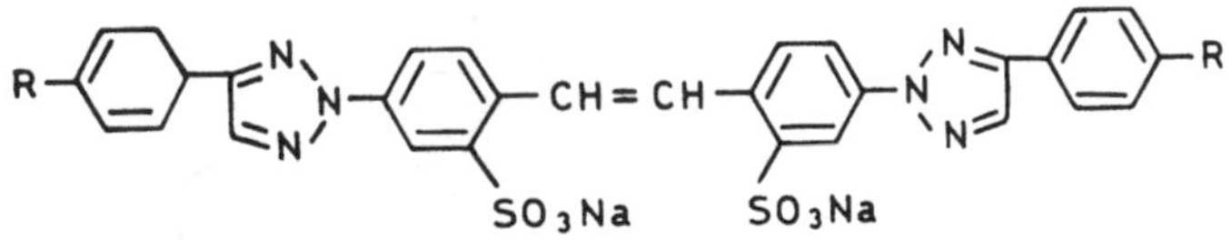

3a: R=H; 3b: R = SO_3Na

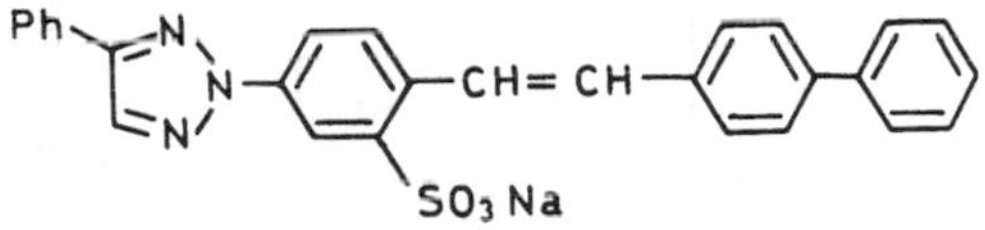

4

Figure 3.8 Some FWAs for cotton.

are also applicable. Polyamide and wool are whitened with anionic FWAs applied in the manner of acid dyes at low pH, but disperse FWAs can also be used for polyamide [2,4,5,9,80,81].

4.2 Whiteners for Cotton and Regenerated Cellulose

DASC Type Whiteners

According to a 1971 estimate [5] over one quarter of all fibers processed in the textile industry are white cottons and regenerated cellulose. Only a negligible proportion of this is not fluorescently whitened. FWAs for cotton also have paramount importance since the majority of FWAs in washing powders are of this type. Most of these FWAs are of the DASC type. Thus, in spite of the great

Table 3.3 Major FWA Structural Types

Letter Designation [7]	[88]	Basic structure	Generic name
DASC			Diaminostilbenedisulfonic acid-cyanuric chloride; 4,4'-bis[(1,3,5-triazin-1-yl)amino]stilbene-2,2'-disulfonic acid
—	TS		4,4'-bis(1,2,3-triazin-2-yl)stilbene-2,2'-disulfonic acid
NTS			Naphthotriazolylstilbene and bis(naphthotriazolyl)stilbene
DSBP			Distyrylbiphenyl; 4,4'-bis(styryl)biphenyl
P	PYZ		Pyrazoline; 1,3-diphenyl-2-pyrazoline

C	COUM	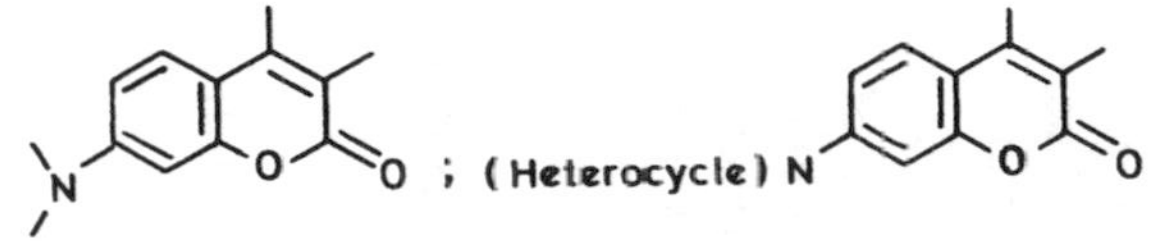	Coumarin

; (Heterocycle) N

BO — Bis(benzoxazol-1-yl)deriv.

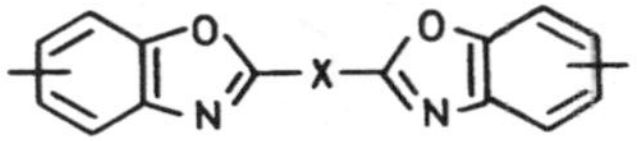

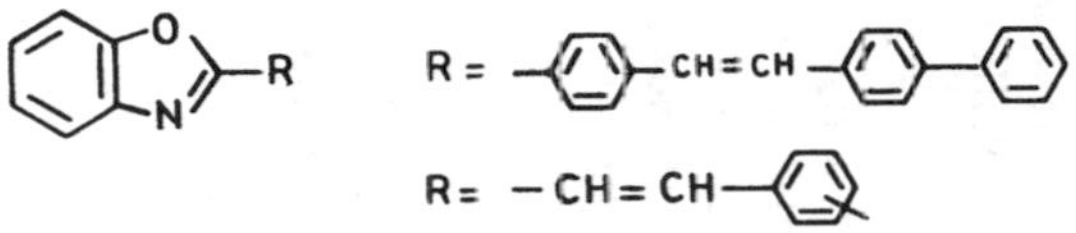

R = — 4-phenyl-4'-(benzoxazol-2-yl)-stilbene

R = — 2-styryl-benzoxazole

BI — Bis(benzimidazol-1-yl)deriv.

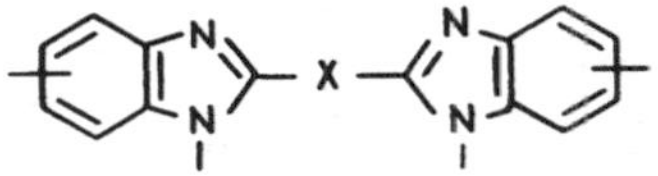

BIS — 4,4'-bis(benzimidazol-1-yl)-stilbene

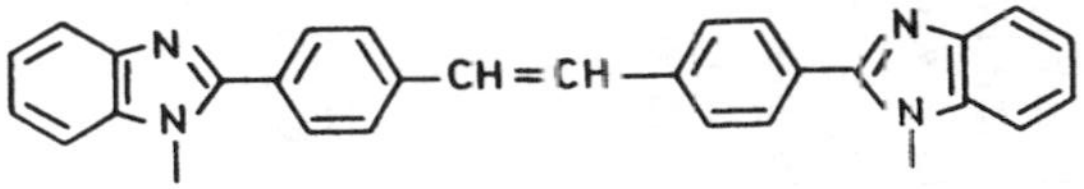

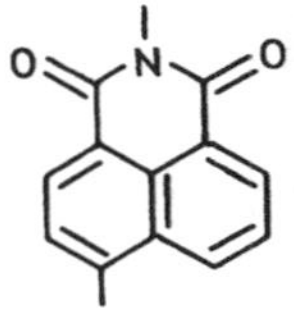

— Naphthalimide

variety of structural types, of which only the major ones are shown in Table 3.3, a significant proportion in common use is of one basic type (DASC). In addition, the FWA of formula 1 (Fig. 3.8), which was in wide use in the 1960s, is still extensively used for cotton in the textile industry today.

The continued popularity of DASC FWAs is due to low price, high efficiency, and the ease with which the synthesis can be modified. They are prepared from cyanuric chloride (CC) which is condensed with 4,4'-diaminostilbene-2,2'-disulfonic acid (DAS) (Fig. 3.9).

One of the active chlorine atoms in CC is condensed with an aromatic amine ($ArNH_2$) such as aniline, at metanilic or sulfanilic acids. The product (2 mols) is then reacted with DAS (1 mol) and finally with a primary or secondary aliphatic amine such as diethanolamine [8]. In a variation of FWA 1, a higher proportion of diethanolamine is used, and less metanilic acid, thus producing an asymmetric molecule [85]. By varying the number of sulfonic acid groups in the aromatic amine $ArNH_2$ from none to two, an FWA may be synthesized containing from two to six sulfonic groups. The number of hydroxyl groups in the aliphatic amine can also be changed (e.g., methylamine or monoethanolamine instead of diethanolamine). Thus the polarity of a DASC whitener, and its solubility in water, can be readily varied within a wide range.

Affinity and Substantivity: The affinity of a dye is the change of chemical potential on dyeing and is thus related to the equilibrium distribution (under exhaust conditions) of the dye (FWA) between the treatment solution and the fiber [90]. If most of the FWA exhausts on the fiber, then it has high affinity.

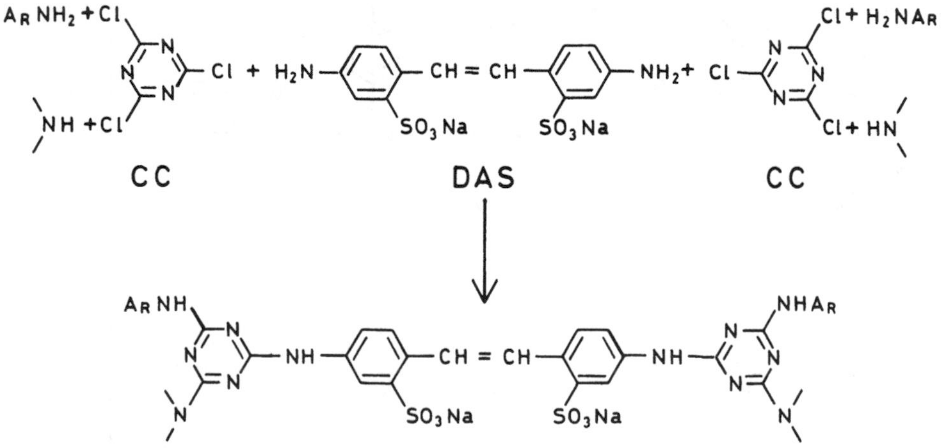

Figure 3.9 Preparation of DASC type FWA.

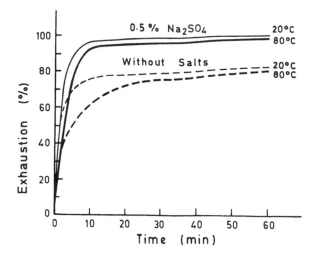

Figure 3.10 Affinity curve for high-affinity FWA (type I). (From Ref. 2.)

Substantivity is related to the heat of dyeing [90]. Since a dyeing process is exothermic, heat of dyeing is negative. Thus raising the temperature reduces the affinity of a dye (FWA), the effect being greater the greater the substantivity. Changes in the structure of an FWA molecule, which affect its polarity, affect both affinity and substantivity.

The basic types of whiteners include:

1. High affinity (type I). When aniline is used in the DASC synthesis (Fig. 3.9), the product has only moderate polarity and has a high affinity and rate of strike for cellulose. Such products are commonly used in washing powders, but for textile applications they suffer from the disadvantage of low water solubility. More important, they have only moderate leveling power and can only be used under conditions where leveling is not likely to be a problem. These type I products have high affinity (Fig. 3.10), about 80% being exhausted on the fiber at equilibrium at 20°C [2,91-93]. Affinity can be increased by adding a salt to the treatment bath, and in the presence of 0.5% w/v concentration of sodium sulfate, exhaustion is complete. The effect of temperature in reducing affinity is small, though leveling is improved. These properties are suited to exhaust application from hot liquors (above 60°C), such as together with a peroxide bleach, followed by a hot wash [92,93].

Type I FWAs can only be used at neutral pH or above. They are particularly suited for whitening regenerated cellulose, which requires FWAs of higher affinity than for cotton.

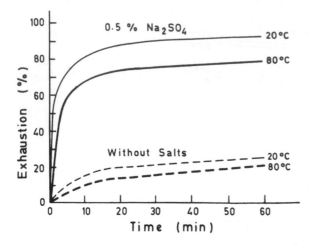

Figure 3.11 Affinity curve for type II FWA. (From Ref. 2.)

2. Low-medium affinity (type II). Introducing an extra two sulfonate groups into the molecule, as in FWA 1 (Fig. 3.8), increases the solubility, reduces the affinity for cotton, and increases the substantivity. Type II FWAs (exemplified by FWA 1) have low affinity in soft water (Fig. 3.11), shown by a slow rate of strike, good leveling power, and a low degree of exhaustion [2,91-93]. They are very sensitive to salt concentration, and, in the presence of 0.5% w/v sodium sulfate, exhaustion of greater than 95% is achieved. Since substantivity is high, the effect of temperature in reducing affinity is higher than with type I FWAs.

The versatility of this type explains its popularity. It can be used in the presence of salts in an exhaust process, with good leveling. On the other hand it is ideally suited to a padding process when used in soft water. In a padding process the cloth is impregnated by being passed rapidly through a trough containing the treatment solution. When a high-affinity FWA is used, which is absorbed rapidly onto the fiber, it is necessary to add continually extra FWA to the treatment solution. However, with low-medium affinity (type II) whitener, there is negligible exhaustion of the FWA onto the cloth during its passage through the padding trough, so that the concentration of FWA does not fall during a long run, causing tailing. Affinity is high in the presence of salts, but the rate of absorption is not high, so this type of FWA can still be used in an anticrease impregnation treatment, where the padding liquors contain salts. FWA 1 is effective in the pH range 5-11, but tends to precipitate from the treatment solution at lower pH. It is particularly suited for use in a pad-batch application, together with a resin treatment or peroxide bleach. Here the cloth is impregnated and then held for a time which may vary from 5 min at 140°C (in resin curing) to 1 day at room temperature (in a cold pad-batch peroxide bleach process). During the holding period FWA penetrates into the fiber, and the fluorescent whiteness develops [18-20].

3. Other types. A third type of FWA has low affinity even in the presence of salts [93]. This type includes the DASC products containing 6 sulfonic groups, prepared using aniline-2,5-disulfonic acid (Fig. 3.9). Due to their high stability to acids, these products are particularly suitable for use in conjunction with resin treatments involving highly acidic conditions [94].

A further type of FWA has medium affinity in soft water and high substantivity. These FWAs are used when there is a risk of unlevelness as in a pack bleaching process. Leveling is assured by raising the temperature, which causes a marked drop in affinity even in the presence of salts [2,92,93].

The disadvantages of DASC type FWAs include:

1. Photochemical instability. This is associated principally with the stilbene double bond, irradiation causing isomerization to the nonfluorescent and nonsubstantive cis isomer [95]. In addition photochemical transformations cause yellowing, or possibly cause the formation of products (e.g., cyclobutane dimers) fluorescing at lower wavelengths which overlap with the absorption band of the trans monomer. Though other stilbene-derived FWAs for cotton undergo such transformations [11,83], many are known (including the DSBP type) which have greater photochemical stability and improved lightfastness (Fig. 3.8, FWAs 2-4).

Once applied to the fabric, where the molecule has reduced mobility, DASC compounds are much more stable to light than when in solution, and they have acceptable lightfastness.

2. Chlorination. The active nitrogen (-NH group) in DASC whiteners causes them to be incompatible with hypochlorite and chlorite bleaching agents. This has led to the development of other (more expensive) FWAs which can be used in conjunction with these bleaches. Once applied to the cloth, high affinity DASC whiteners are quite stable to hypochlorite bleach, but are destroyed by chlorite. It is possible that the -NH group is also involved in lightfastness, since stilbene FWAs without this group usually also have improved light stability.

3. Other interfering agents. All anionic FWAs are incompatible with heavy metal ions such as those of copper, iron, manganese, and zinc [96]. If these are likely to be present, a chelating agent should be added. Resin finishes containing zinc salts and nitrates as crosslinking catalysts are known to cause rapid yellowing of goods finished with DASC whiteners, unless washed off very well after the treatment.

Other Structural Types for Cellulosics

Some hypochlorite bleach-stable FWAs (FWAs 2-4) for cotton and regenerated cellulose are shown in Fig. 3.8 [2,8,80,81]. FWA 2 is
completely compatible with hypochlorite, but due to reduced affinity
at high temperatures is not recommended for an exhaustion process
above 60°C. FWA 3b is particularly recommended for use together
with a cross-linking resin treatment. Clearly the basic structural
types of FWAs 2-4 can be modified by substitution, as in DASC whiteners, to obtain compounds of varying affinity and water solubility.
These products are more expensive than those of the DASC type.

4.3 The Application of FWAs to Cotton

Treatment of cotton with FWAs is rarely carried out as a separate
process; it is usually incorporated together with some other finishing
stage [2,5,9,20,32,91-94,97-102]. For the highest white effects,
good chemical bleaching is essential.

Figure 3.12 [100] shows the operations along with which whiteners can be incorporated and the type of FWA suited to each case.
The majority of cotton whiteners are not compatible with hypochlorite

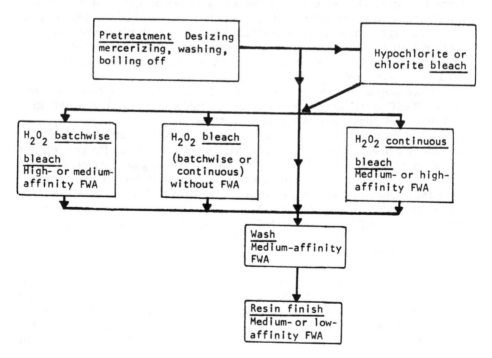

Figure 3.12 The application of FWAs to cotton. (From Ref. 100.)

or chlorite, and if these bleaching agents are to be used, whitening must be carried out at a later stage. However, some FWAs for cotton (such as FWAs 3 and 4, Fig. 3.8) are stable to hypochlorite after they have been applied to the cloth and can be used before a hypochlorite bleach, while FWA 2 can be used along with hypochlorite. All FWAs may be used together with hydrogen peroxide. Preliminary whitening before bleaching, followed by a supplementary whitening at a later stage, gives high, level whiteness [97]. In principle, whitening can be combined with all wet pretreatment processes, with the possible exception of desizing. Similarly, FWA can be applied with the bleach and then later with a resin finish.

Batchwise bleaching and whitening with hydrogen peroxide, as an exhaust process with a high liquor ratio, needs a high-affinity whitener or one with medium affinity together with sodium sulfate. For pad-batch whitening, where the FWA has ample time to penetrate into the fiber, a medium-affinity FWA is preferred, thus avoiding problems of leveling and tailing. For whitening together with continuous peroxide bleaching in the J box, in rope form, or with open-width cloth, either a high- or medium-affinity FWA is suitable. It may be applied together with the bleaching agent both at the impregnation stage and in the curved section of the J. The advantage of a medium-affinity FWA is ease of operation, but best results are obtained with a high-affinity type, avoiding tailing by using a small trough to which liquor is being added continuously in order to maintain a constant level and concentration. A constant running speed without stoppages must also be maintained to ensure evenness. High-affinity FWAs are faster to washing. When wet-on-wet impregnation is used, entailing extra-high-strength bleaching liquors, type II FWAs should be used since the high-affinity types are not completely stable to the stronger bleach. Type I FWAs are sensitive to acidity, and if a subsequent moist resin finish under highly acid conditions is contemplated (see later), a type II FWA is preferable at the bleaching stage. However, high-affinity FWAs are not affected by the normal pad-dry-cure-heat cross-linking process.

If peroxide bleaching has been carried out without whitening, FWAs can be incorporated in the washing-off stage (Fig. 3.12). Here there is a definite danger of tailing and unlevelness with a high-affinity type, and the use of a type II whitener is preferred. It is applied at 40-50°C in the last box of the washing range.

FWAs are often applied together with a resin finish [93,94], sometimes (e.g., with cotton shirtings) in order to supplement a whitener applied at an earlier stage. The FWA is added to the padding batch together with a resin, and the high salt content and low pH dictate the use of a medium- to low-affinity type. Type II FWAs can only be used if the pH is not below 5. For acid cross-linking processes, involving holding the treated cloth for 18-24 hr at room temperature and about pH 1, FWAs of low affinity must be used [32, 93,94]. Fillers, softeners, and shading dyes may also be added to the resin bath, but cationic additives including fabric softeners, should not be used.

4.4 The Whitening of Polyamide

Structural Types

All the general structural types which are suggested with cellulose are also effective in textile applications with the less polar polyamide fibers [2,9,30,32,101,103-105]. These include DASC whiteners of both high and medium affinity, e.g., FWA 1 (Fig. 3.8) as well as FWA 5 (Fig. 3.13), triazole-stilbene disulfonic acids such as FWA 3a and DSBP whiteners, particularly FWA 2 (Fig. 3.8). Figure 3.13 shows some structures recommended particularly for polyamide. FWA 5 is less effective than FWA 2 [30]. Pyrazoline compounds, including FWAs 6a, 6b, are effective with polyamide [106] as is also the naphthotriazolestilbene derivative 7. These compounds all contain sulfonic acid groups and with polyamide act as acid dyes. In addition a group of neutral compounds can be applied to polyamide, exemplified by FWAs 8-10. They are applied as aqueous dispersions and are thus "disperse dye" types. They are also effective in washing powders. Coumarin whiteners (e.g., FWA 10) have poor lightfastness. Benzoxazole derivatives (Table 3.3) are also used for the whitening of polyamide textiles [13].

Textile Application

Acid Types: The sulfonic acid derivatives are not applied in the manner of direct dyes as in the whitening of cotton, at neutral or higher pH, but usually at pH 4-4.5. Though commercial polyamide contains an excess of carboxylic end groups over amino groups, it is the latter which operate in the fixation of an FWA on the substrate [17]. The nature of the fixation is not considered to be mainly salt formation, though ionic interactions are important. The low pH is required in order to ensure the electrical neutrality of the fiber. Slightly more stringent conditions (higher temperature, lower pH) are required for nylon 6 than for nylon 66.

As with cellulosics, fluorescent whitening is combined with some other process, particularly with hydrosulfite bleaching, heat setting, or high-temperature finishing. When combining reductive bleach with a medium-affinity FWA, weak acid (e.g., acetic or formic) is added to pH 4-4.5. However, with a high-affinity type this may not always be possible, and the acidity due to the hydrosulfite itself (pH 5.5-6.0) is used. For exhaustion of the FWA onto the fiber, 20-30 min at 95°C are required.

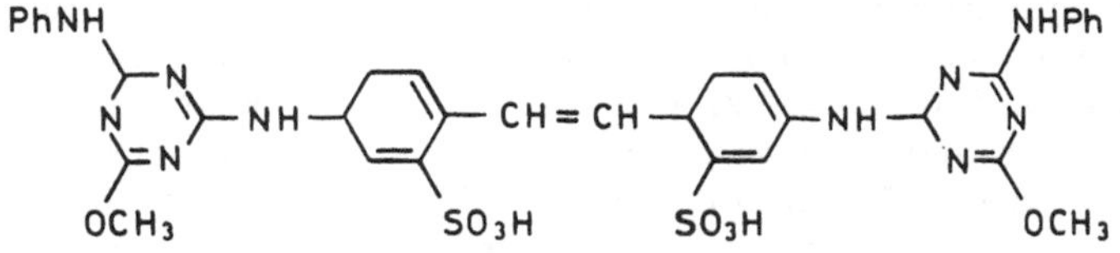

5

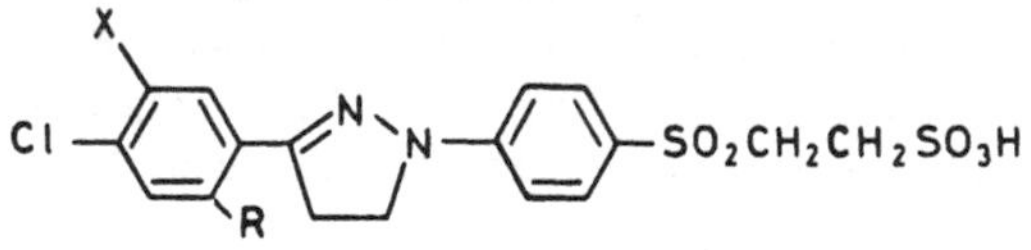

6a : X = H, R = H   6b : X = Cl, R = CH_3

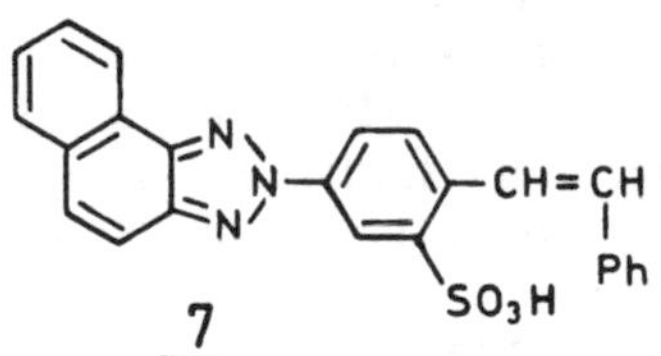

7

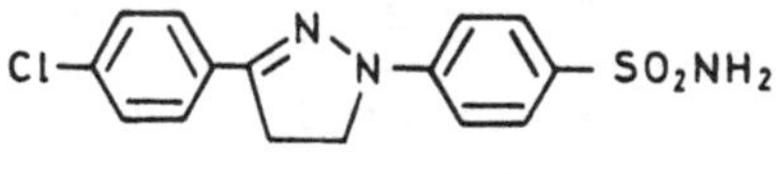

8

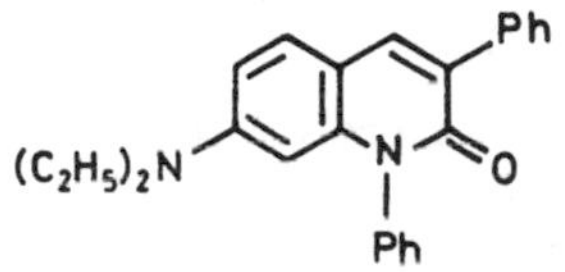

9

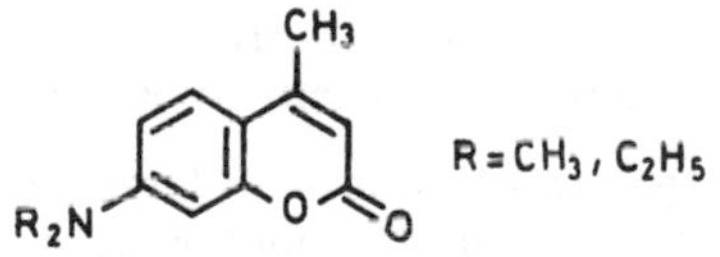

10

Figure 3.13 Some FWAs for polyamide.

Padding Processes: Fluorescent whiteners for polyamide can be incorporated into a semicontinuous or continuous padding process [105]. In one such process (hydrosetting) the cloth is padded with FWA, sodium dithionite, and detergent at 90°C and then heated at 120-130°C for 15-20 min (130°C for nylon 66). While bleaching takes place, the FWA diffuses into the fiber and the fluorescent whiteness is developed. In the pad-roll process, the impregnated cloth is held for 1-2 hr at 90°C. In a continuous pad-steam process, steaming is for 3 min at 105°C and for shorter periods at higher temperature. High-affinity FWAs are used, under slightly acid conditions.

Acid-Shock Treatment: The FWA can be incorporated into the preliminary continuous pad-scour stage whose main purpose is the removal of sizes and oils [30,99,101,103]. The open-width cloth is
impregnated with scouring detergent and FWA at above 50°C. A
high concentration (15 g/liter) of high-affinity FWA is used. The
cloth then passes through a series of hot wash boxes at 90°C, acidified to pH 3.5-4.0, which may also contain FWA in low concentration.
The cloth may be allowed to pile up between padder and wash boxes
to improve dirfusion into the fiber. The process is more suitable
for nylon 6. For nylon 66 padding is at 80°C followed by washing at
pH 3.5, 90°C. This treatment is usually followed by heat setting.

Pad-Bake (Thermosol) Treatment: The thermosol (pad-bake) process is most commonly used for polyester and involves heating for a
short time at high temperature, which also achieves heat setting.
During this heat treatment the FWA diffuses into the fiber forming
a solid solution. The process may also be combined with finishing
agents, such as resins and softeners. The cloth is first impregnated with FWA and other finishing agents together with acid to pH
3.5-4.0. FWAs 3a, 3b, 6a, 6b, and 7 (Figs. 3.8, 3.13) are among
those recommended for this process [106]. Heating is for 10-30 sec
at 190°C for nylon 6 and at 200°C for nylon 66.

Thermosol whitening is less suited to polyamide than to polyester,
because of slower penetration into the fiber. Improved results are
obtained if an agent is added to the padding bath which prevents
the fiber from drying out too rapidly [106,107]. An alternative is
the dip-thermosol process, where the fabric is held in the treatment
solution for 2-4 min at 85-90°C before baking [101].

Disperse Whiteners: These (e.g., FWAs 8-10) are effective at pH 7
and above in the pad-roll, pad-steam, and thermosol processes [103-
105]. Pyrazoline compounds particularly are used.

Blends with Other Fibers: Polyamide-cotton, viscose, or wool can be
whitened using a high-affinity FWA at pH slightly below 7, and in the
presence of hydrosulfite.

Mass-Whitening of Polyamide: Polyamide, like other thermoplastic fibers including polyester and polyacrylonitrile, can be mass-whitened
[13,30,31]. This is achieved by adding FWA during the polymer
manufacturing process or during the melt spinning stage.

4.5 Polyester and Polyester–Cotton Blends

Structural Types

Polyester, together with cellulose acetates, belong to a class of nonpolar fibers which require completely different methods of whitening
compared to the polar natural fibers. With nonpolar fibers, the FWA

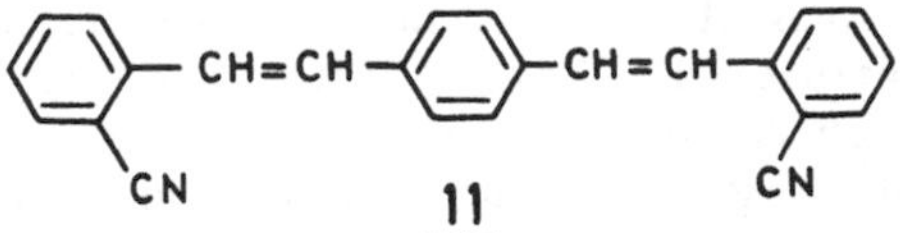

11

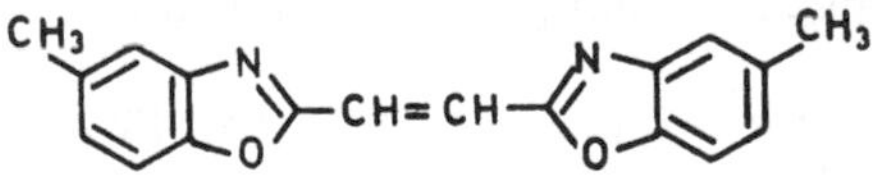

12 a

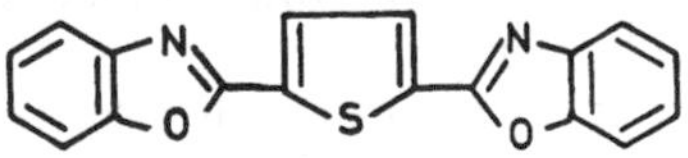

12 b

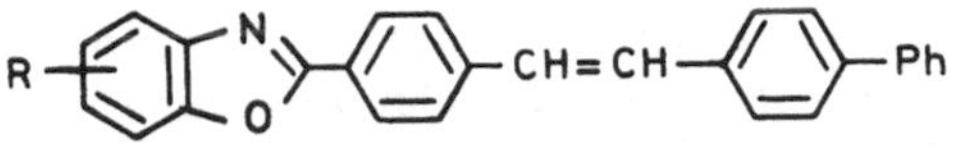

13 R = H, alkyl

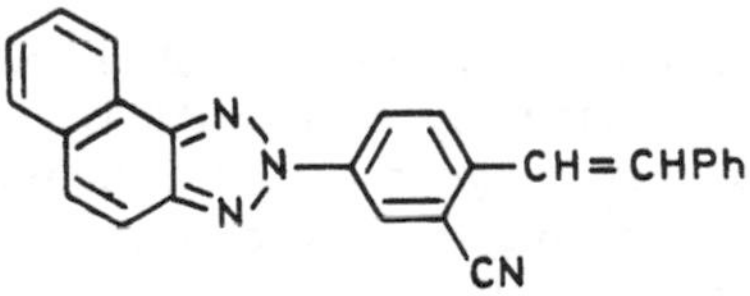

14

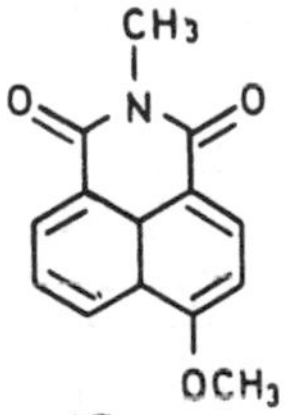

15

Figure 3.14 Some FWAs for polyester.

diffuses into the polymer substrate at high temperature, forming a solid solution, and for this purpose nonpolar whiteners are needed.

Figure 3.14 shows examples of some typical chemical structures. These include the styrylstilbene compound (11), bisbenzoxazoles (e.g., 12a, b), 4-phenylbenzoxazyl stilbenes (e.g., 13), naphtho-triazolyl-stilbenes (e.g., 14), and naphthalimides (e.g., 15).

Methods of Application

These compounds are usually applied as fine aqueous dispersions and are thus exactly equivalent to the disperse dyes used for dying polyester. Methods of application include:

Pad-bake (thermosol) and absorption thermofix (AF)
Pad-steam
High-temperature exhaust, or at lower temperature with a carrier

Pad-Bake (Thermosol) Method [2,9,24,25,30,32,93,100,101,109,111-113]: This method can be used in conjunction with the padding-on of finishes such as antistatic, softening, and soil-release agents. The baking also heat-sets the fabric. After padding and drying, baking is typically for 20 sec at 190-200°C. If temperatures above 200°C are used, yellowing may occur. After the FWA has been applied by padding and drying, the cloth has a yellowish-green tinge, since the whitener has not yet penetrated into the fiber. Only on baking is the full white effect developed. Ironing also brings up the whiteness.

In the absorption-thermofix (AF) process [23,101,110], the FWA is flocculated onto the fiber at 50-80°C in the presence of long-chain polyoxyethylene esters. After drying, whiteness is developed at 170-200°C. This process is particularly suitable for combining, in the absorption stage, with a chlorite bleach at 85-90°C for 30 min. Polyester whiteners are usually stable to chlorite.

An alternative method for developing the whiteness, after padding and drying at 140°C, is by steaming. This requires 10-15 min at 100°C or 5 min at 120-130°C. Separate heat-setting of the fabric is required.

Exhaust Methods: The best white effects on polyester are obtained by a high temperature exhaust process, preferably at 120-130°C for up to 30 min. It is usual to add a dispersing agent to the exhaust bath, and the addition of a carrier may improve the whitening. The process can be combined with a softening or antistatic treatment (or with sodium chlorite bleaching below 120°C). Whitening can also be carried out at temperatures up to 100°C in the presence of carriers, but significantly lower whiteness levels are achieved [23,110].

Household Washing of Polyester

Fluorescent whitening during the washing process is intended to replace slight whiteness losses during washing. However, no FWA has been introduced into the market which is efficient for polyester under household washing conditions, usually at 60°C and below. Thus a high level of fastness to washing and drying in the sun is particularly important for FWAs used for textile application to polyester. It has been shown [114] that one commonly used polyester FWA has little or no whitening effect on the polyester component in a polyester-cotton blend under washing conditions. Graying of polyester-cotton blends, an effect usually attributed to dirt redeposition, may also be due to loss of FWA after repeated washing. FWAs for polyester effective at 60°C have been claimed in patents [115].

Polyester-Cotton Blends

No FWA is substantive to both these fibers, and it is necessary to whiten them separately when they are blended [24,93,100,109,111-113]. The polyester component is usually mass-whitened before blending

[13,30,31], but benefits from further, topical, whitening.

Usually the polyester component is whitened first, followed by chemical bleaching and finally whitening of the cotton. This order is preferred since, during the stage of polyester whitening, the polyester FWA is also deposited on the cotton and, since it does not penetrate, the effect is to cause discoloration. With the above order, of working, the polyester FWA is washed off the cotton during subsequent processing. In addition, if the polyester is whitened by a pad-bake process at 190°C or above, there is a slight yellowing of the cotton, which may be rectified by subsequent chemical bleaching.

A continuous process [100] involves pad-bake whitening and setting of the polyester, followed by chlorite or hypochlorite bleaching of the polyester (if required). Then the cotton is bleached in a peroxide bath. Whitening of the cotton may be carried out at this stage, but preferably together with a final resin finish. Polyester may also be whitened together with chlorite bleaching. If the polyester does not need whitening, whitening of the cottom may be carried out at any stage, including during a combined scouring and peroxide bleaching process.

An alternative method of whitening involves the combined whitening of polyester-cotton blends using a mixture of FWA for cotton and FWA for polyester, incorporated into a final-stage resin cure at 140°C during 5 min. The fabric then benefits from a final wash which removes unfixed polyester FWA.

4.6 Cellulose Acetates

Cellulose acetate (diacetate) and cellulose triacetate are nonpolar fibers which are whitened in a manner similar to fully synthetic fibers, with a disperse whitener [9,116-118]. Those used for polyamide, in particular pyrazolines similar to FWA 8 (Fig. 3.13), are effective. The naphthalimide FWA 15 (Fig. 3.14) is also recommended, and bis-(benzoxazol-2-yl) derivatives (Table 3.3) are also used.

Acetate is usually exhaust-whitened at 80°C or below and at pH 5-7 during 20-60 min. Triacetate is similarly whitened, at the boil, but can also be treated by a pad-bake process combined with heat setting (e.g., 30-60 sec at 170-190°C).

4.7 Polyacrylonitrile Fibers

The properties of pure polyacrylonitrile, including its dyeing and whitening properties, are improved by the incorporation of up to 15% of copolymers [119]. These copolymers contain acidic (sulfonic, sulfate, or carboxylic) groups [120] enabling the acrylic fiber to be dyed or whitened by cationic (basic) dyes or whiteners which contain a quaternary ammonium group. However, acrylic fibers can also be whitened with nonionic disperse-type whiteners.

Some 25 different polyacrylonitrile fibers of commercial importance exist [121], containing various proportions of anionic groups. In addition, there exists a range of "modacrylic" fibers containing between 35 and 85% of polyacrylonitrile, each of which has its individual dyeing characteristics. However, very little of these latter fibers are made into white fabrics.

Cationic FWAs are fixed on the fiber by diffusing inside and forming a salt with the acid groups. Their mechanism of action is thus essentially one of ion exchange, which is in contrast to the case with nylon and wool [17]. Whitening to maximum whiteness with a typical cationic FWA (up to 2% on the fiber) will not saturate more than 10% of the anionic sites [121]. As with other fibers, the rate of whitening (dyeing) is slow below the glass transition point, which in the case of polyacrylonitrile is 70-80°C.

Structural Types

Many of the major structural types, not including DASC, have been adapted for whitening polyacrylonitrile fibers (Fig. 3.15). Examples of water-soluble cationic whiteners are the pyrazoline 16, the bis-benzoxazole(cyanine) 17, the naphthalimide 18, and a group of mixed benzimidazole-benzofuran salts exemplified by 19 [122]. Disperse-type whiteners include the naphthalimide 15, the pyrazoline 20, the 7-aminocoumarin derivatives 10 and 21, and bisbenzimidazoles (e.g., 22). All of these (except 15) can be quaternized to give cationic types, while conversely, FWAs 16-18 can be used unquaternized as disperse whiteners.

Modes of Application

Whitening is commonly by an exhaust technique in a slightly acid bath [2,5,9,97,101,121-123]. Many acrylic fibers have high natural whiteness and do not need chemical bleaching. If this is necessary, it is common to bleach in the staple yarn form, by an exhaust process, using sodium chlorite. This can be combined with fluorescent whitening, typical conditions being 30-45 min at the boil in the presence of weak acids (oxalic, acetic, formic) at pH 3-4 [5]. A nonionic detergent and a leveling agent are added. However, some FWAs, including the pyrazoline types, are not stable to chlorite. FWAs 15, 18, 19, 21, and 22 exemplify types which are chlorite stable.

FWAs can also be applied in a bath together with a softening agent at pH 4-6. A rapid, one-stage bleaching, whitening, softening, and differential shrinking (bulking) process for staple yarn has

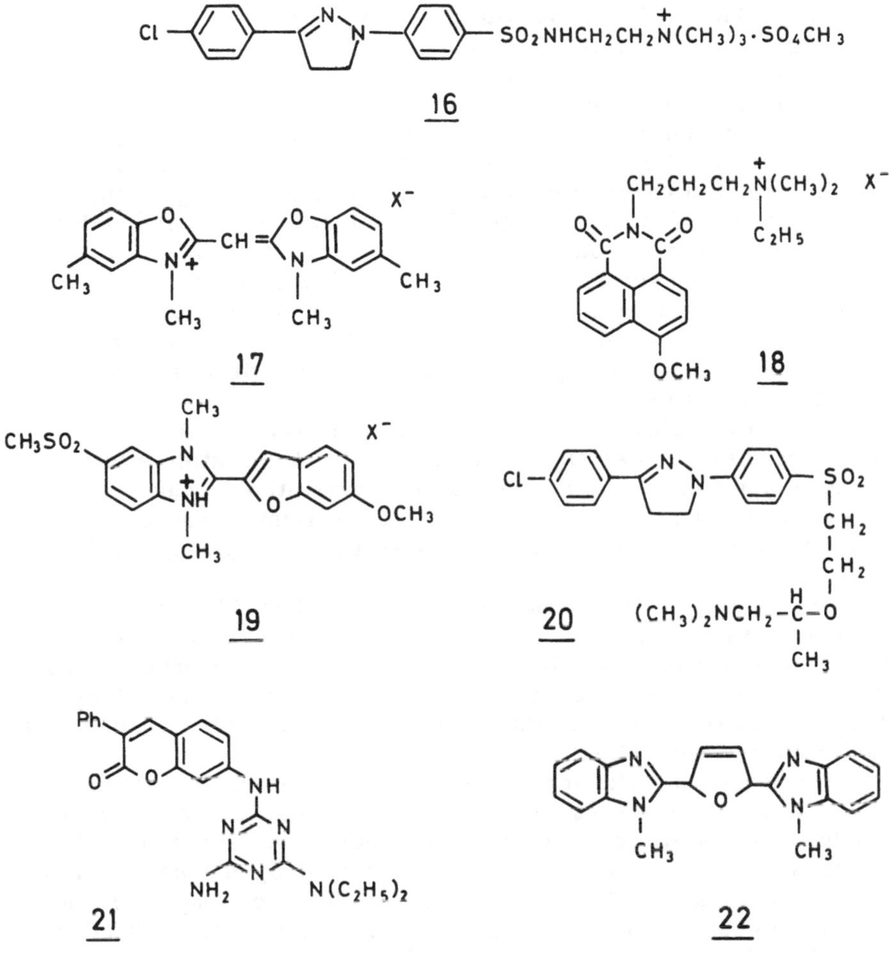

$Cl-\langle\bigcirc\rangle-\langle{}^{N-N}_{\quad}\rangle-N-\langle\bigcirc\rangle-SO_2NHCH_2CH_2\overset{+}{N}(CH_3)_3\cdot SO_4CH_3$

16

17

18

19

20

21

22

Figure 3.15 Some FWAs for polyacrylonitrile.

been developed [101]. Disperse-type whiteners can also be applied by a continuous pad acid-shock treatment, as for polyamide [121].

4.8 Wool

Photoyellowing of Wool

The most effective process of the chemical bleaching of wool is with peroxide, but the product is not light stable and soon yellows. Subsequent reduction-bleaching, usually with stabilized dithionite (hydrosulfite) preparations, improves the light stability, so that peroxide bleaching followed by hydrosulfite bleaching is advised. However, yellowing is still faster than without prior bleaching [124,125].

FWAs sensitize the photochemical degradation and yellowing of wool, particularly peroxide-bleached wool in the wet state [11,126-138]. Peroxide-bleached wool containing FWA, if hung to dry in strong sunlight after washing, may suffer irreparable yellowing [129]. A similar problem occurs with silk [130]. Photoyellowing in the wet state is due to oxidation of tryptophan (and to a lesser extent histidine and methionine) residues by singlet oxygen [130-132]. However, a different mechanism, possibly involving cleavage of the protein backbone, is operative in wool with normal moisture content, where the effect is much less noticeable [128,130]. Yellowing in the presence of FWA is slightly slower if thiourea dioxide is used instead of hydrosulfite [133].

The above yellowing is more serious than that caused by the light instability of FWAs themselves [11,128]. This latter effect is particularly noticeable in 7-aminocoumarin derivatives, which are not recommended now for wool. Of the commercially available FWAs, those of the DASC type are said to cause the least photoyellowing [129].

The amount of wool bleached was estimated in 1971 to be about 10% of total world consumption [5]. This was equivalent to about 150,000 tons a year (clean basis). However, this proportion of world wool consumption has fallen during recent years and today (1981) is probably less than 3% (excluding Russia). By no means all bleached wool is treated with FWA.

The technical difficulties involved in the fluorescent whitening of wool have led to a considerable amount of research on the subject. Avoiding the use of peroxide bleaching would give only a partial solution to the problem. Nor is the problem solved by not using fluorescent whiteners during textile finishing; unwhitened damp wool still yellows seriously. Many washing powders contain FWAs effective with cellulosics and nylon, particularly high-affinity DASC types, and these are slightly substantive to wool. Thus the wool may pick up FWA during household washing and will yellow while being dried. The problem is not limited to white woolen garments; colors on wool, particularly pastel shades, can lose their purity due to yellowing of the background as a result of repeated household washing. Because of this, acrylics are now preferred to wool for babywear.

Protection against Photoyellowing

Partial protection, during shop storage, may be achieved by wrapping a garment in PVC film containing a benzotriazole UV abosrber. Another partial solution, used with carpets which are seldom washed, is to leave residual hydrosulfite on the fabric. Other oxygen

scavengers are also applied. A method has been developed for reducing the yellowing of fluorescently whitened wool by treatment
with thiourea-formaldehyde resin [127-132,134]. The sulfur compound apparently acts by quenching singlet oxygen and excited states
of tryptophan residues [128-132]. The disadvantage of this treatment, apart from the cost is that it has only moderate washfastness.
It would not be practicable to incorporate the resin into washing powders intended for the washing of wool. Other sulfur compounds
(including thioglycolic acid and its salts) retard photoyellowing, but
all have disadvantages [129-131].

Surface whitening has been tried, involving coating the wool
fibers with a polymer containing the FWA [11,128,135]. This treatment does confer appreciably better stability to UV light, particularly in the wet state, but suffers from a number of disadvantages [128].
A different approach involves the use of UV absorbers, 2-hydroxy-
phenylbenzotriazole derivatives showing some promise [128,131,136-
138]. These compounds effectively retard yellowing caused by UV
irradiation, but those tested so far themselves cause yellowing in the
presence of visible sunlight, unless incorporated into a nonyellowing
polymer surface coating. Though UV absorbers would not be used
with fluorescently whitened wool, because of quenching of the fluorescence, they may be beneficial when applied to chemically bleached
wool together with pastel shades.

Chemical Types

Wool contains basic groups and is usually whitened by acid dye-type
FWAs, similar to those used for polyamide. As with nylon, salt formation is not considered to play a major role in dyeing of wool [17,
139]. Acid-stable DASC products, as used with cotton, are recommended [123-124]. Whiteners used for polyamide (Fig. 3.13) are also
effective, including FWAs 5 and 6. The NTS whitener 7 is slightly
effective on wool, as is FWA 2 (Fig. 3.8).

Textile Application Methods

A problem specific to wool (apart from photoyellowing) is that it absorbs in the UV region. Thus relatively large amounts are required
in order to achieve optimum effects [140]. Wool cannot stand alkaline conditions above 50°C, but can be heated to 100°C at pH 5.5.

Fluorescent whitening is carried out exclusively together with a
reductive bleaching agent (usually after a preliminary peroxide bleach)
in weakly acid medium [2,9,97,124,126]. The acidity (pH 5.5) is
provided by the hydrosulfite preparations used for reductive bleaching. With medium-affinity whiteners, the pH is adjusted to 4 with
acetic acid. Specially activated reductive bleaching agents are claimed to allow one-step bleaching and whitening, without preliminary

peroxide bleaching, in a semicontinuous pad-batch process. This involves treatment at 95°-103°C during 20-30 min [124]. FWA is incorporated into many commercial hydrosulfite preparations.

Wool Blends

In wool-polyester blends [113], best lightfastness is achieved by whitening only the polyester component, which is done before chemical bleaching of the wool. The wool may then be whitened, if required, together with the reductive bleach. Wool-polyacrylonitrile can be whitened at pH 3.5 in a reductive bleach bath using either a mixture of FWAs or an FWA of the disperse pyrazoline type (e.g., FWAs 8 or 20) which is substantive to both components. Similarly with wool-polyamide, wool-cotton, and wool-regenerated cellulose, many FWAs are substantive to both components [124, 126,141]. Wool-polyacrlyonitrile can also be whitened together with an acid peroxide bleach.

5. ANALYSIS OF FWAs

Thin-layer chromatography has been used widely for both qualitative [142-146] and quantitative [147-151] analysis of FWAs. High-pressure liquid chromatography has also been used [152,153]. Bibliographies cover publications on the evaluation of FWAs up to 1973 [1,6a].

FWA concentration on cloth can be determined from reflectance measurements in the UV absorption band [33] and by using Eq. (2). Concentration on cloth can also be evaluated from fluorescent emission measurements, using Eq. (1). The logarithmic relationship of Eq. (1) holds for concentrations met with in practice. This method is applicable to the end-use evaluation of an FWA for cotton, using a standard FWA for comparison. The FWA to be tested is applied to bleached cotton under strictly standard conditions, using for instance those specified for preparing the Ciba-Geigy textile white scale [13,44]. An average of at least 16 radiance factor measurements on the whitened cloth should be taken, using filter 11 of the Elrepho photometer. The cloth is calibrated using a standard FWA over a range of concentrations.

6. TOXICOLOGICAL PROPERTIES

Due to their wide use, particularly in washing powders, the toxicological aspects of FWAs have been extensively investigated [9,20,88, 155,156]. The stress has been laid on FWAs used in washing powders, but those used in textile finishing have also been tested. Fears of photocarcinogenic and mutagenic activity were shown by repeated and careful experiments to be without basis [1,88]. One FWA was found to be allergenic, and is not now in commercial use [88]. Two FWAs, including FWA 2, were shown to cause marked to extreme mucous membrane irritation when in high concentration [1,88,89]. Apart from this, no adverse effects can at present be attributed to FWAs.

It may be noted that, in spite of the intimate contact which textile FWAs have with the skin, manufacturers have not disclosed the chemical structures of their brand-name products, though they have indeed done this [7] in relation to many of the FWAs used in washing powders.

REFERENCES

1. F. Coulston, F. Korte, Eds.; R. Anliker, and G. Müller, Guest Eds. *Fluorescent Whitening Agents, Environmental Quality and Safety*, Suppl Vol. IV, Georg Thiem Publ., Stuttgart (1975).
2. A. K. Sarkar, *Fluorescent Whitening Agents*. Merrow Publishing Co., Watford, England (1971).
3. R. Griesser, *Physical Principles of Whiteness Improvement*, Booklet 6010, Ciba-Geigy, Basle (1976).
4. R. Zweidler and H. Hefti, *Kirk-Othmer Encyclopedia of Chemical Technology*, 3rd Ed., Vol. 4. (M. Grayson and D. Eckroth, Eds.). Wiley, New York, pp. 213-226 (1978).
5. H. Hefti, in [1], pp. 51-58.
6a. L. E. Weeks, J. L. Staubly, W. A. Millsaps, and F. G. Villaume, *Bibliographical Abstracts on Evaluation of Fluorescent Whitening Agents*, 1929-1968, STP 507, American Society for Testing and Materials, Philadelphia (1972).
6b. ASTM D459-79.
7. *List of Fluorescent Whitening Agents for the Soap and Detergent Industry*, DS 53A, American Society for Testing and Materials, Philadelphia (1972).
8. H. Gold, *The Chemistry of Synthetic Dyes*, Vol. 5 (K. Venkataraman, Ed.). Academic Press, New York, pp. 535-679 (1971).
9. R. Williamson, *Fluorescent Brightening Agents*, Elsevier North-Holland, New York (1980).
10. T. Föster, *Ann. Physics (Leipzig)* 2, 55 (1948); *Disc. Faraday Soc.* 27, 7 (1959); H. Morawetz, *Pure Applied Chem.* 52, 277 (1980); C. A. Parker, *Photoluminescence of Solutions*, Elsevier, Amsterdam (1968).
11. I. H. Leaver, *Aust. J. Chem.* 30, 87 (1977); 32, 1961 (1979); I. H. Leaver, B. Milligan and L. A. Holt, *Aust. J. Chem.* 29, 437 (1976).
12. A. Reiser. L. J. Leyshon, A. Saunders, M. V. Mijovic, A. Bright, and J. Bogie, *J. Am. Chem. Soc.* 94, 2414 (1972).

13. *Ciba Review 1973/1*, Ciba-Geigy, Basle, 1973.
14. K. H. Drexhage, *Topics in Applied Physics*, Vol. 1, *Dye Lasers*, 2nd Ed. (F. P. Schafer, Ed.). Springer Verlag, Berlin, Chap. 4 (1977); K. H. Drexhage, *Laser Focus 9(3)*, 35 (1973).
15. R. Humphrey Baker, M. Gratzel, and R. Steiger, *J. Am. Chem. Soc. 102*, 848 (1980); H. Gusten and R. Meisner, *J. Photochem. 21*, 53 (1983).
16. T. Föster, *Fluoreszenz Organische Verbindungen*, Vardenhoeck and Ruprech, Gottingen, p. 109 (1951).
17. R. H. Peters, *Textile Chemistry*, Vol. III, *The Physical Chemistry of Dyeing*, Elsevier, Amsterdam, Chap. 26 (1975).
18. R. Levene and S. Magrizo, *Textile Res. J. 51*, 559 (1981).
19. W. Schifferle and R. Keller, *SVF Fachorgan 17*, 193 (1962); *J. Soc. Dyers Colour. 78*, 359 (1962).
20. B. J. O'Hare, *Chem. Ind.* 1220 (1966).
21. M. R. Padye and S. S. Padye, *Ind. J. Technol. 5*, 357 (1967).
22. I. Soljacic and K. Weber, *Textilveredl. 14*, 97 (1979).
23. H. Theidel, *Bayer Farben Revue 29*, 64 (1978).
24. H. Hefti, *Textilveredl. 4*, 94 (1969).
25. Anon., *Ciba Review 1963/3*, 42.
26. J. H. Nobbs, D. I. Bower, and I. M. Ward, *J. Polym. Sci., Polym. Phys. Edn. 17*, 259 (1979); J. H. Nobbs, D. I. Bower, I. M. Ward, and D. Palterson, *Polymer 15*, 287 (1974).
27. A. Wagner, *Bayer Farben Review, Special Ed. 7E*, 2 (1965).
28. T. L. Dawson and J. C. Todd, *J. Soc. Dyers Colour. 95*, 417 (1979).
29. C. Eckhardt and R. Von Rutte, in [1], pp. 59-64.
30. C. Eckhardt and H. Hefti, *J. Soc. Dyers Colour. 87*, 365 (1971).
31. A. Wieber, in [1], pp. 65-82.
32. R. Williamson, *Int. Dyer 151*, 359, 408 (1977).
33. G. Anders, in [1], pp. 83-93.
34. R. Griesser, *Soap Cosmetics Chem. Spec. 53(1)*, 54 (1977).
35. R. Griesser, *Textil Praxis Int. 32(8)*, 31; *32(9)*, 24; *32(10)*, 32 (1977).
36. S. N. Glarum and S. E. Penner, *Am. Dyestuff Rep. 43*, 310 (1954).
37. E. Allen, *Am. Dyestuff Rep. 48(14)*, 27 (1959).
38. E. Allen, *J. Opt. Soc. Am. 54*, 506 (1964); *J. Color Appearance 1(5)*, 28 (1972); *Appl. Optics 12*, 289 (1973).
39. T. H. Morton, *J. Soc. Dyers Colour. 79*, 238 (1968).
40. G. G. Taylor, *J. Soc. Dyers Colour. 71*, 697 (1955).
41. R. Sève, *Die Farbe 26(1/2)*, 89 (1977).
42. B. Berglund and A. S. Stenius, *Die Farbe 26(1/2)*, 17 (1977).
43. G. Anders, *J. Soc. Dyers Colour. 84*, 125 (1968); G. Anders, *Textilveredlung 3*, 116 (1968); W. Bernard and B. Schmidt, *Textilbetreib 98(1)*, 38 (1980).

44. *Ciba-Geigy White Scales*, Publication 2406, Ciba-Geigy, Ltd., Basle (1973).
45 E. Ganz, *Appl. Optics 15*, 2039 (1976).
46. G Anders and C. Daul, *J. Am. Oil Chem. Soc. 48*, 80 (1971).
47. E. Ganz, *Appl. Optics 18*, 2963 (1979).
48. S. V. Vaeck, *J. Soc. Dyers Colour. 95*, 262 (1979); *J. Soc. Dyers Colour. 96*, 501 (1980).
49. R. Levene and A. Knoll, *J. Soc. Dyers Colour. 94*, 144 (1978).
50. G. Anders and E. Ganz, *Appl. Optics 18*, 1067 (1979).
51. A. Berger and D. Strocka, *Appl. Optics 12*, 338 (1973); K. Takohama, H. Sobagaki, and Y. Nayatani, *Die Farbe 28*, 249, 264 (1979); S. T. Henderson and B. Holstead, *Color Res. Appl. 6*, 212 (1980).
52. I. Nimeroff, *Colorimetry*, Monograph 104, National Bureau of Standards, Washington, D.C. (1968).
53. F. Grum, S. Saunders, and T. Whightman, *Tappi 53*, 1264 (1970).
54. A. Berger and D. Strocka, *Appl. Optics 14*, 726, (1975).
55. P. S. Stensby, *J. Color Appearance 2(1)*, 39 (1973).
56. A. Berger and A. Brockes, *Bayer Farben Revue, Special Ed. 3E* (1966).
57. F. Grum, F. Witzel, and P. S. Stensby, *J. Opt. Soc. Am. 64*, 210 (1974).
58. R. Hunter and W. Schramm, *J. Opt. Soc. Am. 59*, 881 (1969).
59. E. Ganz, *Appl. Optics 18*, 1073 (1979); E. Ganz and R. Griesser, *Appl. Optics 20*, 1395 (1981).
60. *Annex to International Standard* ISO 2470-1977 (E).
61. A. S. Stenius, *J. Opt. Soc. Am. 65*, 213 (1975).
62. P. S. Stensby, *Soap Chem. Specialties 43(5)*, 84 (1967); P. S. Stensby, *Detergency Theory and Test Methods*, Part III (W. G. Cutler and R. C. Davis, Eds.). Marcel Dekker, New York, p. 801 (1980).
63. G. Wyszecki and W. Stiles, *Color Science*, John Wiley, New York (1963); D. B. Judd and G. Wyszecki, *Color in Business, Science and Industry*, 3rd Ed., John Wiley, New York (1975).
64. J. Cegarra, J. Ribe, O. Videl, and F. Fernandez, *J. Text. Inst. 67*, 5 (1976).
65. R. S. Hunter, *J. Opt. Soc. Am. 50*, 44 (1960).
66. ASTM Test Method E-313; AATCC Test Method 110-1979.
67. A. Berger, *Die Farbe 8*, 187 (1959).
68. K. J. Nieuwenhuis, *J. Am. Oil Chem. Soc. 45*, 37 (1968).
69. B. K. Swenholt, F. Grum, and R. F. Witzel, *Color Res. Appl. 3*, 141 (1978).
70. F. Grum and J. M. Patek, *Tappi 48*, 357 (1965); RC 332, *Tappi 49*, 167A (1966).

71. S. V. Vaeck, *Ann. Sci. Textiles Belges 1*, 95 (1966).
72. R. Thielert and G. Schliemann, *J. Opt. Soc. Am. 63*, 1607 (1973).
73. R. Levene, A Knoll, and S. Magrizo, *J. Soc. Dyers Colour. 96*, 623 (1980).
74. E. Ganz, *J. Color Appearance 1(5)*, 33 (1972).
75. R. Griesser, *Methoden und Einsatzmöglichkeiten der Farbmetrischen Weissbewertung von Textilen*, Booklet 9134D, Ciba-Geigy, Basle (1980); *Rev. Prog. Color. Rel. Topics 11*, 26 (1981).
76. G. Anders, *Textilveredl. 9*, 10 (1974).
77. D. H. Alman and F. W. Billmeyer, *Color Res. Applications 1*, 141 (1976).
78. A. S. Stenius, *Die Farbe 26(1/2)*, 63 (1977).
79. R. Zweidler, *Textilveredl. 4*, 76 (1969).
80. H. Gold, in [1], pp. 25-46.
81. A Dolars, C. W. Schelhamer, and J. Schroeder, *Angew. Chem. Int. Ed. 14*, 665 (1975).
82. A. E. Siegrist, *J. Am. Oil Chem. Soc. 55*, 114 (1978).
83. R. Anliker, 5th Int. Symp. Colour, Basle, 24-28 Sept., 1973.
84. T. Rubel, *Optical Brighteners. Technology and Applications*, Noyes Data Corp., Park Ridge, N. J. (1972).
85. D. Barton and H. Davidson, *Rev. Prog. Color. Rel. Topics 5*, 3 (1974).
86. *Reports on the Progress of Applied Chemistry*, Society of Chemistry and Industry, London (1968-1975).
87. M. Morf, in [1], pp. 283-307.
88. C. Gloxhuber and H. Bloching, *Clin. Texicol. 13(2)*, 171 (1978).
89. P. Thomann and L. Kruger, in [1], pp. 193-198.
90. T. Vickerstaff, *The Physical Chemistry of Dyeing*, Oliver and Boyd, London, p. 102 (1954); I. D. Rattee and M. Breuer, *The Physical Chemistry of Dye Absorption*, Academic Press, New York, pp. 171-195 (1974).
91. W. Schölerman, *Bayer Farben Revue, Special Ed. 7E* 7 (1965).
92. H. Mantz and A. Bruggerman, *SVF Fachorgen 19(6)*, 426 (1964).
93. W. Schürings, *Textilveredl. 15*, 236 (1980).
94. H. Mantz, *Bayer Farben Revue 23*, 28 (1973).
95. W. T. Weller, *J. Soc. Dyers Colour. 95*, 187 (1979).
96. I. Soljacic and R. Cunko, *Melliand Textilber. 60*, 1032 (1979); *Textile Asia 11*, 112 (1980).
97. H. Zaiser, *Bayer Farben Revue, Special Ed. 7E*, 36 (1965).
98. G. Rösch, *Textilbetreib 91(5)*, 55 (1973).
99. H. B. Partlow, *Am. Dyestuff Rep. 68(3)*, 64 (1979).
100. J. Mazenauer, in *Ciba Review 1973/3*, Ciba Geigy, Basle, p. 36 (1973).
101. W. Guth, *Melliand Textilber. 8*, 715 (1980); W. Guth and A. Bruggerman, *Textilveredl. 16*, 96 (1981); Anon., *Bayer Farben Revue, Special Ed. 16* (1977).

102. W. Schürings, *Textilveredl.* 5, 15 (1970).
103. G. Rösch, *Textilbetreib* 91(6), 59 (1973).
104. J. Weihsback, *Int. Dyer* 144, 179 (1970).
105. H. Nick, *Bayer Farben Revue* 15, 44 (1965).
106. British Patent 1,556,207 (1979); *Int. Dyer* 153, 561 (1979).
107. German Patents 2,403,308 (1973); 2,235,073 (1971).
108. H. Mantz, *Bayer Farben Revue, Special Ed.* 7E, 28 (1965).
109. G. Rösch, *Textilbetreib* 91(7/8), 54 (1973).
110. Anon., *Textile Asia* 10(10), 20 (1979).
111. M. Liebiger, *Dyers Digest* 7(3), 2 (1978).
112. C. Scherk and R. Jenny, *SVF Fachorgen* 20, 333 (1965);
     U. Kirner, *Melliand Textilber.* 49, 1081 (1968).
113. *Bayer Farben Revue, Special Ed.* 20 (1982).
114. S. V. Vaeck, *Textilveredl.* 14, 528 (1979).
115. German Patent 2,817,931 (1978); U.S. Patent 3,570,866 (1969).
116. G. Rösch, *Textilbetreib* 91(11), 56 (1973).
117. C. Frommelt, *Bayer Farben Revue* 18, 48 (1979).
118. E. A. Kleinheidt, *Bayer Farben Revue* 22, 53 (1972).
119. G. E. Ham, *Text. Res. J.* 24, 604 (1954).
120. B. Kramish, *J. Soc. Dyers Colour.* 73, 85 (1957).
121. G. Rösch, *Textilbetreib* 91(10), 53 (1973).
122. R. Anliker, H. Hefti, A. Rauchle, and H. Schläpfer, *Textilveredl.* 11, 369 (1976).
123. A. Bruggermann, *Bayer Farben Revue, Special Ed.* 7E, 16 (1965).
124. C. Frommelt, *Bayer Farben Revue* 25, 80 (1975); BASF Process Data Sheet CTE-054e (1978).
125. B. Milligan and D. J. Tucker, *Text. Res. J.* 32, 634 (1962).
126. W. Schurings, *Textilveredl.* 3, 542 (1968).
127. A. Bendak, *Am. Dyestuff Rep.* 62(1), 46 (1973).
128. B. Milligan, Proc. 6th Quinquenial Int. Wool Text. Res. Conf., Pretoria, S. Africa, Vol. 4, 167 (1980).
129. G. C. Ramsay, *Wool Science Rev.* 39, 27 (1970); L. A. Holt, B. Milligan, and L. J. Wolfram, *Text. Res. J.* 44, 846 (1974); M. G. King, *J. Text. Inst.* 62, 251 (1971).
130. I. H. Leaver, *Photochem. Photobiol.* 27, 451 (1978).
131. J. A. Maclaren and B. Milligan, *Wool Science: The Chemical Reactivity of the Wool Fibre*, Science Press, Marrickville, NSW., Australia (1981).
132. I. H. Leaver, *Text. Res. J.* 48, 610 (1971); F. G. Lennox, M. G. King, I. H. Leaver, G. C. Ramsay, and W. E. Savige, *Applied Polym. Symp.* 18, 353 (1971).
133. L A. Holt and B. Milligan, *J. Text. Inst.* 69, 117 (1980).
134. B. Milligan and D. J. Tucker, *Text. Res. J.* 34, 681 (1964).
135. L. A. Holt, B. Milligan, and L. J. Wolfram, *Text. Res. J.* 45, 257 (1975); B. Milligan, L. A. Holt, and I. H. Leaver, Proc. 5th Int. Wool Text. Res. Conf., Aachen, Vol III, 607 (1975).

136. P. J. Waters and N. A. Evans, *Text. Res. J. 48*, 251 (1978).
137. P. J. Waters, N. A. Evans, L. A. Holt, and B Milligan, *Proc. 6th Quinquenial Int. Wool Text. Res. Conf.*, Pretoria, S. Africa, Vol. 4, pp. 195-204 (1980).
138. I. H. Leaver, P. J. Waters, and N. A. Evans, *J. Polym. Sci. Polym. Chem. Ed. 17*, 1531 (1979); N. A. Evans and P. J. Waters, *Text. Res. J. 51*, 432 (1981).
139. G. A. Gilbert and E. K. Rideal, *Proc. Soc., Ser. A. 182*, 335 (1944); G. A. Gilbert, *Proc. R. Soc., Ser. A. 183*, 167 (1944).
140. M. G. King, *Appl. Optics 14*, 1627 (1975).
141. C. Frommelt, *Bayer Farben Revue 27*, 48 (1977).
142. J. C. Brown, *J. Soc. Dyers Colour. 80*, 185 (1964).
143. K. Figge, *Fette Seifen Anstrichm. 70*, 680 (1968).
144. H. Thiedel and G. Schmitz, *J. Chromatog. 27*, 413 (1967). A. H. Lawrence and D. Ducharme, *J. Chromatog. 194*, 434, (1980).
145. J. Lanter, *J. Soc. Dyers Colour. 82*, 125 (1966).
146. H. Thiedel, in [1], pp. 104-110.
147. J. Schulze, T. Polcaro, and P. Stensby, *Soap Cosmet. Chem. Spec. 50*, 45 (1974).
148. H. Bloching, W. Holtmann, and M. Otten., *Seifen-Ole-Fette-Wachse 105*, 33 (1979).
149. J. B. F. Lloyd, *J. Forens. Sci. Soc. 17*, 145 (1977); A. Abe and H. Yoshimi, *Water Res. 13*, 1111 (1979).
150. H. Thiedel, in [1], pp. 111-114.
151. G. Anders, in [1], pp. 104-110, 143-156.
152. D. Kirkpatrick, *J. Chromatog. 139*, 168 (1977); *121*, 153 (1976).
153. B. P. McPherson and N. Omelczenko, *J. Am. Oil Chem. Soc. 57*, 388 (1980).
154. A. Nakea, M. Morita, and M. Yamanaka, *Bunschi Kagaku 29*, 69 (1980), *Chem. Abs. 92*, 165618V (1980).
155. *Fluorescent Whitening Agents, MVC-Report 2*, Proceedings of a Symposium held at the Royal Institute of Technology, Stockholm, Sweden (1973).
156. A. W. Burg, M. W. Rohovsky, and C. J. Kensler, *Crit. Rev. Envir. Control 7*, 91 (1977).

4

WOOL BLEACHING

RAPHAEL LEVENE / Israel Fiber Institute, Jerusalem, Israel

I. INTRODUCTION

Wool is bleached to a good cream color. The brilliant whites achieved with cotton and synthetics cannot be approached with wool, and no attempt is made to do so. Fully bleached wool is slightly whiter than unbleached scoured cotton. The light cream color of bleached wool is accepted as aesthetically pleasing by consumers who appreciate its superb and unique advantages, such as comfort, warmth, and durability.

According to a 1971 survey [1] the proportion of wool which was then bleached was about 10% of total world consumption. Thus about 150,000 tons of wool (clean basis) were then processed by bleaching. However today (1981) probably less than 3% of world consumption, excluding Russia, passes through a bleaching stage. There is a predominance of wool bleaching in the Far East, and negligible amounts of bleached wool, other than in blends, are produced in some Western Countries.

Nevertheless, the bleaching of wool remains an important subject. Bleached wool is still produced as such in many countries, and bleaching together with other finishing processes is common. Carried out together with dyeing, bleaching improves the brightness of the colors, particularly pastels, and extends the gamut of shades which can be achieved [2,3]. Mild bleaching during scouring slightly lightens the color of raw wool [3]. Bleaching improves the ground color of fabric before printing and after chlorination [4]. An additional purpose in all these cases is to counteract yellowing which occurs during dyeing, scouring printing, or chlorination. In addition, wool blends are bleached in considerable and increasing amounts [5].

Wool bleaching before dyeing is not necessary, except when bright pastel shades are required. However, all wool, whether it is to be bleached or dyed, must first be scoured. In a sense this is itself a bleaching process, in that it increases the reflectance of the gray wool. Commercial scouring processes leave a small amount of tenaciously held dirt on the fiber, and part of the whiteness improvement during bleaching is due to removal of this residue [6].

The decline in market demand for bleached wool is partly due to the high levels of whiteness obtained with cotton and man-made fibers. As against these, the whiteness which can be achieved with wool is limited by its sensitivity to chemical attack, particularly in an alkaline medium.

Thus, sodium hypochlorite severely weakens and discolors the fiber. Bleaching with hydrogen peroxide must be performed under very mild alkaline conditions; optimum conditions as used for bleaching cotton rapidly destroy wool. Sodium chlorite does not give an effective bleach on wool [7], while the sulfur compounds used as reductive bleaching agents have only a relatively mild effect.

In addition bleached wool, particularly after treatment with hydrogen peroxide, has low color-fastness to light. Though wool is bleached by the blue component of sunlight, it undergoes photochemical degradation accompanied by yellowing when subjected to irradiation by the ultraviolet (UV) component [8-12]. This process is sensitized by the presence of a fluorescent whitening agent (FWA), often incorporated into a bleaching process or possibly taken up from a washing powder during laundering. With fluorescently whitened wool, even visible light causes yellowing, and though protection is afforded by dyeing to deep shades, the problem exists with pastel shades, as well as with undyed goods.

The amount of wool bleached also depends on current fashions. The increased popularity of berber carpets has emphasized the use of wool's natural colors. In many markets wool blends have become more important, particularly for carpets [13] and suitings, and there has been a move away from printed wool.

The nature of the natural yellow-pigmentation of wool is not known. The color at least partly develops on the sheep's back and during storage, in the presence of alkaline suint, and aided by moisture, heat, and impaired ventilation [13-17]. Such conditions occur in India and induce an unscourable bright yellow coloration (canary yellow) [18]. Regions of canary yellow occur in many wools and have been shown to coincide with regions of high alkalinity [13-17]. It is possible that the yellow stain results from aldol polymerization of aldehydes (e.g., glycolaldehyde) metabolically formed in the suint [15, 16].

Another cause of color might be photoyellowing [8-11,19,20], though dirt and grease, and shielding by the wool itself, must give a large measure of protection against this, except at the tips. Photoyellowing of damp wool may be a result of the formation of aminophenols which polymerize to deeply colored polymers related to melamine [19,20]. It has been suggested that, in addition to the presence of yellow chromophores formed by the degradative effect of heat, light, moisture, and alkalinity, wool keratin may have a natural yellowness resulting from the existance of a low-lying, semiconducting layer of mobile electrons [21].

In addition to the above factors, wool is further yellowed by scouring, particularly under alkaline conditions [22,23]. Thus the pigments in wool which is to be bleached are derived from various sources. They are not necessarily only melanoproteins, similar in structure to those found in dark wools and human hair [25,26]. Yellowing caused by alkali is generally more easily removed than that caused by heat [27].

The whiter the initial color of a raw wool, the whiter it tends to be after bleaching [28]. When considering a raw wool for bleaching, obviously one of good color will be chosen. Wool can be very darkly pigmented, but breeding through the ages has produced sheep with fleeces of a cream color [25]. Australian Merino wools have good color, but South African Cape Merino is whiter. The whitest wools are from Southern Argentina, this being attributed to the cool climate [29,30].

Wool is degraded and yellowed under alkaline conditions (pH >9.5) and temperatures above 50-55°C. However, it will withstand heating to 100°C (and slightly above) under weakly acid conditions (pH 2.5-5.5) [3]. Most wool bleaching is with hydrogen peroxide at pH 8-9 and at temperatures below 55°C. A further improvement of color is obtained by subsequent reductive bleaching with sodium dithionite (hydrosulfite, $Na_2S_2O_4$) stabilized against oxidation. This two-stage oxidation-reduction "full bleach" or "combination bleach" gives the best whites available [4,32-35]. Peroxide bleaching is also carried out at slightly acid pH which inflicts less damage on the fiber. This allows higher working temperatures (around 80°C), and thus more rapid processing, but even in the presence of activators [4,24], the whiteness attained is less than that with alkaline peroxide [32]. Rapid bleaching is also possible at pH 8/70°C in the presence of a stabilizer [29].

Sodium formaldehydesulfoxylate (hydroxymethyl sulfinate) and the zinc salt have been used for wool bleaching [3,6,36-39], but are now recommended mainly for stripping dyed wool, for bleaching heavily colored fibers, or for improving the color of raw wool during scouring [6,34,36-40]. These products are prepared by the action of formaldehyde on zinc dithionite [36,40]. Small amounts of wool are bleached with a fairly recently introduced commercial product [32, 34,37,41,42] which is stated to be zinc hydroxymethyl sulfinate formulated togather with activators [34,41]. This is claimed to give, in one rapid operation, a degree of whiteness approaching that achieved with combination bleaching [4,34,42-44] and with less damage to the wool.

Bleaching may be applied at any stage of wool wet processing from scouring onwards. Batchwise processes are most usual, through continuous processes have been developed. Previous accounts of wool bleaching appeared more than ten years ago [25,33,36,45-48], though more recent practical guides have been published [4,29,34, 41,49], and two recent reviews [50,51].

2. DAMAGE DURING BLEACHING

2.1 Structure of Wool

Wool is a protein fiber, about 85% of it being composed of keratin [52-54]. Though the nonkeratinous proteins of wool are far from being unimportant, the main chemical behavior can be described in terms of polypeptide keratin chains which are cross-linked by cystine disulfide bridges (Fig. 4.1). Chemical damage to the fiber is caused mainly, either by breaking of these disulfide bridges or by hydrolytic splitting of the polyamide chain [10,36,45,48,52-58].

A wool fiber is a complex, composite structure [10,52-54,59] consisting of a thin, hard outer cuticle or sheath with a high degree of

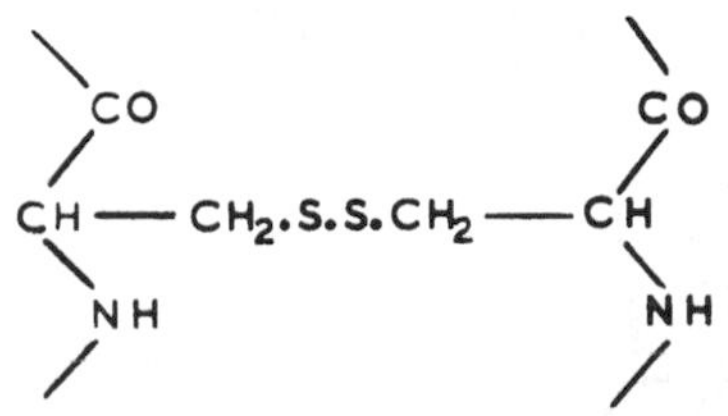

$$\overset{\displaystyle\diagdown}{\underset{\diagup}{CH}}\overset{\displaystyle\diagup CO}{\underset{\diagdown NH}{}}\!-\!CH_2.S.S.CH_2\!-\!\overset{\displaystyle\diagup}{\underset{\diagdown}{CH}}\overset{\displaystyle\diagdown CO}{\underset{\diagup NH}{}}$$

Figure 4.1 Cross-linking by cystine disulfide residues.

cystine cross-linking (the exo- and epicuticles), which is separated by the endocuticular layer from a microfibular cortex (Fig. 4.2) [52, 58,59]. The nonkeratin proteins (wool gelatins) reside mainly in the cuticle and in the cell membranes of the cortex.

2.2 Effect of Acid and Alkaline Hydrogen Peroxide

Nonkeratinous proteins have a low degree of cross-linking and are thus easily swelled. They constitute therefore the "weak points" of the fiber, being attacked preferentially by an aqueous medium particularly at pH < 3 [52,53]. Once dissolved out, though they constitute only a small proportion of the fiber, the composite struc- ture as a whole is weakened and also made more prone to further attack. Preferential hydrolytic attack within the polypeptide chain is adjacent to electron-withdrawing groups, particularly aspartic acid residues. These are found to a large extent in the nonkeratinous proteins [49]. Similarly, attack on cystine disulfide is preferentially in the region of polar side chains [10,60].

Keratinous protein is resistant to dilute mineral acids and even more so to carboxylic acids [36,58]. Weak acids cause no significant degree of breaking of either disulfide or peptide links, even at the boil [31]. In contrast even weakly basic conditions cause fiber da- mage from both the above causes, but mainly from the breaking of di- sulfide links [10,45,52,56,61]. Alkaline fiber damage shows itself in yellowing, loss of strength, a fall in abrasion resistance, and an in- creased tendency to felting. Since yellowing occurs above 55°C and pH 9, peroxide bleaching under such conditions cannot give optimum whiteness [62].

Disulfide links are oxidatively split by alkaline hydrogen perox- ide and cysteic acid residues (I) are formed [10,45,56,63-65]:

$$\overset{|}{\underset{|}{CH}}CH_2-S-S-CH_2-\overset{|}{\underset{|}{CH}} \rightarrow \ \rightarrow \ \overset{|}{\underset{|}{CH}}CH_2SO_3H \qquad (I)$$

The oxidation apparently proceeds via intermediate sulfoxide compounds, formed without $-S-S-$ fission [45]. Cysteic acid residues are also

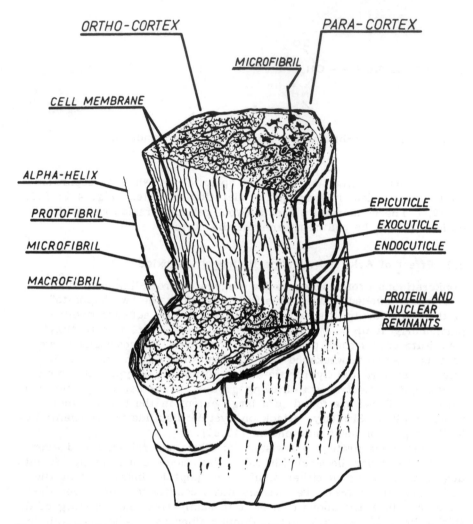

ORTHO-CORTEX

PARA-CORTEX

MICROFIBRIL

CELL MEMBRANE

ALPHA-HELIX

PROTOFIBRIL

MICROFIBRIL

MACROFIBRIL

EPICUTICLE

EXOCUTICLE

ENDOCUTICLE

PROTEIN AND
NUCLEAR
REMNANTS

Figure 4.2 Structure of a wool fiber (diagrammatic). (From Ref. 59.)

produced by alkali [56,66], by acid chlorination, and by photooxidation. The presence of these electron-withdrawing groups in the polypeptide accelerates hydrolytic chain splitting, even more so than do
aspartic acid residues [52,56]. Though acid chlorination oxidatively
attacks mainly the cuticle, alkaline hydrogen peroxide oxidation and
photooxidation have been shown to occur within the fiber [67].
     Damage to the fiber can thus be cumulative, that caused at an
early stage of processing making the wool more prone to degradation
at a later stage. Chlorinated wool and wool which has suffered

phototendering become more susceptible to degradative attack during bleaching [9,68]. This effect is very marked in wool which has
undergone a shrink-resist treatment with permonosulfuric acid [69].
An alkaline scouring process not only yellows the fiber [22-24], but
also enhances yellowing on subsequent bleaching [70]. Conversely,
excessive damage caused by an alkaline bleach can accelerate yellowing during steaming [28]. This latter effect is caused by wool damage [27], part of which is damage to the cuticular sheath [45,58],
and also by residual alkali remaining on the fiber [71].

Alkaline peroxide rupturing of disulfide bonds is accompanied
by increased ability of the fiber to swell [45,72], a fall in both wet
and dry fiber strength and in abrasion resistance [45,57,73,74].
There is a fairly good linear relationship between fall in fiber strength
and loss of disulfide sulfur [45,74]. Loss of strength in fact is not
excessive; the removal of as much as half of the cystine content
causes a fall of only 35% in tensile strength. Maximum stability of
wool fibers in aqueous boiling medium is at pH 3-4 (Fig. 4.3) [31,
52,75]. The effect of alkali and acid is exacerbated by the presence
of sodium sulfate and other salts. Abrasion resistance and cystine
content also are related in a fairly linear manner [31]. Minimum loss
of abrasion resistance is at pH 4-5 (Fig. 4.4) [31]. It is noteworthy
that loss of abrasion resistance is more sensitive to acid conditions
(below pH 3) than it is to alkali.

Hydrogen peroxide alone causes cleavage of peptide bonds; a
3% hydrogen peroxide solution will dissolve wool completely in three

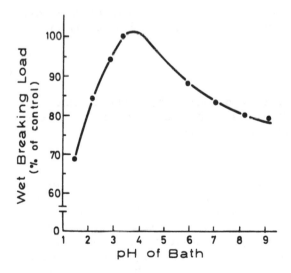

Figure 4.3 Effect of pH at the boil on the wet breaking load of
yarns. (From Ref. 31.)

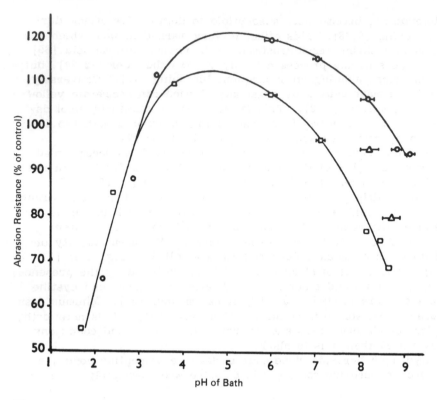

Figure 4.4 Effect of pH at the boil on abrasion resistance of yarns. ○, no salt; △, 1.5 g sodium sulfate per liter; □, 9.2 g sodium sulfate per liter. (From Ref. 31.)

days at 60°C [45,76]. Attack on wool keratin by acidic peroxide is however not fast [63], and if free thiol groups are present (cysteine residues), there is an increase of cross-linking:

$$2RSH \xrightarrow{O} R-S-S-R$$

It is common practice to protect wool from damage during bleaching in acid baths by the addition of formaldehyde. Rupturing of disulfide links is balanced by the formation of methylene bridges [45, 52,77]. Formaldehyde is present in a widely used commercial preparation added as an activator during peroxide bleaching under weakly acid conditions [4,32,34]. It is also present during reductive bleaching at pH 2-3 with an activated hydroxymethyl sulfinate preparation [4,32,34,37,41,42]. Excessive cross-linking will cause embrittlement of the fiber, shown by loss in extensibility and abrasion resistance [52].

2.3 Reductive Bleaching Agents

Reducing agents attack the cystine disulfide bonds. Sulfitolysis of these bonds by bisulfite ions in acid solution is a well-studied reaction [10,45,56]:

$$\begin{array}{ccc} | & | & HSO_3^- & | & | \\ CHCH_2S-SCH_2CH & \rightleftharpoons & CHCH_2SH & + & CHCH_2SSO_3^- \\ | & | & | & | \end{array}$$

$$\qquad\qquad\qquad II \qquad\qquad III$$

This reversible reaction proceeds to the right to the maximum extent at pH 3-4, producing cysteine (II) and cysteine-S-sulfonate (III) residues. The most common reductive bleaching agent is sodium dithionite ($Na_2S_2O_4$), which reacts similarly to bisulfite [78] and which also reduces disulfide links nonreversibly [35,56]. Sodium and zinc hydroxymethyl sulfinate, which are sometimes used for wool bleaching, are only active at 80°C and above, and at pH 3 and below (formic acid). Both dithionite and hydroxymethyl sulfinates cause significant fiber damage [10,78]. Slightly acid solutions of sodium dithionite, even when stabilized, contain bisulfite ions formed by disproportionation:

$$2 S_2O_4^= + H_2O \rightarrow S_2O_3^= + 2 HSO_3^-$$

and these are also formed by aerial oxidation to sodium pyrosulfite (metabisulfite, $Na_2S_2O_5$). Bisulfite, sulfite, and pyrosulfite ions are in aqueous equilibrium [79]. Acidic solutions of hydroxymethyl sulfinate salts at 80°C also contain dithionite ions [80]. Sulfitolysis of wool is inhibited by anionic surfactants [81].

2.4 Assessment of Wool Damage

Physical Tests

Reference has already been made to physical tests for assessing the extent of damage which has been caused by a bleaching process. Relevant standards are listed in Table 4.1. One of the most important and most generally applied tests is the measurement of tensile strength and elongation [31,77,84,85]. Other test involve the measurement of abrasion resistance [31,72,77,84-86] and the cubex washability test for area shrinking and felting [85,88]. Several dyeing tests have been devised. Kiton red G (CI Acid Red 1) penetrates the fiber and dyes it only where the outer surface has been damaged [36,45,89-91]. Uptake of methylene blue under standard conditions

Table 4.1 Standard Physical and Chemical Tests for Bleached Wool

Test	ASTM test method [82]	IWTO [83] or ISO test method
Physical tests		
Tensile strength (bundle)	D-1294-79; D2524-79 [82a]	5-73
Breaking load and elongation	D-1682-64 [82b]	—
Abrasion resistance	D-1175-80 [82b]	—
Feltability	D-3599-77 [82a]	—
Shrinkage (cubex)	D-1284-76 [82b]	ISO 38/2 N 79
Yellowness index	D-1925-70 [82c]; E313-73 [82d]	—
Light fastness	AATCC16B,C - 1977 [82a]	ISO R105/I, Pt II; R105/V, Pt II.
Chemical tests		
pH of aqueous extract	D-2165-78 [82a]	2
Alkali solubility	D-1283-80 [82a]	4-60
Cystine + cysteine content	—	15-66
Cysteic acid content	—	23-70

is dependent on cysteic acid content and also indicates the extent of damage [36,58,59,90,91].

The tendency of bleached woolen cloth to yellow, particularly when fluorescently whitened, has often been rated by standard light-fastness tests (see Table 4.1). Since these tests stipulate irradiation of cloth of normal moisture content, while yellowing is most serious with wet cloth, the ratings are of limited significance.

Chemical Tests

Many chemical tests have been devised for assessing wool damage. Those which have been standardized are listed in Table 4.1. The alkaline pH of an aqueous extract is not a measure of wool damage, but a value above 9.5 [71] can cause damage in later processing.

Routine testing for detecting excessive damage resulting from bleaching is most readily carried out by determining the solubility in alkali (0.1 N NaOH) under strictly standardized conditions [36, 45,46,48,58,63,73,85,89,91]. The alkali ruptures the disulfide bonds, and the amount of polypeptide then dissolved is greater the shorter the chains. Values of 12-17% are normal for scoured and combed wool. A value of 20-24% shows acceptable damage after bleaching [89], and greater than 30% is excessive [63,71,89.91,92]. Fiber breaking load and tensile strength fall fairly linearly with increasing alkali solubility, until a fall of about 20% in the initial value is reached [45,62, 71]. Fall in abrasion resistance also correlates well with increasing alkali solubility [87]. The test is insensitive to damage caused by strong alkali [45,72], but not to alkaline conditions used in bleaching.

Fall in cystine + cysteine content indicates damage from oxidative, reductive, or alkaline treatment. The standard test (Table 4.1) involves acid hydrolysis and colorimetric determination of the cystine, together with a minor proportion of cysteine, in the hydrolyzate [10, 57,71]. This measure shows a good linear proportionality against wet tensile strength [31,62,71] and alkali solubility [62,72] for a particular sample of wool bleached with alkaline peroxide.

Determination of the cysteic acid content involves total hydrolysis and ion exchange or electrophoretic separation [10]. The test indicates the degree to which disulfide bonds have been oxidized, which also includes oxidation to intermediate stages which precede the final stage of formation of cysteic acid residues accompanied by breaking of cross-links [45,63-65]. Excellent linear correlations have been demonstrated for cysteic acid content against cystine content [62,92], alkali solubility [62,92], and wet strength [62].

The alkali solubility has thus been shown to correlate well with cystine and cysteic acid content, fiber strength, and abrasion resistance for a particular sample of wool treated with alkaline peroxide. This justifies its use as a routine procedure for assessing damage suffered by alkaline bleached wool.

Less is known about the relationships which the above chemical parameters bear to physical damage caused by reductive bleaching. Treatment with sodium bisulfite causes an increase in alkali solubility and a reduction in cystine content [37,93]. Similar results are shown by sodium dithionite; in concentrations above 5 g/liter, sodium dithionite causes a marked increase in alkaline solubility, but is safe at low concentrations [35]. An increase of 8% in alkali solubility is normal for bleaching with an activated hydroxymethyl sulfinic acid salt [44].

3. THE MEASUREMENT OF WOOL YELLOWNESS

3.1 Yellowness and Whiteness Indices

No less important than the amount of damage which a bleaching process has caused to wool is its efficacy in reducing the yellowness, or increasing the whiteness, of the substrate. The assessment of the whiteness of textile substrates is discussed elsewhere in this book, together with an explanation of the colorimetric terms and symbols essential to the understanding of the present discussion [94]. The yellowness or whiteness of wool before and after bleaching is usually assessed instrumentally and expressed numerically using a yellowness or whiteness formula.

Table 4.2 shows a selection of some of the simpler formulae which have been used for wool. They are of two types:

1. Yellowness indices (Formulae 1-5). Values of these fall as ideal white (A = G = B = 100) is approached; a lower yellowness index represents a whiter sample. These indices are not usually used with any other textile substrate other than wool.

2. Whiteness formulae (Formulae 6-10). Values calculated with these formulae increase with increasing whiteness and approach 100 as ideal white is approached. These formulae are commonly used for other textile substrates [94] except for Formula 6 which is a measure of paper brightness. The formulae of Table 4.2 have been based on tristimulus photometer reflectance readings (A,G,B) as well as on these replaced by tristimulus values (X,Y,Z).

Other more complicated whiteness formulae have been used [2, 95,96]. Yellowness has also been defined as the color saturation in Hunter's $L\alpha\beta$ color space [2,97,107], and this has been shown to relate in a linear manner to the yellowness index of Formula 2 [107]. Color saturation in ANLAB space has also been proposed as a measure of yellowness [108].

It is essential that a yellowness or whiteness index should correlate well with the results of visual assessment, at least to the extent of giving the same ranking order. Consider a series of wool samples which have been bleached under varying conditions so as to

Table 4.2 Yellowness and Whiteness Indices

Formula No.	Name [reference]	Formula[a]	References to Users
1	Yellowness index [95]	$YI = 100(X-Z)/Y$	[2,95,96]
2	ASTM yellowness index [82c,97]	$YI = 100(A-B/G$ $= 100(1.28X-1.06Z)/Y$	[2,22,96,98,101]
3	Yellowness index [102]	$YI = 100 \ (R_{650} - R_{425})/R_{550}$	[27,39,103,104]
4	Yellowness index [82d,97]	$YI = 100(Y-Z)/Y$	[2]
5	Jacquemart [105]	$YI = [(100-Y)^2 + k(X-Z)^2]^{1/2}$ $k = 1.0, \ 5.5$	[2,23,62,95,96,104]
6	TAPPI T452 [106]	$W = R_{457}$	[4,7,27,43,44]
7	Berger [94,95]	$W = Y + 3(B-A); \ W = Y + 3(Z-X)$	[2,37,95,96]
8	Stephansen [94,95]	$W = 2B-A; \ W = 2Z-X$	[2,95,96]
9	Taube [82d,94,95]	$W = 4B-3G; \ W = 4Z-3Y$	[95,96]
10	Croes [94,95]	$W = G + B - A; \ W = Y + Z - X$	[2,95,96]

[a]YI = yellowness index; W = whiteness; R_{650}, R_{425}, R_{550}, R_{457} = reflectance at these wavelengths (nm) measured using a narrow wavelength band filter.

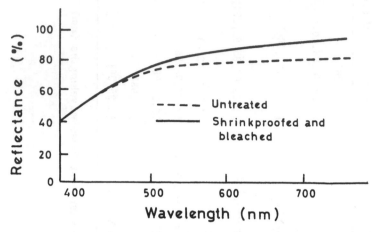

Figure 4.5 Reflectance spectra of two wool fabrics. (From Ref. 39.)

compare the efficacy of various bleaching techniques or to define
the optimum conditions of bleaching. If the instrumentally assessed ranking of the bleached substrates differs from that obtained
by visual ranking, the conclusions drawn from the investigation
are questionable. The visual assessment must be the arbiter and
not a semi-empirical instrument measure.

    Another point is demonstrated by Fig. 4.5 [39]. The upper
line represents a sample which is brighter than that represented
by the broken line, although it appears yellower and has a higher
yellowness index. This former sample, being brighter, can, nevertheless, be dyed to brighter shades. Incidently, though the samples differ in whiteness, their reflectances at 457 nm (Formula 6)
are practically the same.

    The degree of correlation between instrumental measurement
and visual assessment has been tested for a wide range for formulae,
including those of Table 4.2 (except Formula 6) [2,95,96]. Though
these investigations have been applied mostly to collections of greasy
or scoured wools, the results are also relevant to bleached substrates.
It is important in such tests to choose a sample set in which the hue,
saturation, and luminosity of the set's components all vary independently and randomly and cover a range representative of values found
in practice. Otherwise, though high linear correlation coefficients
may be found between predictions for a particular formula as against
visual assessments, the results will not be applicable to a different
collection of wools [109]. This is demonstrated by the results of an
early investigation [95] which could not be confirmed when tested by
others [2,110].

The weighting to be given to each of the three instrumental parameters which define the color of undyed wool has yet to be determined. The yellowness indices of Formulae 1 and 2 give undue weight, in rating whiteness, to substrates with a greenish hue (Fig. 4.5) and probably do not weight luminance correctly [2]. With bleached karakul wools, perceived color has been shown to correlate well with a number of instrumental measures, including ΔL^* and ΔE in CIE L^* a^* b^* space and also Formulae 5 and 10 of Table 4.2 [96].

The above discussion demonstrates the care which should be taken in applying the formulae of Table 4.2. When a series of bleaching operations with one agent are carried out under standard conditions using samples of the same batch of wool, the effect of changing various parameters (e.g., time, temperature, concentration) may probably be estimated correctly using a yellowness index, or a linear whiteness formula (7-10); the bleached samples will not differ in hue, while saturation and luminance will probably be interdependent. However, when different bleaching agents are being compared, resulting in samples of differing hue [32], as also when bleaching in conjunction with other finishing processes, the results of applying the formulae of Table 4.2 should be treated with caution.

3.2 Obtainable Volume

Increased whiteness of a wool substrate allows colors which are of lower saturation (pastel shades) and which have higher luminance (brighter colors). This has been expressed quantitatively as the obtainable volume (O.V.), that is, the maximum volume of colors obtainable in CIE (x,y,Y) and other color spaces on a wool substrate [2,111]. O.V. has been recommended as an alternative measure of whiteness [111]. The calculation of O.V. is complicated, but it has been shown to correlate well with the average of the tristimulus values $(X + Y + Z)/3$ of the substrate for a varied range of scoured wool samples.

4. THE PRACTICE OF WOOL BLEACHING

4.1 History

Wool was the first fiber to be chemically bleached, by a process known as stoving [25,36,51,59,112-114]. The damp, slightly alkaline wool was hung in a closed stove in an atmosphere of sulfur dioxide, obtained by burning sulfur. The method was used in Roman times [112,114] and is referred to in Apuleius' "The Golden Ass" [113]. Sulfur dioxide poisoning was one of the many hazards to which the dyer was subjected.

Only the surface of the wool was bleached, and the effect was not fast. Nevertheless, the method was cheap and the whiteness obtained was at least as good as that achieved with linen and cotton which, up to the end of the eighteenth century, were sunlight bleached by "crofting." Only when hypochlorite was introduced for cotton at the beginning of the nineteenth century did wool bleachers seek improvements.

Stoving has continued in common use well into the twentieth century [29,36]. Burning sulfur was replaced by sulfur dioxide from cylinders [36,115]; an aqueous solution of sulfur dioxide was also used or a solution of sodium bisulfite at pH 5 [36]. After the introduction of stabilized hydrogen peroxide in 1926, its use for bleaching wool became the preferred method.

4.2 Bleaching with Hydrogen Peroxide [51]

Stabilization

The oxidizing and bleaching ability of hydrogen peroxide solutions increases with increasing alkalinity to a maximum in the region of pH 11.5 (45,89]. Such a high alkalinity, particularly in the presence of an oxidizing agent, would rapidly destroy wool, so the bleaching solution must be buffered to a maximum of pH 9.5. Phosphate buffers are commonly used for this purpose, particularly tetrasodium pyrophosphate (TSPP), but also trisodium polyphosphate. Apart from the buffering action, there is a stabilizing effect on the peroxide, counteracting its alkali-accelerated decomposition to water and oxygen. A commercial preparation, in use for many years, contains TSPP together with sodium oxalate. Solium silicate similarly acts as a buffer and stabilizer. Alkaline earth silicates, including colloidal magnesium silicate, are particularly effective [89,116]. These alkaline buffers may be added to the bleaching bath in sufficient concentration to achieve pH 8.0-9.5, but more usually they are added at a lower concentration together with ammonia.

The stabilizing effect of TSPP is adversely affected by water-hardness (i.e., Ca and Mg salts). The latter are precipitated from solution, which gives the fiber a harsh feel. The effect is also reduced above 60°C and pH 10, probably as a result of hydrolysis to trisodium phosphate. In contrast, water hardness does not harm sodium silicate, and the stabilizing effect is even improved by Mg ions. However, bleaching with silicates may give the fiber a harsh handle [36,89,116-118].

Stabilizers act, at least in part, by complexing heavy metal (Fe, Cu, Mn) ions, which catalytically accelerate the decomposition of hydrogen peroxide [32,45,119-121]. The presence of these ions also seriously accelerates fiber degradation [58]. It is advantageous to add a sequestering agent, such as EDTA, to the bleaching bath

[36,46,122]. Organic stabilizing agents are also marketed, including a phosphonic acid derivative [123], complexing agents [87,116,124], soluble proteins and protein hydrolysates and surfactants including protein fatty acid condensates [73,89,123,126].

Steeping Methods

Wool is conventionally steeped in a bath containing from 7.5 to 30 ml of 35% w/w hydrogen peroxide per liter (1-4 vol.) together with a wetting agent. Liquor ratios vary from 15:1 to 50:1, but are usually between 20:1 to 30:1. With a bath fitted with good circulation, as used with loose wool in a package machine, ratios between 5:1 to 10:1 can be used. The goods are usually entered cold, the temperature raised to 40-50°C, and treatment continued for 3-6 hr. Alternatively the temperature is allowed to cool down from 55°C to about 38°C overnight [36]. Under these conditions wool damage is kept within acceptable limits (Figs. 4.6 and 4.7) [33,126]; more severe conditions cause excessive degradation with only marginal, and eventually regressive, improvement in whiteness. Optimal conditions have been determined using multiple regression factorial analysis [62,92,127]. Maximum whiteness was obtained after 4 hr at 45°C with 3 vol. peroxide and phosphate buffer, and after 1.12 hr with silicate.

Similar levels of whiteness, with less fiber damage, can be achieved if potassium persulfate (12-15% on weight of fiber) is added as an activator [32,89]. The bath must then not be heated, and bleaching continues for an extended period at room temperature. In

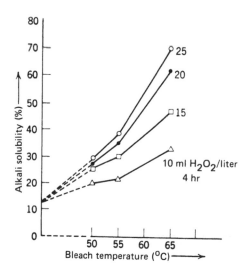

Figure 4.6 Relationship between alkali solubility and temperature for various concentration of 30% hydrogen peroxide. (From Refs. 33 and 126.)

Russia, at least up to recent years, sodium silicate in low concentration waa used, together with ammonia and preferably in the presence of a softening agent [128].

A standing bath can be reused many times, until it becomes fouled, the peroxide and alkali content being adjusted for losses before starting each batch [36,46,129]. Wool proteins, dissolved during the repeated bleaching process, stabilize the peroxide by complexing trace concentrations of copper [25,121]. After alkaline bleaching, the wool should be rinsed with dilute acid. This is particularly important if subsequent heat treatment is contemplated.

High-Temperature Rapid Peroxide Bleaching

Commercial bleaching bath auxiliaries are available which contain peroxide activators or stabilizers which allow higher temperatures and thus quicker bleaching [116]. With one such stabilizer, bleaching is for 20-40 min at 70-80°C. The pH is set at 7.8-8.0 using trisodium polyphosphate [29,130]. Bleaching under weakly acid conditions is fairly common practice. The advantages are minimal damage to the wool, allowing higher temperatures and thus rapid bleaching. However, though the majority of such bleaching is in the presence of a commercial activater [4,34,41,131-133], lower whiteness levels are obtained [32]. This activating preparation contains a complexing agent, formaldehyde and mono- and polycarboxylic acid salts, chiefly

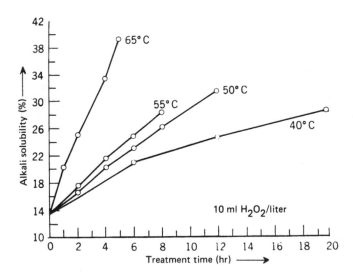

Figure 4.7 Relationship between alkali solubility and treatment time at various temperatures, with a fixed concentration of 30% hydrogen peroxide. (From Refs. 33 and 126.)

formates. Through it has been claimed that performic acid is formed, apparently this is not so [32,116,134]. Treatment is typically
at pH 5.5 for 45-60 min at 80°C or for 2-3 hr at 65°C. The bleached goods tend to have a pink hue which is removed by subsequent
reductive bleaching with hydrosulfite.

Pad-Batch Methods

Cold pad-batch processes give lower levels of whiteness but save
energy and work at a greatly reduced volume. A process has been
used, particularly for loose carpet wool, in which the wool is padded
in the last bowl of the scouring train [25,33,36,46,47,49,135] or back-
washing unit [25] and bleaching continued during subsequent drying
and storing or processing in the dry state. In a similar process,
wool impregnated in the scouring train at pH 3-5 is held for several
days in a damp state [25,36,136]. With blanket wool the cloth is
then in a particularly favorable condition for raising [25]. In a commercially established process using cold pad-batch dyeing equipment,
wool tops are impregnated with peroxide in the presence of formic
acid and isopropanol and stored cold for two days [137]. Damp wool
must be protected from sunlight, and there is also a danger of bacterial growth [25,58,138].

With the acid-peroxide activator, padding-bath concentrations of
up to 15 vol. peroxide may be used for cold-pad-batch bleaching.
Piece goods are not suited to this process since they may show an
uneven effect due to a temperature gradient caused by spontaneous
internal warming of the batch. A warm pad-batch procedure is suggested, in which the impregnated wool is stored in insulated bins
[133].

Continuous Processes with Peroxide

Continuous processes for wool are justified when large quantities are
to be treated, as with tops. A pad-steam process has been described
in which cloth or combed wool is padded with 6-8 vol. peroxide at
pH 9 (phosphate stabilizer) and then steamed for 8-11 min at 102°C
[139]. Lower whiteness levels are obtained with padding at pH 5,
though wool damage is reduced. The acid-peroxide activator is
readily adaptable to such a process. Padding is with 3-15 vol. peroxide at pH 5-5.5, and steaming is for 10-20 min at 100°C [4,131,
133]. This is suitable for loose wool, slubbing, hanks, and piece
goods. Another pad-steam process uses a stabilizer which allows
steaming at 85°C/pH 7.8-8 [29,130].

4.3 Reductive and Combination Bleaching

Sodium Dithionite

The major proportion of wool reductive bleaching is with sodium dithionite (hydrosulfite). Such solutions are unstable, and improvement

in bleaching performance is obtained by adding a stabilizer, which buffers the solution to between pH 5.5 up to near neutral pH. Phosphates have been used for this purpose, but nonphosphate buffers are also used [34,140]. Commercial preparations usually also contain a complexing agent. The bleach obtained with hydrosulfite is superior to that given by bisulfite, but less than that obtained with peroxide [35,37,141].

Hydrosulfite [2-5 g/liter) is best added to the bath when it has been heated to the working temperature (45-65°C). The goods are then introduced together with a surfactant, the bath is covered, and bleaching is continued for 1-2 hr. Longer periods may also be used in a bath left to cool. Residual odors may be prevented by adding peroxide to the last rinse bath. Bleaching with hydrosulfite is cheaper than with peroxide, but the bath cannot be replenished and reused. Hydrosulfite is not suitable for pad-batch processes, since it is too rapidly oxidized by air.

Combination Bleaching (Oxidizing-Reducing System)

Reductive bleaching is rarely practiced alone, but only in order to further improve whiteness after peroxide bleaching (combination, full, or double bleaching) [33,126]. A conventional combination bleach involves treatment with warm alkaline peroxide followed by a second treatment with dithionite. It is common practice to include a fluorescent whitening agent with the second (reductive) stage, of a combination bleach. Alternatively a third, reductive, stage is sometimes added, incorporating a fluorescent whitener, this in spite of the risk of photosensitized yellowing which exists with FWA treated wool, particularly when wet. Many commercial hydrosulfite preparations are formulated together with FWA.

Activated Sulfinic Acid Derivative

This product, zinc hydroxymethyl sulfinate formulated together with activators [34,41], is much more stable than is hydrosulfite, allowing higher temperatures and shorter times for bleaching [4,34,37, 42-44]. It is claimed to give, in one stage, and without preliminary peroxide treatment, a level of whiteness approaching that achieved with combination bleaching [4]. Typical conditions for bleaching in a long liquor are 30 min at 80°C and pH 2.6-3.0 (formic acid). There is a tendency for the bleached goods to have a harsh feel and a sulfurous odour, so that after-rinsing with a dilute solution of peroxide containing a softening agent may be necessary. Unlike hydrosulfite, this preparation can be used in a cold pad-batch process (15 hr at room temperature) and is also adaptable to a continuous pad-steam procedure (20 min steaming at 102°C).

4.4 Other Bleaching Processes

Thiourea Dioxide

Thiourea has been introduced as an alternative to hydrosulfite in textile applications such as vat dyeing and reduction clearing [142], and has also been proposed for reductive wool bleaching [77,142, 143]. For this purpose it may be as effective as sodium dithionite [142,144], and when used together with FWA gives a marginally better wool photostability [144]. It is claimed that efficient bleaching is achieved when as little as 0.5% (on weight of goods) is used [142]. However, optimal conditions for efficient bleaching with this reagent without excessive fiber damage have not been delineated. It is not itself a reducing agent, but is hydrolyzed in solution by hot dilute sodium hydroxide to urea and sodium sulfinate (Na_2SO_2). The latter has strong reducing properties. Thiourea dioxide is stable to hydrolysis by dilute acid and is thus relatively ineffective at low pH.

Photobleaching

Though wool is yellowed by ultraviolet light, it is bleached (in the absence of FWA) by blue light from which the UV component has been filtered. Intense irradiation with blue light has been suggested as a means of bleaching wool, particularly raw wool [10,11,145,146]. This is potentially adaptable to a rapid, continuous, dry process [145], but the economic advantages over chemical bleaching are not clear [146]. Photobleaching is accelerated in the presence of zinc thioglycolate and other reducing agents [8,19,98,147].

Mordant Bleaching of Dark Wools

Heavily pigmented (e.g., karakul) wools and dark cross-bred wools for carpets and blankets have been peroxide-bleached after mordanting with ferrous sulfate [5,25,29,36,40,45,96,116,121,129,148]. The wool is first treated with ferrous sulfate and sodium dithionite (the latter to prevent formation of ferric ions at this stage), usually in the presence of formaldehyde. The iron salt is absorbed preferentially by the melanoprotein pigment, causing localized activation of the peroxide. The latter is then applied after thorough washing. Iron-catalyzed degradation is concentrated on pigmented protein, and damage to unpigmented protein is also reduced by the formaldehyde. In this way light-colored material for dyeing can be obtained from heavily pigmented wools.

Solvent Bleaching

Alkaline peroxide bleaching in water-perchloroethylene or -trichloroethylene emulsions has been proposed [149,150]. Similar results are obtained as from the conventional method, but in shorter times and using less peroxide. The process can be combined with fluorescent whitening. Permonosulfuric acid can replace hydrogen peroxide in solvent bleaching, but there is no concommitent shrink-proofing [161].

Other Bleaching Agents

Potassium permanganate has occasionally been used for wool, as a replacement for hydrogen peroxide [29,33]. Sodium chlorite in acid solution has been tested, followed by reductive bleaching [25,77, 129,152-154]. Though a degree of shrink-resistance is conferred to the wool, the level of whiteness obtained is low [154]. Chlorite will bleach urine-stained wool [153], but similar results can be achieved using peroxide [32]. Similarly, peracetic acid has been suggested for wool bleaching [48,155], but better results are achieved using peroxide and with less damage [156]. Sodium pyrophosphate perhydrate has been proposed as a wool bleaching agent [157]. An interesting innovation involves peroxide bleaching in the presence of a protein cleavage enzyme, such as Esparase [158,159]. This also confers a measure of shrink-resistance [158].

4.5 Bleaching in Combination with Other Processes

Scouring

Peroxide pad-batch bleaching, with padding in the last bowl of a scouring train or back-washing unit, has already been described (Sec. 4.2, discussion on Pad-Batch Methods). Continuous bleaching of raw wool is also carried out together with scouring. This involves the addition of sodium bisulfite [3,160,161], sodium dithionite or, occasionally, hydrogen peroxide to the last bowl. The wool, after leaving the bowl, is then rinsed and dried. Since immersion is for 4 min or less, at 60°C, only a slight bleaching effect can be achieved. With poorly colored wools a worthwhile whiteness improvement is obtained with sodium bisulfite, but the effect with wools of good color is hardly noticeable [3,161]. However, bisulfite neutralizes alkalinity, thus preparing the wool for subsequent heating and dyeing processes [161].

Dyeing

Bleaching together with dyeing is possibly the major bleaching operation applied to pure wool. The bleaching action on the wool ground permits the achievement of bright, pastel shades and also, since dyeing is the most damaging of all the aqueous finishing processes applied to wool [52], counteracts the accompanying yellowing of the background. Combining bleaching with dyeing, rather than carrying out a separate preliminary treatment, gives more level, more reproducible, and more controllable results, as well as saving the expense of an additional stage.

Hydrogen peroxide may be used, either with or without the acid-peroxide activator [4,133,162]. Peroxide can be used only with acid dyes and should only be added 15-20 min before the end of the dyeing process, so as to avoid possible adverse effects on some of these dyes. Rate of dyeing is enhanced with fibers which have been damaged by peroxide bleaching [10,164]. Bisulfite also increases the rate of dyeing with wool [10,165]. A number of mild reducing bleaching-preparations are marketed for incorporating into dye-baths [4,29,38,48,51,166]. These are based on bisulfite and hydroxylamine sulfate among other compounds, and include a sequestering agent. They may be used at pH < 7 and are compatible with acid and pre-metalized dyes, though they are not recommended for use with chrome dyestuffs.

Several cold-pad-batch dyeing processes have been developed together with bleaching. An additional factor here is that cold dyeing causes less yellowing of the ground [137]. Wool tops may be dyed with reactive dyes at weakly acid pH in the presence of urea and hydrogen peroxide [137,167]. Dyeing is also possible with acid dyes in the absence of urea, together with the acid peroxide activator [4,131,133]. Sodium metabisulfite is often incorporated into the padding liquor in cold pad-batch dyeing with certain reactive dyes in the presence of urea [29,137,168]. In addition to improving the depth or dyeing, this also produces brighter shades [137,168].

Printing

Before printing, wool is usually made more receptive to dyes by chlorination in weakly acid solution using dichlorocyanuric acid. This is followed by treatment with an antichlor, usually sodium bisulfite or metabisulfite. In addition to neutralizing excess chlorine on the goods, this counters the yellowing which accompanies chlorination. A clearer background is obtained by using instead a more powerful reductive bleaching agent, such as sodium hydroxymethyl sulfinate [39], activated zinc hydroxymethyl sulfinate [4], or stabilized dithionite [29,140].

Shrink-Resist and Anti-Felt Treatment

Chlorination of wool, either as an anti-felt treatment or to prepare it for a resin "Superwash" (e.g., "Hercosett") finish, also requires an antichlor after-treatment with bisulfite. This may be replaced, before dyeing, by a more powerful reductive bleach as above, or it may be followed by a combination bleach. It is possible to combine several stages by adding the antichlor, together with a mild reductive bleaching preparation, to the dye-bath for pastel shades [169].

Bleaching is possible before chlorination, for example with loose wool, but excessive fiber damage will result unless the subsequent chlorine dosage is reduced [170]. It is thus usual to bleach after chlorination [4,169]. "Superwash"-treated wool may be bleached with peroxide, or, for best whites, using a combination bleach [170].

An alkaline peroxide bleach may overdamage wool which has been chlorinated [4,153]. With "Hercosett" wool, peroxide bleaching is slightly slower and less effective than with normal wool [90,92]. "Synthappret BAP" resin treated wool may be bleached before or after treatment. However, peroxide bleaching before a shrink-resist treatment generally causes less damage than with the reverse sequence. A peroxide bleach also improves the effectiveness of most subsequent shrink-resist treatments, particularly if a silicate stabilizer is used [69]. Wool which has undergone a shrink-resist treatment with an acid bromate solution [171,172] can subsequently be bleached with stabilized sodium dithionite [173]. This treatment replaces one with thiosulfate, also removing excess bromine.

Other Finishing Processes

Introduction of sulfur dioxide into a steam heat-setting chamber counteracts yellowing [174]. Wool which has been flame-proofed by Zr-treatment may be bleached reductively, but Ti-treated wool is not suitable for bleaching [175]. Mothproofing treatments can also be carried out together with bleaching [166,176].

4.6 The Bleaching of Wool Blends

Blends of wool with other fibers have attained increasing importance in the United States [5]. Generally, in Western countries, although pure wool has maintained its overall strength [177], blends with polyester and PAN have become more popular.

The conditions for bleaching wool are essentially those for pure wool; the other component is never more sensitive to damage than is wool, except to a marginal extent. However, wool damage in blends is less serious than with pure wool, since the other component dilutes the wool and supports it. Thus a polyester blend containing only a small proportion of wool can be treated with alkaline peroixde at the boil, in combination with scouring [5]. In all types of wool blends, conventional combination bleaching (Sec. 4.3, discussion on "Combination Bleaching") gives the best effect [41].

Wool-Cotton [4,41,48,154,178,179]

These blends are best bleached using conventional alkaline peroxide conditions. This may be followed by reductive bleaching with dithionite. Sulfinic acid derivatives at low pH should not be used, since this damages the cotton. A cold pad-batch process has been developed using alkaline peroxide in the presence of a sodium silicate-based stabilizer [178,179]. A continuous pad-steam process, with steaming for 15 min in the presence of a tetrasodium pyrophosphate-based stabilizer, is less effective.

Polyester-Wool [4,5,41,48,180]

Bleaching here is restricted to wool, which is the minor component, and conventional combination bleaching gives the best results. Rapid peroxide bleaching (Sec. 4.2, discussion on "High-Temperature Rapid Peroxide Bleaching") is recommended, e.g., for 1 hr at 70°C at slightly acid pH [180]. If simultaneous fluorescent whitening of the polyester is desired, a carrier and FWA are included and treatment continued for a further half-hour at 98°C. Even shorter bleaching times are possible in the presence of activated zinc hydroxymethyl sulfinate allowing continuous operation [42]. This may be in a long liquor, for 10 min at 105°C or in a pad-steam process for 15-30 min at pH 4-4.2 [4,42,43]. However, a preliminary trial is essential, since neither all wool nor all polyester is compatible. The long-liquor process is also suitable for bleaching the wool component before it is dyed. When dyeing the polyester component, staining of the wool by disperse dyes is removed by a reductive scour with stabilized dithionite. With selected chrome dyes this is possible after dyeing of the wool [5]. The higher the temperature of dyeing the less the wool is stained.

Wool Acrylic Blends

An acrylic-wool mixture is usually peroxide bleached at pH 8.5-9 [29,41]. Bleaching at pH 5.5 in the presence of acid peroxide activator may be used, possibly followed by reduction bleaching, and treatment with sulfinic acid derivatives is also possible [41]. None of these procedures is particularly effective for the PAN component, but the high whiteness of the latter generally makes bleaching unnecessary. Brighter shades of dyeing can be obtained if, after dyeing the acrylic component with a cationic dye, a zinc hydroxymethyl sulfinate wool reduction clearing and bleaching treatment is applied [27].

Other Mixtures

Conventional alkaline bleaching of wool-polyamide, at pH 8.5-9.0 and 50°C, is recommended only in the presence of an antioxidant protective agent which reduces attack of peroxy compounds on polyamide fiber [41]. The latter is not necessary in a rapid-bleach procedure at pH 5.5 [41], or when the blend contains less than 25% polyamide [181]. Low-temperature bleaching can also be used [182]. All reduction procedures are suitable [43,44].

Care should be taken with an alkaline bleaching process for a wool mixture containing acetate or triacetate. Acetate particularly may be damaged if, at 50°C, a pH of 8.5 is exceeded. An activated reductive bleach is particularly recommended for this blend [41].

ACKNOWLEDGEMENTS

The author wishes to thank the International Wool Secretariat Technical Center at Ilkley for supplying up-to-date information on wool bleaching. He is particularly grateful to Mr. Irwin Seltzer, IWS, Israel Office, for supplying information and for a critical and constructive reading of the manuscript. Thanks are also due to the South African Wool Textile Research Institute and to the Wool Research Organisation of New Zealand for sending copies of their reports, and to authors who sent copies of their articles. He is indebted to manufacturers who sent him technical literature on request.

REFERENCES

1. H. Hefti, *Fluorescent Whitening Agents, Environmental Quality and Safety*, Supp. Vol. IV (F. Coulston and F. Korte, Eds.). Georg Thieme Publ. Stuttgart, pp. 51-58 (1975).
2. M. G. King, *Wool Sci. Rev.* 49, 2 (1974); 50, 14 (1974).
3. J. L. Hoare, *J. Text. Inst.* 65, 402 (1974).
4. BASF, *Processes of Bleaching Wool, Alone and in Blends with Other Fibres*, Process Data Sheet CTE-054e (1978).
5. M. Drewniak, *Am. Dyestuff Rep.* 68(6), 45 (1979).
6. J. L. Hoare, *WRONZ Rep. No.* 47 (1978).
7. F. A. Barkhuysen and R. A. Leigh, *SAWTRI Tech. Rep. No. 364* (1977).
8. G. C. Ramsay, *Wool Sci. Rev.* 39, 27 (1970).
9. C. H. Nicholls, *Developments in Polymer Photochemistry-1.* (N. S. Allen Ed.). Applied Science Publ., London, pp. 125-144 (1980).
10. J. A. Maclaren and B. Milligan, *Wool Science: The Chemical Reactivity of the Wool Fibre*, Science Press, Marrickville, NSW, Australia (1981).
11. A Bendak, *Am. Dyestuff Rep.* 62(1), 46 (1973).
12. R. Levene and M. Lewin, this Volume, "The Fluorescent Whitening of Textiles," Sec. 4.8 and loc. cit.
13. D. A. Ross, *Wool 7*, 37 (1980-81); *World Text. Abs.* 1981/5732 (1981).
14. J. L. Hoare, *J. Text. Inst.* 62, 110 (1971); *WRONZ Comm. No. 2* (1968).
15. F. G. Lennox and M. G. King, *Text. Res. J.* 38, 754 (1968).

16. J. L. Hoare, *WRONZ Rep. No. 46* (1978).
17. J. L. Hoare, and R. G. Steward, *J. Text. Inst. 62*, 455 (1971).
18. V. S. Kenkare, V. C. Gupte, and S. Ramachandran, *Wool Woollens India 11(1-2)*, 41 (1974); *11(3)*, 15 (1974); *11(5)*, 19 (1974); V. S. Gupte and S. Ramachandran, *Wool Woollens India 10(10)*, 2 (1973); M. Singh, Q. Sheikh, and N. Ahmed., *J. Text. Inst. 66*, 231 (1971).
19. D. M. Lewis, *P. D. Report No. 156*, IWS, Ilkley (1973).
20. F. G. Lennox, M. G. King, I. H. Leaver, G. C. Ramsay, and W. E. Savige, *Appl. Polym. Symp. 18*, 353 (1971).
21. J. L. Hoare, *J. Text. Inst. 65*, 503 (1974).
22. D. W. Turpie and E. Gee, *SAWTRI Tech. Rep. No. 288* (1976).
23. H.-J. Meiswinkel, G. Blankenburg, and H. Zahn., *Text. Praxis Int. 36*, 403 (1981).
24. V. Kopke, *J. Text. Inst. 61*, 388 (1970); *Wool Sci. Rev. 38*, 10 (1970).
25. L. Chesner, *Wool Sci. Rev. 30*, 16 (1964); *31*, 1 (1964).
26. G. Nitschke, *Faserforsch. Textiltech. 7*, 401 (1956).
27. G. P. Norton and C. H. Nicholls, 3rd Int. Wool Text. Res. Conf. (CIRTEL), III, 108 (1965); *J. Text. Inst. 55*, T462 (1964); *Text. Res. J. 37*, 1031 (1967); *J. Text. Inst. 51*, T1183 (1960).
28. J. Cegarra, J. Ribé, and J. Gacen, *J. Soc. Dyers Colour. 85*, 147 (1969).
29. *Wool, A Sandoz Manual*, Sandoz Ltd., Basle, Switzerland (1979).
30. J. Gacen, J. Cegarra, and M. Caro, *Tientex 44(10)*, 11 (1979).
31. R. V. Peryman, *J. Soc. Dyers Colour. 70*, 83 (1954).
32. W. Selb, *Text. Praxis Int. 36*, 163, 559 + XXV, 691 + XVII, 867 (1981).
33. H. E Millson and W. von Bergen, in *Wool Handbook*, 3rd Ed., Vol. 2, (W. von Bergen, Ed.), Interscience Publishers, New York, pp. 763-774 (1970).
34. C. Frommelt, *Bayer Farben Revue 25*, 80 (1975).
35. N. J. J. van Rensburg, *Text. Res. J. 38*, 318 (1968).
36. E. R. Trotman, *Textile Scouring and Bleaching*, Griffin, London (1968).
37. N. J. J. van Rensburg, *SAWTRI Tech. Rep. No. 162* (1972).
38. J. L. Hoare, R. G. Stewart, and C. E. Brown, *WRONZ Comm. No. 9* (1971).
39. G. A. Swanepoel and J. Becker, *SAWTRI Tech. Rep. No. 105* (1968).
40. B. M. Baum, J. H. Finley, J. H. Blumbergs, E. J. Elliot, F. Scholer, and H. F. Wooten, in *Kirk-Othmer Encyclopedia of Chemical Technology*, 3rd Ed., Vol. 3. (M. Grayson and D. Echroth, Eds.). Wiley Interscience, New York, pp. 938-958 (1978).

41. C. Frommelt, *Bayer Farben Revue 27*, 46 (1977).
42. BASF, Techn. Leaflet M 5431e (1975).
43. O. Schmidt, *Textilind 73*, 588 (1971); *Colourage 24(18A)*, 55 (1977).
44. O. Schmidt, *ITB Dyeing, Printing, Finishing (World Eng. Ed.)* 4, 371 (1972).
45. R. H. Peters, *Textile Chemistry*, Vol. II, *Impurities in Fibres; Purification of Fibres*, Elsevier Publishing Co., Amsterdam, pp. 276-321 (1967).
46. *Interox Textile Bleaching Manual*, Section 3 (1979).
47. T. E. Bell, in *Encyclopedia of Polymer Science and Technology*, 1st ed., Vol. 2. (H. F. Marks, N. G. Gaylord, and N. M. Bikales, Eds.), Wiley Interscience, New York, pp. 438-467 (1968).
48. W. Schuring, *Textilveredl. 3*, 542 (1968); A. Korde, *Text. Dyer Printer 1(6)*, 42, *1(7)*, 43, *1(8)*, 44 (1968); G. Rosch, *Spinner Weber Textilveredl. 84*, 628, 760 (1966).
49. H. Heiz, *Deutsche Farbe Kalender 83*, 124 (1979); Anon. *SVF Lehrgang Textilveredl.*, No. 12, H025 (1975).
50. J. Cegarra and J. Gacen, *Wool S i. Rev. 59*, 2 (1983).
51. M. White, *Text. Prog. 13(2)*, 8 (1983).
52. H. Baumann, *Fibrous Proteins: Scientific, Industrial and Medical Aspects*, Vol. 1. (D. A. D. Parry and L. K. Creamer, Eds.). Academic Press, London, pp. 299-370 (1979).
53. H. Zahn and P. Kusch, *Melliand Textilber. 62*, 59 (1981); Eng. Ed., *1/81*, 75 (1981).
54. J. H. Bradbury, 5th Int. Wool Text. Res. Conf., *I*, 439 (1975).
55. A. Robson, 5th Int. Wool Text. Res. Conf. *I*, 137 (1975).
56. W. H. Leon, *Text. Progr. 7(1)*, 1 (1975).
57. W. Schefer, *Textilveredl. 15*, 484 (1980).
58. E. R. Trotman, *Dyeing and Chemical Technology of Textile Fibres*, 5th Ed., Griffin, London (1975).
59. M. G. Dobb, F. R. Johnston, J. A. Nott, L. Oster, J. Sikorski, and W. S. Simpson, *J. Text. Res. 52*, T153 (1961).
60. H. Lindley and R. W. Cranston, *Biochem. J. 139*, 515 (1974).
61. R. S. Asquith, Ed., *Chemistry of Natural Protein Fibers*, Plenum Press, New York (1977).
62. J. Cegarra, J. Gacen, and M. Caro, *J. Soc. Dyers Colour. 94*, 85 (1978); J. Cegarra in *Reports of the Seventh Joint Session of the Aachon Textile Research Institute*, Vol. 85 (K. Ziegler, Ed.). German Wool Research Institute and Aachen Technical College, Aachen, pp. 66-84 (1981).
63. A. L. Smith and M. Harris, *J. Res. Nat. Bur. Standards 16*, 301, 309 (1936); *17*, 97, 577 (1936); *18*, 623 (1937); *Am. Dyestuff Rep. 25*, 180P (1936).
64. R. Consdon and A. H. Gordon, *Biochem. J. 46*, 8 (1950).

65. J. Nachtigal and C. Robbins, *Text. Res. J. 40*, 454 (1970); B. J. Sweetman, J. Eager, J. A. Maclaven, and W. E. Savage, 3rd. Int. Wool Text. Res. Conf. *II*, 62 (1965).

66. P. Miro, J. J. Garcia-Dominguez, and J. L. Parra, *J. Soc. Dyers Colour. 85*, 407 (1969).

67. J. M. Marzinkowski and H. Baumann, 6th Int. Wool Text. Res. Conf. *II*, 411 (1980).

68. W. S. Simpson and C. T. Page, 6th Int. Wool Text. Res. Conf. *V* 183 (1980); I. L. Weatherall, 5th Int. Wool Text. Res., Conf. *II*, 580 (1975).

69. L. Benisek and M. J. Palin, *J. Soc. Dyers Colour. 99*, 154 (1983).

70. J. Cegarra, J. Gacen, and M. Caro, *Text. Chem. Col. 6*, 83 (1974); H. J. Henning, *Z. ges. Textilind. 69*, 237 (1969).

71. H. Zahn, *Wool Sci. Rev. 32*, 1 (1967); H. Zahn, H. J. Henning, and G. Blankenburg, *Text. Inst. Ind. 8*, 125 (1970).

72. J. C. Brown, *J. Soc. Dyers Colour. 75*, 11 (1959).

73. P. Alexander and F. Hudson, *Wool, its Chemistry and Physics*, 2nd Ed. Chapman and Hall, London (1963).

74. E. Elod, H. H. Novotny, and H. Zahn, *Kolliod-Z. 93*, 50 (1940); *100*, 283 (1942).

75. H. Baumann and B. Potting, *Textilveredl. 13*, 74 (1978).

76. E. Elod, H. Novotny, and H. Zahn, *Melliand Textilber. 21*, 617 (1940); *23*, 313 (1942).

77. A. I. Valko, B. Bitter, and L. Gagnon, *Am. Dyestuff Rep. 58(2)*, (1969).

78. G. Valk, 3rd Int. Wool Text. Res. Conf. *2*, 375 (1965); C. S. Lowe, A. C. Lloyd, and A. C. Smith, *Am. Dyestuff Rep. 30*, 81 (1941); A. E. Brown and M. Harris, *Ind. Eng. Chem. 40*, 316 (1948); *Am. Dyestuff Rep. 36*, 316 (1947).

79. R. M. Golding, *J. Chem. Soc.* 3711 (1960).

80. R. G. Tinker, T. P. Gordon and W. H. Corcoran, *Inorg. Chem. 3*, 1407 (1964); L. A. Pokhilko, N. E. Bulusheva, and A. V. Senakhov, *Referat. zhur.* 12B6 (June 1981); *J. Soc. Dyers Colour. 97*, 551 (1981).

81. J. L. Parra, J. Garcia-Dominguez, M. R. Infante, and J. M. Garcia, *J. Text. Inst. 68*, 191 (1977).

82. *Annual Book of ASTM Standards*, American Society for Testing and Materials, Philadelphia (1981).

82a. In [82], Part 33.

82b. In [82], Part 32.

82c. In [82], Part 35.

82d. In [82], Part 46.

83. IWTO Specifications, *Standards for Testing Wool*, IWS, Ilkley (1980).

84. W. Garner, *Textile Laboratory Manual*, Vol. 5, 3rd Ed., Elsevier, New York, 1967; F. F. Kobayash and W. von Bergen, *Wool*

Handbook, 3rd. Ed., Vol. 2. (W. von Bergen, Ed.). Interscience, New York, Chap. 16 (1970).

85. H. J. Henning and C. Mezingue, *Text. Praxis, Int. 35(1)*, 5+XX; *35(2)*, 179+XXIII; *35(3)*, 298+XVIII (1980).

86. S. L. Anderson, *Wool Sci. Rev. 32*, 16 (1967).

87. K. Lees and F. F. Elsworth, Int. Wool Text. Res. Conf. *E*, 425 (1955); *J. Soc. Dyers Colour. 68*, 207 (1952).

88. H. J. Henning, *Wool Sci. Rev. 38*, 35 (1970).

89. Interox, *A Bleacher's Handbook* (1979).

90. W. Schefer, *Textilveredl. 13*, 55 (1978).

91. R. Merkle, *Textil-Rundsch. 19*, 655 (1964); *20*, 7, 51 (1965).

92. J. Gacen, J. Cegarra, M. Caro, and L. Aizpurua, *J. Soc. Dyers Colour. 95*, 389 (1979).

93. H. Zahn, *Textil-Rundsch. 19*, 573 (1964).

94. R. Levene and M. Lewin, in this volume, "The Fluorescent Whitening of Textiles," Sec. 3.

95. J. Cegarra, J. Ribé, D. Vidal, and J. F. Fernandez, *J. Text. Inst. 67*, 5 (1976); *IWTO Report No. 15*, Monaco (June 1972).

96. D. C. Teasdale and A. Bereck, *Text. Res. J. 51*, 541 (1981).

97. R. S. Hunter, *N.B.S. Circular* C 429 (1942).

98. M. G. King, *J. Text. Inst. 62*, 251 (1971).

99. K. J. Whiteley, M. J. Clark, S. J. Welsman, and J. H. Stanton, *J. Text. Inst. 71*, 117 (1980); M. J. Clark and K. J. Whiteley, *J. Text. Inst. 69*, 121 (1978).

100. B. Milligan and D. J. Tucker, *Text. Res. J. 34*, 681 (1964).

101. J. L. Hoare and B. Thompson, *J. Text. Inst. 65*, 281 (1974).

102. G. P. Norton and C. H. Nicholls, *J. Text. Inst. 51*, T1183 (1960).

103. J. Cegarra and J. Gacen, *J. Soc. Dyers Colour. 91*, 61 (1975).

104. J. Cegarra, J. Gacen, and J. Ribé, *J. Soc. Dyers Colour. 84*, 457 (1968).

105. J. C. Jacquemart, *Bull. Inst. Text. Fr. 111*, 249 (1964); *103*, 115 (1962); P. Ponchel and N. Minh Man, *Bull. Inst. Text. Fr. 136*, 377 (1968).

106. Techn. Assocn. Pulp Paper Ind., TAPPI, *Testing Methods*.

107. M. G. King, *J. Text. Inst. 61*, 513 (1970).

108. K. McLaren, *J. Text. Inst. 62*, 453 (1971).

109. R. Levene, *J. Text. Inst. 74*, 223 (1983).

110. M. J. Hammersley and B. Thompson, *IWTO Report No. 14*, Basle (June 1976); Paris (Dec. 1976).

111. E. Coates, M. G. King, and B. Rigg, *J. Text. Inst. 65*, 34 (1974); *63*, 573 (1972).

112. J. Tann, *Text. History 1*, 158 (1969); *World Text. Abs.* 1970/606 (1970).

113. G. Anders, R. Anliker, and G. J. Veenemans, *The Secret of Whiteness*, Ciba-Giegy (1972); *Ciba J.* Winter 1965/66.

114. R. A. Peel, *J. Soc. Dyers Colour.* 83, 281 (1967).
115. C. F. Ward, *J. Soc. Dyers Colour.* 52, 445 (1939).
116. W. Ney, *Text. Praxis Int.* 29, 1392, 1552 (1974).
117. J. Cegarra, J. Ribé, and J. Gacen, *J. Soc. Dyers Colour.* 80, 123 (1964); 83, 189 (1967).
118. H. Baier, *SVF Fachorgen* 16, 72 (1961).
119. W. D. Nicoll and A. F. Smith, *Ind. Eng. Chem.* 47, 2548 (1955).
120. J. A. Gascoine, *J. Text. Inst.* 53, p. 422 (1962).
121. L. Chesner and G. C. Woodford, *J. Soc. Dyers Colour.* 74, 531 (1958).
122. J. Cegarra, J. Gacen, and J. Maillo, *Tientex* 42, 83, 411 (1977).
123. N. J. J. van Rensberg, *SAWTRI Tech. Rep. No.* 151 (1971).
124. O. Oldenroth, *Seifen Ole, Fette. Wachse* 93, 371 (1967); H. Gyslins, A. Rauchlc, and H. Fenke, *Textilverdl.* 1, 315 (1966).
125. M. G. King, *J. Text. Inst.* 62, 251 (1971).
126. K. Ziegler, *Text. Praxis Int.* 17, 376 (1962).
127. M. J. Palin, D. C. Teasdale, and L. Benisek, *J. Soc. Dyers Colour.* 99, 261 (1983).
128. Y. Kobliakova, personal communication.
129. L. Chesner, *J. Soc. Dyers Colour.* 79, 139 (1963).
130. Sandoz, Leaflet TPa 24.
131. O. Schmidt, *Text. Praxis Int.* 28, 164 (1973); *Z. ges. Textilind.* 69, 401 (1967); 66, 849 (1964); W. von Breitenstein, *Text. Praxis Int.* 23, 186 (1968).
132. K. Reinke, *Text. Praxis Int.* 28, 633 (1973).
133. BASF Techn. Leaflet, M5756e (1981).
134. P. Groitsch, Ph.D. Thesis, U. Stuttgart (1970).
135. E. Shanley, H. O. Kauffmann, and W. H. Kibbel, *Am. Dyestuff Rep.* 40, 1 (1951); B. K. Easton, *Text. Chem. Color 1,* 592 (1969); *Can. Text. J.* 78, 32 (1961).
136. W. S. Wood and K. W. Richmond, *J. Soc. Dyers Colour.* 68, 337 (1952); R. Steidl, *Melliand Textilber.* 45, 300 (1964); C. S. Whewell, A. Charlesworth, and R. L. Kitchin, *J. Text. Inst.* 40, P769 (1949).
137. J. F. Graham, R. R. D. Holt, and D. M. Lewis, *Colourage* 26(12), 48 (1979); J. F. Graham and R. R. D. Holt, *Colourage* 25(10), 43 (1978).
138. G. Nitschke, *Z. ges. Textilind.* 71, 788 (1969).
139. J. Cegarra and J. Gacen, *J. Soc. Dyers Colour.* 91, 61 (1975). *Appl. Polym. Symp.* 18, 607 (1971); *Textilveredl.* 11, 390 (1976); J. Cegarra, J. Ribé, and J. Gacen, *J. Soc. Dyers Colour.* 83, 189 (1967); J. Cegarra, J. Gacen, and J. Ribé, *J. Soc. Dyers Colour.* 84, 457 (1968).
140. BASF, Techn. Leaflet M 5400c (1974).

141. G. Valk, *Melliand Textilber. 1*, 18 (1965).
142. M. Weiss, *Am. Dyestuff Rep. 67(8)*, 35 (1975); *67(9)*, 72 (1978); *Can. Text. J. 97*, 47 (1980).
143. S. Imai and T. Hanyu, *Chem. Abs. 84*, P166,187j (1976); *84*, P152, 108 (1976); O. Ohura, *Chem. Abs. 76*, 4790z (1972).
144. L. A. Holt and B. Milligan, *J. Text. Inst. 50*, 117 (1980).
145. H. F. Launer, *Text. Res. J. 41*, 211, 311 (1971); *36*, 606 (1966); *35*, 395, 813 (1965).
146. C. Garrow, E. P. Luede, and C. M. Roxburgh, *Text. Inst. Ind. 9*, 286 (1971).
147. J. E. Tucker, *Text. Res. J. 39*, 830 (1969).
148. A. Bereck, H. Zahn, and S. Schwartz, *Text. Praxis Int. 37*, 621+xv (1982).
149. N. J. J. van Rensburg, *SAWTRI Tech. Rep. Nos. 187, 197* (1973).
150. G. Reinert, *Ciba-Geigy Rev. 1972/2*, 37 (1972); M. A. White, *Wool Sci. Rev. 53*, 50 (1977).
151. N. J. J. van Rensburg and S. G. Scanes, *SAWTRI Bull. 5(3)*, 14 (1971).
152. C. Schirlé and J. Meybeck, *Bull. Inst. Text. Fr. 45*, 31 (1954); *46*, 29 (1954); C. Earland and K. G. Johnson, *Text. Res. J. 22*, 591 (1952); R. Merkle, *Textil-Rundsch. 20*, 131 (1965).
153. S. Serapimoff, R. Valtschewa, and T. Mahljanska, *Melliand Textilber. 44*, 53 (1963); *46*, 170 (1965); *53*, 197 (1972).
154. F. A. Barkhuysen and R. A. Leigh, *SAWTRI Tech. Rep. No. 364* (1977).
155. E. Ludwig and E. Engel, *Melliand Textilber 8*, 903 (1967); A. M. Khan and S. M. Shah, *Chem. Abs. 92*, 112134 (1980).
156. O. Schmidt, *Melliand Textilber. 49*, 316 (1968).
157. I. S. Samuseva, S. V. Zakharova, and A. F. Shiskina, *Chem. Abs. 78*, 31268 (1973).
158. Mitsubishi Gas Co., *Chem. Abs. 95*, P134,336j (1981).
159. N. Nandurka and T. R. Rao, *Wool Woollens India 14(4)*, 11 (1977).
160. R. G. Stewart, *Am. Dyestuff Rep. 55*, 361 (1966); *Proc. Int. Conf. Wool Scouring, Brno, 1st*, 77 (1971).
161. J. L. Hoare, *WRONZ Rep. No. 47* (1978).
162. O. Schmidt, *Z. ges. Textilind. 67*, 723 (1968); *Textil Praxis Int. 23*, 675 (1968); K. Reinke, *Textil Praxis Int. 26*, 744 (1971); *Textilveredl. 3*, 555 (1968); P. Senner, D. Ulmer, and J. Renner, *Z. ges. Textilind. 68*, 763, 858 (1966).
163. O. Schmidt and R. Reinke, *Wirkerei, Strickerei Technik. 18*, 591 (1968).
164. D. R. Lemin and T. Vickerstaff, *Symposium on Fibrous Proteins*, Society of Dyers and Colourists, Bradford, p. 129, (1946); T. Vickerstaff, *The Physical Chemistry of Dyeing*, Oliver and Boyd, London, p. 427 (1954).

165. H. Meichelbeck and H. Knittel, *Z. ges. Textilind.* 72, 379 (1970).
166. K. Schaffner, *Textilveredl.* 1, 334 (1966).
167. Z. A. Mamykina, A. V. Berovskaya, M. N. Serebrennikova, G. P. Kozhnova, and V. G. Berezin, *Texstil. Promyshlennost.* 40(5), 56 (1980).
168. D. M. Lewis, *Wool Sci. Rev.* 50, 22 (1974).
169. BASF, Techn. Leaflet 1889e (1980).
170. K. R. F. Cockett, *Wool Sci. Rev.* 56, 20 (1980).
171. M. Lewin, U. S. Patents 2,923,596 (1959); 3,106,440 (1963).
172. M. Lewin and M. Avrahami, 3rd Int. Wool Text. Res. Conf. *II*, 270 (1965).
173. M. Lewin and J. Weintraub, *Israel Fiber Inst. Rep. No. C44* (1968).
174. P. W. Butcher and D. A. Ross, *WRONZ Comm. No. 23* (1974).
175. H. Heinz, *Textilveredl.* 16, 53 (1981).
176. G. Holler, *Handbach der Textilhilfsmittel.* (C. Chwala and V. Anger, Eds.). Verlag Chemie, Weinheim, pp. 829-842 (1977).
177. G. Laxer, *Am. Dyestuff Rep.* 70(6), 41 (1981).
178. R. A. Leigh, *SAWTRI Tech. Rep. No. 282* (1976).
179. C. J. Lupton, *Am. Dyestuff Rep.* 67(9), 35 (1978).
180. K. Turschmann, *Textilbetreib.* 93(4), 49 (1979); Anon., *Bayer Farben Revue, Special Edn.* 20 (1982).
181. S. Shaw and W. S. Wilson, *J. Soc. Dyers Colour.* 71, 857 (1955).
182. *Du Pont Bulletin No. 259.*

INDEX

9 780367 451820